Progress in Mathematics
Volume 302

Paolo Mastrolia
Marco Rigoli
Alberto G. Setti

Yamabe-type Equations on Complete, Noncompact Manifolds

 Birkhäuser

Paolo Mastrolia
Dipartimento di Matematica
Università degli Studi die Milano
Milano, Italy

Marco Rigoli
Dipartimento di Matematica
Università degli Studi die Milano
Milano, Italy

Alberto G. Setti
Dipartimento di Fisica e Matematica
Università dell'Insubria
Como, Italy

ISBN 978-3-0348-0375-5   ISBN 978-3-0348-0376-2 (eBook)
DOI 10.1007/978-3-0348-0376-2
Springer Basel Heidelberg New York Dordrecht London

Library of Congress Control Number: 2012944419

Mathematics Subject Classification (2010): Primary: 53C21, 58-02; Secondary: 35J60, 35B45, 58J50

Printed on acid-free paper

Springer Basel AG is part of Springer Science+Business Media (www.birkhauser-science.com)

Contents

Introduction   1

1   **Some Riemannian Geometry**   7
   1.1   Preliminaries   7
     1.1.1   Moving frames and the first structure equations   8
     1.1.2   Covariant derivative of tensor fields   10
     1.1.3   Meaning of the first structure equations   12
     1.1.4   Curvature: the second structure equations   14
     1.1.5   Einstein manifolds and Schur's Theorem   16
   1.2   Comparison theorems   18
     1.2.1   Ricci identities   18
     1.2.2   Cut locus and regularity of the distance function   21
     1.2.3   The Laplacian comparison theorem   22
     1.2.4   The Bishop-Gromov comparison theorem   28
     1.2.5   The Hessian comparison theorem   31
   1.3   Some formulas for immersed submanifolds   32

2   **Pointwise conformal metrics**   37
   2.1   The Yamabe equation   37
     2.1.1   The derivation of the Yamabe equation   37
     2.1.2   The Kazdan-Warner obstruction   40
     2.1.3   The Weyl and Cotton tensors   43
   2.2   Some applications in the compact case   49
     2.2.1   A rigidity result of Obata   49
     2.2.2   A result by M. F. Bidaut-Véron and L. Véron   57
     2.2.3   A version of Theorem 2.12 on manifolds with boundary . .   62
     2.2.4   A rigidity result of Escobar   67

3   **General nonexistence results**   73
   3.1   Some spectral considerations   74
     3.1.1   The main nonexistence result   78
   3.2   The endpoint case $K = -1$ and the Poisson equation   93

v

 3.3 A refined version of Theorem 3.2 98

4 A priori estimates   105
 4.1 Estimates from below 105
 4.2 Estimates from above 111
 4.3 Sharpness of the previous results 115
 4.4 Some further estimates 117
 4.5 Nonexistence results for the Yamabe problem 121

5 Uniqueness   127
 5.1 A sharp integral condition 127
 5.2 A remark on the asymptotic behaviour of solutions: examples in
      $\mathbb{R}^m$ and $\mathbb{H}^m$ 130
 5.3 Uniqueness via the weak maximum principle 132
      5.3.1 A useful form of the weak maximum principle 133
      5.3.2 A comparison result 140
      5.3.3 Uniqueness of ground states 143
 5.4 Some geometric applications and further uniqueness 146
      5.4.1 Conformal diffeomorphisms 146
      5.4.2 Uniqueness for the Yamabe problem 148
      5.4.3 An L^∞ a priori estimate 149

6 Existence   157
 6.1 A general procedure 158
      6.1.1 Another comparison result 158
      6.1.2 More basic spectral theory and a result of Li, Tam and Yang 158
      6.1.3 Two useful lemmas 162
      6.1.4 Existence of a maximal solution 165
 6.2 Subsolutions and existence 166
      6.2.1 Existence with $\lambda_1^L(M) < 0$ 166
      6.2.2 $\lambda_1^L(M) < 0$: some sufficient conditions 170
      6.2.3 A more general case 177
 6.3 Global sub- and supersolutions 180
 6.4 The case of the Yamabe problem 188
 6.5 Appendix: the Monotone Iteration Scheme 191

7 Some special cases   197
 7.1 A nonexistence result 197
      7.1.1 A Rellich-Pohozaev formula 207
      7.1.2 A nonexistence result for hyperbolic space 211
      7.1.3 An integral obstruction 219
 7.2 Special symmetries and existence 221
 7.3 The case of Euclidean space and further results 227
      7.3.1 A linear comparison result 227

Contents

    7.3.2   Back to Corollary 5.8 229

    7.3.3   The Euclidean space 231

Bibliography   **239**

List of Symbols   **247**

Index   **253**

Introduction

This book originates from a graduate course given at the University of Milan in 2007.

Our goal is twofold: first, to present a self-contained introduction to the geometric and analytic aspects of the Yamabe problem on a complete noncompact Riemannian manifold, treating existence, nonexistence, uniqueness and *a priori* estimates of the solutions. Secondly, we intend to describe in a way accessible to the nonspecialist a range of methods and techniques that can be successfully applied to more general nonlinear equations which arise in applications.

The classical Yamabe problem concerns the possibility of pointwise conformally deforming a metric of scalar curvature $S(x)$ on the manifold M to a new metric with prescribed scalar curvature $K(x)$. In the case where K is constant it is a natural higher dimensional generalization of the Poincaré–Köbe Uniformization Theorem for Riemann surfaces and can be seen as a way to select a privileged metric on the manifold.

If $\langle\,,\,\rangle$ is the original metric of the Riemannian manifold M and we denote with $\widetilde{\langle\,,\,\rangle} = \varphi^2 \langle\,,\,\rangle$, $\varphi > 0$, a conformally deformed metric, then the two scalar curvatures $S(x)$ and $\widetilde{S}(x)$ are related by the equation

$$\varphi^2 \widetilde{S}(x) = S(x) - 2(m-1)\frac{\Delta\varphi}{\varphi} - (m-1)(m-4)\frac{|\nabla\varphi|^2}{\varphi^2}$$

(see equation (2.7) in Chapter 2), where Laplacian, gradient, and norm are those of the metric $\langle\,,\,\rangle$. In the case where the dimension m of the manifold is greater than or equal to three, it is useful to set

$$\varphi = u^{\frac{2}{m-2}}$$

so that the above equation takes the form

$$\widetilde{S} u^{\frac{m+2}{m-2}} = Su - 4\frac{m-1}{m-2}\Delta u.$$

Thus the *Yamabe problem* amounts to finding a positive solution u of the familiar *Yamabe equation*

$$c_m \Delta u - Su + K u^{\frac{m+2}{m-2}} = 0, \tag{1}$$

where $c_m = 4\frac{m-1}{m-2}$ and $K = \tilde{S}$, the prescribed scalar curvature of the conformally deformed metric. If M is compact and K is constant, after an initial attempted solution by H. Yamabe [Yam60], the problem was solved thanks to efforts of N. Trudinger [Tru68], T. Aubin [Aub76] and R. Schoen [Sch84] (see the nice survey paper by J.M. Lee and T.H. Parker, [LP87], for a complete and self-contained treatment). The solution was obtained using variational methods, and one of the main analytic difficulties stems from the fact that $\frac{m+2}{m-2}$ is the critical exponent for the Sobolev embedding $W^{1,2} \hookrightarrow L^{\frac{2m}{m-2}}$.

A natural generalization of the classical Yamabe problem is the case where K is nonconstant and/or M is noncompact. In this direction we mention the pioneering work of J. L. Kazdan and F. W. Warner, [KW74a], [KW74b], [KW75a], [KW75b]. It should also be mentioned that even the classical Yamabe problem of deforming the metric to one of constant scalar curvature in the noncompact setting is in general not solvable, as first shown by Z. R. Jin, [Jin88].

The Yamabe problem for noncompact manifolds with variable prescribed curvature is the subject of the present monograph. Indeed, we describe methods which allows us to consider the more general *Yamabe-type equations* (resp. inequalities) of the form

$$\Delta u + a(x)u - b(x)u^\sigma = 0 \quad (\text{resp.} \geq 0) \tag{2}$$

where $\sigma > 1$, and we study nonexistence, *a priori* estimates, uniqueness and existence.

Equations of the form (2) and still more general differential inequalities of the form

$$u\Delta u + a(x)u^2 - b(x)u^{\sigma+1} \geq -A|\nabla u|^2 \tag{3}$$

arise in complex analysis (e.g. in the study of the structure of complete Kähler manifolds, [LY90], [Li90] and [LR96], in the Schwarz Lemma for the ratio of volume elements of Kähler manifolds of the same dimension, [Gri76], in the study of pluriharmonic functions on a Kähler manifold, [PRS08]), in the study of harmonic maps with bounded dilation ([EL78] and [PRS08] Chapter 8), in the classification of locally conformally flat manifolds ([PRS07]), in the study of Yang-Mills fields, and in population dynamics, to quote only a few examples.

Existence and nonexistence of positive solutions of (2) clearly depend on the geometry of the underlying manifold, typically encoded by curvature or volume growth of geodesic balls, on properties of the coefficients (typically the relative signs of the coefficients $a(x)$ and $b(x)$) and their asymptotic behavior and on the mutual interplay of the two. This interplay can be taken into account in terms of the relative asymptotic behavior of the coefficients versus the geometry at infinity of the manifold or, at a deeper level, in terms of spectral properties of Schrödinger operators naturally associated to the equation.

From the geometrical interpretation of the equation, it is natural to expect it will be easier to have existence when a and b are "close" enough, for instance they have the same sign, while it will be more difficult to have existence (and therefore it will be easier to prove nonexistence) when a and b are farther apart, for instance when they have opposite sign. This expectation is confirmed by both the existence and the nonexistence results that we will describe.

The geometry of the manifold also plays a natural role in the uniqueness results as well in the *a priori* estimates on the solutions. The latter have a particular geometric interest since they are responsible for the completeness/noncompleteness of the deformed metric.

As mentioned above, we use a variety of techniques adapted to the geometric situation at hand in which the lack of compactness and of symmetry and homogeneity prevents the use of more standard tools typical of compact situations or of the Euclidean setting.

In particular, for existence we will essentially use the method of *sub-super solutions*, [Ama76], [Sat73]. Nonexistence will be obtained using *Liouville-type results* which in turn are obtained using either integral formulas or a method based on the coupling of the supposed solution of the Yamabe-type inequality with that of an appropriate Schrödinger-type inequality associated to it, in a manner reminiscent of the classical generalized maximum principle. Uniqueness will be obtained using variants of the *weak maximum principle* (see, e.g., [PRS05b]) and of clever integration by parts arguments. Finally, *a priori* estimates will be typically obtained using an elaboration of the old idea of the proof of the Schwarz's Lemma by L. H. Ahlfors, [Ahl38], which is at the heart of the maximum principle.

The book is divided into seven chapters.

In the first chapter we give a quick review of Riemannian geometry using the method of *moving frames*. While we assume basic knowledge of Riemannian geometry, several computations will be carried out in full detail in order to acquaint the reader with notation and formalism. We concentrate on derivation of the symmetry properties of the curvature tensors together with a number of other identities that will be repeatedly used in the sequel. In particular, we will describe the commutation rules for covariant derivatives up to fourth order. Then we describe comparison results for the Laplacian of the Riemannian distance function, and for the volume of geodesic balls in terms of lower bounds for the Ricci curvature. We point out that our treatment, which follows that of [PRS05b], does not use Jacobi fields.

In Chapter 2 we first derive equations for the change of curvature tensors under a conformal change of the metric and introduce the Yamabe equation. As a side product of our computations we obtain decomposition of the Riemann curvature tensor in its irreducible components and we exhibit the conformal invariance of the Weyl tensor. Then, we briefly consider the case where M is compact to illustrate the interplay between geometry and analysis, with a few illuminating examples such as the Kazdan-Warner obstruction, a result of Obata on Einstein manifolds and a far-reaching "generalization" due to Véron-Véron, through which we prove further results of Escobar. Along the way we give a detailed proof, which inspires to Petersen's treatise [Pet06a], of a famous rigidity result of Obata. The goal is also to give some geometrical feeling on the subject that will enable us to proceed with the noncompact case: the case of the rest of our investigation.

The core of the monograph begins with Chapter 3, devoted to nonexistence results. As mentioned above, since our methods apply to more general situations which have a wide range of applications, we consider in fact differential inequalities of the form (2) and (3). We describe several nonexistence results; in most of them we assume that u satisfies suitable integrability conditions, that $b(x)$ is nonnegative and that there exists a positive solution φ to the differential inequality

$$\Delta\varphi + Ha(x)\varphi \leq -K\frac{|\nabla\varphi|^2}{\varphi}$$

with H, K parameters satisfying $H > 0$, $K > -1$. Note that in the special case where $K = 0$ the latter condition amounts to the fact that the bottom of the spectrum of the operator $-\Delta - Ha(x)$ is nonnegative. Since $-\Delta$ is a nonnegative operator, the condition is trivially satisfied if $a(x) \leq 0$ on M and may be interpreted as a measure of smallness in a spectral sense of the positive part of $a(x)$. This agrees with the heuristic intuition on the effect of the relative signs of $a(x)$ and $b(x)$ on the existence of solutions. The existence of the positive function φ enters the proof in two different ways. In Theorem 3.2 one uses the functions φ to obtain an integral inequality involving u and its gradient from which one concludes that u is constant, and therefore necessarily identically zero. In a second group of results, the function φ is combined with the solution u to give rise to a diffusion-type differential inequality for which we prove a Liouville theorem. This yields the desired triviality. We also show that when σ is greater than or equal to the critical exponent $(m+2)/(m-2)$, then, by performing an appropriate change of the metric and of the solution, the nonexistence results can be improved to allow even some controlled negativity of the coefficient $b(x)$.

Chapter 4 is devoted to establishing *a priori* upper and lower estimates for the asymptotic behavior of solutions of the differential inequalities

$$\Delta u + a(x)u - b(x)u^\sigma \geq 0, \quad \text{resp.} \quad \Delta u + a(x)u - b(x)u^\sigma \leq 0,$$

under assumptions on $a(x)$ and $b(x)$ related to an assumed radial lower bound for the Ricci curvature. As briefly mentioned above, the results are obtained by applying Alhfors's old idea, namely, one considers an auxiliary function defined in terms of the solution u which by construction attains an extremum, and applies the usual maximum principle. Clearly, the heart of the method consists in finding the best auxiliary function for the problem at hand. We exhibit examples showing that our estimates are essentially sharp. Some further estimates, which cannot be obtained with the previous method, are provided by direct comparison with the aid of the maximum principle (see section 4.3). The chapter ends with some nonexistence results for the Yamabe problem, which complement those described in Chapter 3.

In Chapter 5 we discuss some uniqueness results for positive solutions of Yamabe-type equations (2). The first, the very general Theorem 5.1, states that if the coefficient $b(x)$ is nonnegative and not identically zero, then two solutions whose difference is L^2-integrable are necessarily the same. Although very general, it is sharp, and, remarkably, the assumption on the L^2-integrability cannot be replaced by an L^p condition with $p > 2$. The result is obtained by means of a clever elementary integral inequality. The second result, Theorem 5.2, follows by a comparison argument which relies on a version of the weak maximum principle (see Theorem 5.3) which is interesting in its own. While in most of the results available in the literature, uniqueness is obtained by requiring that solutions have a rather precisely determined asymptotic behavior, our result applies to solutions whose behavior at infinity is specified in a much less stringent manner, see (5.13); moreover, the conclusion is reached assuming only conditions on the volume growth of the manifold. Counterexamples show the sharpness of each result. The chapter ends with a geometric application to the group of conformal diffeomorphisms of a complete manifold and to the uniqueness of solutions of the geometric Yamabe problem.

Chapter 6 deals with existence results for Yamabe-type equations (2) on the complete, noncompact, Riemannian manifold M. The main tool is the *monotone iteration scheme* in various forms, and we give a rather detailed description of it in the appendix at the end of the chapter. The application of the scheme in this context goes back to W. M. Ni, [Ni82], in the Euclidean setting and to P. Aviles and R. C. McOwen, [AM85] and [AM88], for noncompact manifolds. After having introduced some preliminary material on spectral theory, and a useful comparison result, the main body of the chapter is then devoted to the construction of (global and local) super- and subsolutions for the problem. In general terms, supersolutions are obtained under assumptions on the sign of $b(x)$ and of the first eigenvalue of $L = \Delta + a(x)$ on appropriate domains. Because of the combination of signs of the coefficients, subsolutions are harder to find. We give a number of sufficient conditions which ensure that such subsolutions do exist: among them the spectral condition $\lambda_1^L(M) < 0$, for which we provide a new sufficient condition contained in Theorem 6.11. Furthermore, we mention Theorem 6.15, in which existence is guaranteed under a very week growth condition on $b(x)$, and also Theorem 6.16, where a further weakening on the condition on the sign of $b(x)$ is balanced by the necessity of imposing a constant negative lower bound on the Ricci curvature. We explicitly note that the assumptions of our existence theorems match those of the nonexistence results in the previous chapters.

In the last chapter, Chapter 7, we consider some particular cases where the symmetry of the geometry allows one to use special techniques and to obtain stronger results. Typically this happens in Euclidean and Hyperbolic spaces, and more generally in the case of *models* (in the sense of R. Greene and H. Wu, [GW79]), or manifolds with special symmetry. The specific feature of models which make the analysis more precise is that the Laplacian of the distance function from the origin is given explicitly, as opposed to the case of a general manifold where only upper and lower bounds may be obtained under suitable curvature assumptions, by means of the the Laplacian and Hessian comparison theorems, and where the possible presence of the cut locus raises additional difficulties.

We describe refined techniques adapted to the situation at hand and obtain results that, as a by-product, show the degree of sharpness of the general theory and methods we have developed dealing with generic complete Riemannian manifolds. It seems worth remarking that, in the specific case of Hyperbolic space, we provide a nonexistence result with the aid of a Rellich-Pohozaev type formula (see Theorem 7.7) and, even more, in Proposition 7.9 we introduce an integral obstruction to the existence of a conformal deformation which is of a different nature with respect to the Kazdan-Warner condition.

Many of the results presented in this monograph have been obtained over the years by the authors jointly with many collaborators. To all of them we wish to extend our thanks and appreciation. In particular we are indebted to S. Pigola and M. Rimoldi who provided us with the proof of Theorem 2.10 in Chapter 2.

Chapter 1

Some Riemannian Geometry

In this chapter we give a quick review of Riemannian geometry using the moving frame formalism. While we assume basic knowledge of Riemannian geometry, several computations will be carried out in full detail in order to acquaint the reader with notation and formalism. After having introduced frame and coframes, we will describe connection and curvature in terms of the connection and curvature forms. Symmetry properties of the curvature tensors will be described in detail and we will derive a number of identities that will be repeatedly used in the sequel. In particular, we will obtain the commutation rules for covariant derivatives up to fourth order. Along the way, we will introduce Einstein manifolds and prove Schur's lemma. Then we introduce basic results for the Riemannian distance function from a fixed reference point $o \in M$, and discuss briefly the cut locus and some of its properties. We will then describe comparison results for the Laplacian of the Riemannian distance function and for the volume of geodesic balls in terms of lower bounds for the Ricci curvature. We point out that our treatment, which follows that of [PRS05b], does not use Jacobi fields. The chapter ends with a brief section on the geometry of immersed submanifolds to fix notation and terminology used in the second part of Chapter 2.

1.1 Preliminaries

Let $(M, \langle \, , \, \rangle)$ be a Riemannian manifold of dimension m with metric $\langle \, , \, \rangle$. The aim of this section is to fix notation and to describe the essential facts of the geometry of $(M, \langle \, , \, \rangle)$ using É. Cartan's formalism. The usefulness of this approach will become apparent in the sequel, and it will prove to be particularly effective in Chapter 2 in the derivation of the formulae which express the change of the Riemannian, Ricci and scalar curvature under a conformal change of the metric.

1.1.1 Moving frames and the first structure equations

Let $p \in M$ and (U, φ) a local chart such that $U \ni p$, with coordinate functions $x^1, \ldots, x^m$, $m = \dim(M)$. If q is a generic point in U we have, at q,

$$\langle\,,\,\rangle = \langle\,,\,\rangle_{ij}\, dx^i \otimes dx^j, \tag{1.1}$$

where dx^i denotes the differential of the function x^i and $\langle\,,\,\rangle_{ij}$ are the (local) components of the metric, defined by $\langle\,,\,\rangle_{ij} = \langle \frac{\partial}{\partial x^i}, \frac{\partial}{\partial x^j} \rangle$. In relation (1.1), and throughout the book, we adopt the Einstein summation convention for repeated indices. Applying in q the Gram-Schmidt orthonormalization process we can find linear combinations of the 1-form dx^k, which we will call θ^i, such that

$$\langle\,,\,\rangle = \delta_{ij}\, \theta^i \otimes \theta^j, \tag{1.2}$$

where δ_{ij} is the Kronecker symbol. Since, as q varies in U, the previous process gives rise to coefficients that are C^∞-functions of q, the set of 1-forms $\{\theta^i\}$, $i = 1, \ldots, m$, define an orthonormal system on U for the metric $\langle\,,\,\rangle$, i.e., a (local) *orthonormal (o.n.) coframe*. We will sometimes write

$$\langle\,,\,\rangle = \sum_{i=1}^{m} (\theta^i)^2$$

instead of (1.2). We also define the (local) *dual orthonormal frame* $\{e_i\}$, $i = 1, \ldots, m$, as the set of m (local) vector fields satisfying, on the open set U,

$$\theta^j(e_i) = \delta_i^j \tag{1.3}$$

(where δ_i^j is just a suggestive way of writing the Kronecker symbol, reflecting the position of the indexes in the pairing of θ^i and e_j). We have the following

Proposition 1.1. *Let* $\{\theta^i\}$ *be a local o.n. coframe on M, defined on an open set U; then there exist unique 1-forms*

$$\{\theta^i_j\}, \quad i, j = 1, \ldots, m$$

on U such that, $\forall\, i, j = 1, \ldots, m$,

$$d\theta^i = -\theta^i_j \wedge \theta^j, \tag{1.4}$$

$$\theta^i_j + \theta^j_i = 0. \tag{1.5}$$

The forms $\{\theta^i_j\}$ *are called the* Levi-Civita connection forms *associated to the o.n. coframe* $\{\theta^i\}$.

Remark. Equations (1.4) are classically known as the *first structure equations*.

Proof. Let us assume the existence of the forms $\{\theta^i_j\}$ satisfying (1.5) and (1.4) and determine their expression. Of course

$$\theta^i_j = a^i_{jk}\theta^k$$

for some $a^i_{jk} \in C^\infty(U)$ (where $C^\infty(U)$ denotes the set of smooth functions defined on the open set U), and (1.5) is equivalent to

$$a^i_{jk} + a^j_{ik} = 0. \tag{1.6}$$

The 2-forms $d\theta^i$ can be written, for some (unique) coefficients $b^i_{jk} \in C^\infty(U)$, as

$$d\theta^i = \frac{1}{2}b^i_{jk}\,\theta^j \wedge \theta^k, \quad b^i_{jk} + b^i_{kj} = 0.$$

Since (1.4) must hold we have:

$$\frac{1}{2}b^i_{jk}\theta^j \wedge \theta^k = -a^i_{jk}\theta^k \wedge \theta^j = a^i_{jk}\theta^j \wedge \theta^k = \frac{1}{2}(a^i_{jk} - a^i_{kj})\theta^j \wedge \theta^k.$$

It follows that

$$b^i_{jk} = a^i_{jk} - a^i_{kj}. \tag{1.7}$$

Cyclic permutations of the indices i, j, k and use of (1.6) and (1.7) yield

$$b^k_{ij} = a^k_{ij} - a^k_{ji} = -a^i_{kj} + a^j_{ki}; \tag{1.8}$$
$$b^j_{ki} = a^j_{ki} - a^j_{ik} = a^j_{ki} + a^i_{jk}. \tag{1.9}$$

Adding (1.7) and (1.9) and subtracting (1.8) we get

$$a^i_{jk} = \frac{1}{2}(b^i_{jk} - b^k_{ij} + b^j_{ki}). \tag{1.10}$$

The previous relation determines the expression of the forms θ^i_j and also proves uniqueness. Now define

$$\theta^i_j = \frac{1}{2}(b^i_{jk} - b^k_{ij} + b^j_{ki})\theta^k, \tag{1.11}$$

where the b^i_{jk}'s satisfy

$$b^i_{jk} + b^i_{kj} = 0.$$

It is clear that

$$a^i_{jk} = \frac{1}{2}\left(b^j_{ik} - b^k_{ji} + b^i_{kj}\right) = -\frac{1}{2}\left(b^i_{jk} - b^k_{ij} + b^j_{ki}\right) = -a^i_{jk},$$

thus (1.6) is met, and then the θ^i_j defined in (1.11) satisfy (1.5); it is also immediate to verify that they satisfy (1.4). $\qquad\square$

1.1.2   Covariant derivative of tensor fields

The Levi-Civita connection forms are the starting point to define a *covariant derivative of tensor fields* in the following way. Following standard notation we denote with T_pM the tangent space at $p \in M$ and with T_p^*M the cotangent space at p.

We recall that a *tensor field T of type (r,s)* is a law that assigns to all points $p \in M$ a multilinear map

$$T_p : \underbrace{T_p^*M \times \cdots \times T_p^*M}_{r \text{ times}} \times \underbrace{T_pM \times \cdots \times T_pM}_{s \text{ times}} \to \mathbb{R}$$

with the usual differentiability request with respect to the variable p (see e.g. [Lee03]). The set of tensor fields of type (r,s) will be denoted by $T_r^s(M)$. Let us begin considering the case of a *vector field X* on M, i.e., a tensor of type $(1,0)$. We denote with $\mathfrak{X}(M)$ the set of all (smooth) vector fields on M.

Let $\{\theta^i\}$ a local o.n. coframe and $\{e_i\}$ the dual frame.

Definition 1.2. *The* covariant derivative *of the vector field X, ∇X, is the tensor field of type $(1,1)$ ∇X defined in the following way: if $X = X^i e_i$,*

$$\nabla X = (dX^i) \otimes e_i + X^i \nabla e_i,$$

having defined

$$\nabla e_i = \theta_i^j \otimes e_j.$$

Setting

$$X_k^i \theta^k = dX^i + X^j \theta_j^i,$$

∇X can be then written as

$$\nabla X = (dX^i + X^j \theta_j^i) \otimes e_i = X_k^i \theta^k \otimes e_i,$$

and X_k^i is said to be the covariant derivative of the coefficient X^i.

If $Y \in \mathfrak{X}(M)$ we define the *covariant derivative of X in the direction of Y* as the vector field

$$\nabla_Y X = \nabla X(Y),$$

which in components reads as

$$\nabla_Y X = X_k^i \theta^k(Y) e_i = X_k^i Y^k e_i.$$

Definition 1.3. *The* divergence *of the vector field $X \in \mathfrak{X}(M)$ is the trace of ∇X, that is,*

$$\operatorname{div} X = \operatorname{tr}(\nabla X) = \langle \nabla_{e_i} X, e_i \rangle = X_i^i. \tag{1.12}$$

Analogously, for a 1-form ω (i.e., a tensor of type $(0,1)$), we have:

Definition 1.4. *The* covariant derivative of the 1-form ω, $\nabla\omega$, *is the tensor field of type* $(0,2)$ *defined in the following way: if* $\omega = \omega_i \theta^i$,

$$\nabla\omega = (d\omega_i) \otimes \theta^i + \omega_i \nabla\theta^i,$$

with

$$\nabla\theta^i = -\theta^i_j \otimes \theta^j.$$

Note that, setting

$$\omega_{ik}\theta^k = d\omega_i - \omega_j\theta^j_i,$$

it follows that

$$\nabla\omega = \omega_{ik}\theta^k \otimes \theta^i.$$

If $Y \in \mathfrak{X}(M)$ we define the *covariant derivative of ω in the direction of Y* as the 1-form

$$\nabla_Y\omega = \nabla\omega(Y),$$

which in components reads as

$$\nabla_Y\omega = \omega_{ik}\theta^k(Y)\theta^i = \omega_{ik}Y^k\theta^i.$$

For a 0-form $f \in C^\infty(M)$ we set

$$\nabla f = df \quad \text{(exterior differential of } f\text{)}.$$

We point out that this notation may give rise to some ambiguity; indeed, in the literature (and also in this book) ∇f often denotes the *gradient* of f, i.e., the vector field dual to the 1-form df: in this case, using standard notation (see for instance [Lee97]), we can write $\nabla f = (df)^\sharp$, where $\sharp$ is the *sharp map* from the cotangent bundle T^*M to the tangent bundle TM defined by

$$\left\langle (df)^\sharp, Y \right\rangle = \langle \nabla f, Y \rangle = df(Y) = Y(f),$$

for all $Y \in \mathfrak{X}(M)$. Note that, in components, setting $df = f_j\theta^j$ for some smooth coefficients f_j, we have $(\nabla f)^i = \delta^{ij}(df)_j = \delta^{ij}f_j = f_i$ (that is, in an orthonormal frame, differential and gradient of a function have the same coefficients with respect to the (dual) basis $\{\theta^i\}$ and $\{e_i\}$).

Finally, ∇ can be extended in a natural way to a generic tensor T, in order to define a connection on each tensor bundle $T^s_r(M)$: this extension of ∇ satisfies the Leibniz rule and some other nice properties, like the commutativity with the trace on any pair of indices (see again [Lee97]).

Although the covariant derivative has been introduced by means of locally defined objects, it is possible to show (using the transformation laws of $\{\theta^i\}$ and $\{\theta^i_j\}$ as the coframe changes) that the new tensor field thus obtained is *globally* defined.

Remark. One can verify that the previous definition matches the "canonical" one usually given in terms of the Koszul formalism (see for example [Lee97], [Pet06a]). Indeed, as we will see before long, the operator ∇ coincides precisely with the Levi-Civita connection associated to the metric $\langle\,,\,\rangle$ of M.

1.1.3 Meaning of the first structure equations

We now want to discuss the geometric meaning of condition (1.5), namely

$$\theta^i_j + \theta^j_i = 0.$$

To this purpose let us compute the covariant derivative of the metric tensor $\langle\,,\,\rangle$. Using the Leibniz rule, and recalling that for every tangent vector $X_p = X(p) \in T_pM$ (with p in the domain of the o.n. coframe $\{\theta^i\}$) it holds that $\nabla_{X_p}\theta^i = -\theta^i_j(X_p)\theta^j$, we have

$$
\begin{aligned}
\nabla_{X_p}\langle\,,\,\rangle &= \nabla_{X_p}(\delta_{ij}\theta^i\theta^j) = \delta_{ij}(\nabla_{X_p}\theta^i \otimes \theta^j + \theta^i \otimes \nabla_{X_p}\theta^j) \\
&= \delta_{ij}(-\theta^i_k(X_p)\theta^k \otimes \theta^j - \theta^j_k(X_p)\theta^i \otimes \theta^k) \\
&= -\theta^i_k(X_p)\theta^k \otimes \theta^i - \theta^i_k(X_p)\theta^i \otimes \theta^k \\
&= -(\theta^i_k + \theta^k_i)(X_p)\theta^k \otimes \theta^i,
\end{aligned}
$$

and therefore

$$\nabla\langle\,,\,\rangle = 0 \ \text{ if and only if } \ \theta^i_j + \theta^j_i = 0,$$

i.e., *condition (1.5) is equivalent to the parallelism of the metric* (in other words: $\forall\, X, Y, Z \in \mathfrak{X}(M),\ X\langle Y, Z\rangle = \langle \nabla_X Y, Z\rangle + \langle Y, \nabla_X Z\rangle$).

On the other hand, the first structure equations (1.4) tell us that the metric is *torsion-free*. Indeed, let X and Y be two vector fields on M and $[X, Y]$ their *Lie bracket*, defined by

$$[X, Y](f) = X(Y(f)) - Y(X(f)), \quad \forall f \in C^\infty(M). \tag{1.13}$$

We claim that condition

$$[X, Y] = \nabla_X Y - \nabla_Y X \quad \forall\, X, Y \in \mathfrak{X}(M) \tag{1.14}$$

is equivalent to the validity of (1.4). Note that the left-hand side of (1.14) is independent of the choice of a metric on M. Since the *torsion* of a generic connection ∇ on M is the $(0,2)$ tensor field $\mathrm{Tor}(X, Y) = \nabla_X Y - \nabla_Y X - [X, Y]$, this justifies the expression "torsion-free" used above. To prove the equivalence, recall that the exterior differential of a 1-form ω is intrinsically defined by

$$d\omega(X, Y) = X(\omega(Y)) - Y(\omega(X)) - \omega([X, Y]);$$

moreover, as a consequence of the definition of covariant derivative,

$$(\nabla_X\omega)(Y) = X(\omega(Y)) - \omega(\nabla_X Y), \tag{1.15}$$

so that

$$X\big(\theta^i(Y)\big) - \theta^i(\nabla_X Y) = (\nabla_X\theta^i)(Y) = -\theta^i_j(X)\theta^j(Y),$$

that is,

$$X\big(\theta^i(Y)\big) + \theta^i_j(X)\theta^j(Y) = \theta^i(\nabla_X Y).$$

Then we compute $\big(d\theta^i + \theta^i_j \wedge \theta^j\big)(X,Y)$, that is

$$
\begin{aligned}
d\theta^i(X,Y) &+ \theta^i_j \wedge \theta^j(X,Y) \\
&= X(\theta^i(Y)) - Y(\theta^i(X)) - \theta^i([X,Y]) + \theta^i_j(X)\theta^j(Y) - \theta^i_j(Y)\theta^j(X) \\
&= X(\theta^i(Y)) + \theta^i_j(X)\theta^j(Y)) - Y(\theta^i(X)) - \theta^i_j(Y)\theta^j(X)) - \theta^i([X,Y]) \\
&= \theta^i(\nabla_X Y - \nabla_Y X - [X,Y]),
\end{aligned}
$$

and the claim follows.

Remark. By the fundamental theorem of Riemannian geometry (see for instance [Lee97] or [Pet06b]), we deduce that the connection ∇ coincides, as we said previously, with the Levi-Civita connection of the metric $\langle\,,\,\rangle$.

We now define the *Lie derivative* of Y in the direction of X to be $\mathcal{L}_X Y = [X,Y]$, so that condition (1.14) can be written in the form

$$\mathcal{L}_X Y = \nabla_X Y - \nabla_Y X. \tag{1.16}$$

Setting also

$$\mathcal{L}_X f = X(f) \tag{1.17}$$

for $f \in C^\infty(M)$, and

$$(\mathcal{L}_X \omega)(Y) = \mathcal{L}_X(\omega(Y)) - \omega(\mathcal{L}_X Y), \tag{1.18}$$

if ω is a 1-form, we can extend $\mathcal{L}_X$ to a generic tensor field requiring $\mathbb{R}$-linearity and the validity of the Leibniz rule (see also [Lee03], [Pet06a]). Using (1.15), we compute the Lie derivative of the metric in the direction of X, $\mathcal{L}_X \langle\,,\,\rangle$ (note that this latter has to be a covariant tensor of order 2, that is, a $(0,2)$-tensor):

$$
\begin{aligned}
(\mathcal{L}_X \langle\,,\,\rangle)(Y,Z) &= ((\mathcal{L}_X \theta^i) \otimes \theta^i + \theta^i \otimes (\mathcal{L}_X \theta^i))(Y,Z) \\
&= \theta^i(Z)(\mathcal{L}_X \theta^i)(Y) + \theta^i(Y)(\mathcal{L}_X \theta^i)(Z) \\
&= \theta^i(Z)\big[\mathcal{L}_X(\theta^i(Y)) - \theta^i(\mathcal{L}_X Y)\big] \\
&\quad + \theta^i(Y)\big[\mathcal{L}_X(\theta^i(Z)) - \theta^i(\mathcal{L}_X Z)\big] \\
&= \theta^i(Z)X(\theta^i(Y)) - \theta^i(Z)\theta^i(\nabla_X Y - \nabla_Y X) \\
&\quad + \theta^i(Y)X(\theta^i(Z)) - \theta^i(Y)\theta^i(\nabla_X Z - \nabla_Z X) \\
&= \theta^i(Z)(\nabla_X \theta^i)(Y) + \theta^i(Y)(\nabla_X \theta^i)(Z) \\
&\quad + \theta^i(Z)\theta^i(\nabla_Y X) + \theta^i(Y)\theta^i(\nabla_Z X) \\
&= (\nabla_X \theta^i \otimes \theta^i + \theta^i \otimes \nabla_X \theta^i)(Y,Z) + \langle \nabla_Y X, Z\rangle + \langle Y, \nabla_Z X\rangle \\
&= (\nabla_X \langle\,,\,\rangle)(Y,Z) + \langle \nabla_Y X, Z\rangle + \langle Y, \nabla_Z X\rangle \\
&= \langle \nabla_Y X, Z\rangle + \langle Y, \nabla_Z X\rangle,
\end{aligned}
$$

where in the last equality we have used the fact that the metric is parallel with respect to the Levi-Civita connection. Thus, we have proved the useful identity

$$(\mathcal{L}_X \langle \, , \, \rangle)(Y, Z) = \langle \nabla_Y X, Z \rangle + \langle Y, \nabla_Z X \rangle \tag{1.19}$$

for all $X, Y, Z \in \mathfrak{X}(M)$, which will be repeatedly used in the sequel. Note that equation (1.19) in components reads as

$$(\mathcal{L}_X \langle \, , \, \rangle)_{ij} = \langle \nabla_{e_i} X, e_j \rangle + \langle e_i, \nabla_{e_j} X \rangle = X_i^j + X_j^i. \tag{1.20}$$

It can be proved that the Lie derivative of Y in the direction of X has the following geometric meaning (see e.g. [Lee03]):

$$(\mathcal{L}_X Y)_p = \left.\frac{d}{dt}\right|_{t=0} (\varphi_{-t})_* Y_{\varphi_t(p)} = \lim_{t \to 0} \frac{(\varphi_{-t})_* Y_{\varphi_t(p)} - Y_p}{t},$$

where φ_t is the local flow generated by X and $(\varphi_t)_*$ is the push-forward. The analogous applies to $\mathcal{L}_X h$, with h a generic tensor field (see also Chapter 2 for the special case of $\mathcal{L}_X \langle \, , \, \rangle$).

1.1.4   Curvature: the second structure equations

We now consider the second structure equations. Let $\{\theta^i\}$ be a local o.n. frame and $\{\theta^i_j\}$ the corresponding Levi-Civita connection forms. The *curvature forms* $\{\Theta^i_j\}$ associated to the coframe are defined through the *second structure equations*,

$$d\theta^i_j = -\theta^i_k \wedge \theta^k_j + \Theta^i_j. \tag{1.21}$$

Obviously the Θ^i_j are 2-forms. Since, according to (1.5), $\theta^i_j + \theta^j_i = 0$, it follows immediately that the Θ^i_j's satisfy the same antisymmetry condition

$$\Theta^i_j + \Theta^j_i = 0. \tag{1.22}$$

Using the basis $\{\theta^i \wedge \theta^j\}$, $1 \leq i < j \leq m$, of the space of skew-symmetric 2-forms $\bigwedge^2(U)$ on the open set U, we may write

$$\Theta^i_j = \frac{1}{2} R^i_{jkt} \theta^k \wedge \theta^t = \sum_{k<t} R^i_{jkt} \theta^k \wedge \theta^t \tag{1.23}$$

for some coefficients $R^i_{jkt} \in C^\infty(U)$ satisfying

$$R^i_{jkt} + R^i_{jtk} = 0, \tag{1.24}$$

while (1.22) implies that

$$R^i_{jkt} + R^j_{ikt} = 0. \tag{1.25}$$

From (1.24) and (1.25) we thus deduce the symmetries

$$R^i_{jkt} = -R^i_{jtk} = -R^j_{ikt}. \tag{1.26}$$

Differentiating the first structure equation, using the second and the properties of the exterior differential we have

$$0 = d(d\theta^i) = -d(\theta^i_j \wedge \theta^j) = -d\theta^i_j \wedge \theta^j + \theta^i_j \wedge d\theta^j$$
$$= \theta^i_k \wedge \theta^k_j \wedge \theta^j - \Theta^i_j \wedge \theta^j - \theta^i_j \wedge \theta^j_k \wedge \theta^k$$
$$= \theta^j \wedge \Theta^i_j,$$

that is

$$\theta^j \wedge \Theta^i_j = 0. \tag{1.27}$$

These identities go under the name of the *first Bianchi identities*. Using (1.23) we get

$$0 = R^i_{jkt}\theta^j \wedge \theta^k \wedge \theta^t = \sum_{1 \le j < k < t \le m} (R^i_{jkt} + R^i_{ktj} + R^i_{tjk})\theta^j \wedge \theta^k \wedge \theta^t,$$

and then we deduce the first Bianchi identities in the classical form

$$R^i_{jkt} + R^i_{ktj} + R^i_{tjk} = 0. \tag{1.28}$$

It is possible also to show that another consequence of (1.26) and (1.28) is the last, important symmetry

$$R^i_{jkt} = R^k_{tij}. \tag{1.29}$$

One can verify that the coefficients R^i_{jkt} gives rise to a (global) $(1,3)$-tensor, called the *Riemann curvature tensor*,

$$\text{Riem} = R^i_{jkt}\theta^k \otimes \theta^t \otimes \theta^j \otimes e_i.$$

We warn the reader that there are a number of different conventions for the Riemann curvature tensor (see the discussion in [Lee97]). We will often use the *curvature tensor* of type $(0,4)$ given by

$$R = R_{ijkt}\theta^i \otimes \theta^j \otimes \theta^k \otimes \theta^t,$$

with $R_{ijkt} = R^i_{jkt}$. Note that we have performed the classical operation of lowering indices using the metric tensor, that is, in our chosen orthonormal frame,

$$R_{ijkt} = \delta_{is}R^s_{jkt} = R^i_{jkt}$$

(compare with the discussion about gradient and differential in section 1.1.2). We complete the description of the symmetries of the curvature tensor noting that, from the first Bianchi identities, we can obtain the *second Bianchi identities* involving the covariant derivatives of the components of the curvature tensor, namely:

$$R_{ijkt,l} + R_{ijtl,k} + R_{ijlk,t} = 0. \tag{1.30}$$

Tracing the curvature tensor on its first and third indices (or, equivalently, on its second and fourth) we obtain the *Ricci tensor*

$$\mathrm{Ric} = R(e_i, \cdot, e_i, \cdot) = R(\cdot, e_i, \cdot, e_i,) = R_{ijik}\theta^j \otimes \theta^k = R_{jk}\theta^j \otimes \theta^k,$$

where $\{e_i\}$ is the dual basis of $\{\theta^j\}$. Note that, according to (1.29), Ric is a symmetric $(0,2)$-tensor; in other words, $\forall\, X, Y \in \mathfrak{X}(M)$

$$\mathrm{Ric}(X,Y) = \mathrm{Ric}(Y,X), \tag{1.31}$$

and in components

$$R_{ij} = R_{ji}. \tag{1.32}$$

The *scalar curvature* S is defined as the trace of Ric, i.e.,

$$S = \mathrm{Ric}(e_i, e_i) = R_{ijij} = R_{ii}.$$

The *sectional curvature of the 2-plane* $\pi \subset T_pM$ spanned by the vectors u and v is defined to be

$$K_p(\pi) = \frac{R(u,v,u,v)}{\langle u,u\rangle\,\langle v,v\rangle - \langle u,v\rangle^2} \in \mathbb{R}.$$

It is not difficult to verify that the right-hand side of the above formula is in fact independent of the chosen basis of π. Clearly, if u v are an orthonormal basis for π, then

$$K_p(\pi) = R(u,v,u,v).$$

We note that a common notation for the sectional curvature of the plane π spanned by u and v is

$$K_p(\pi) = \mathrm{Sect}(u \wedge v).$$

1.1.5   Einstein manifolds and Schur's Theorem

Definition 1.5. *The manifold* $(M, \langle\,,\,\rangle)$, $\dim(M) = m \geq 2$, *is said to be* Einstein *if*

$$\mathrm{Ric} = \lambda\,\langle\,,\,\rangle, \tag{1.33}$$

for some $\lambda \in \mathbb{R}$.

Using the moving frame formalism we now show that, if the dimension of the manifold is greater than or equal to 3 and equation (1.33) holds for some $\lambda \in C^\infty(M)$, where $C^\infty(M)$ is the set of smooth functions defined on M, then

λ is automatically constant. Indeed, first note that, tracing equation (1.33), we immediately obtain

$$\lambda = \frac{S}{m}. \tag{1.34}$$

We now trace the second Bianchi identities (1.30) with respect to the indices i and l to get

$$R_{ijkt,i} + R_{ijti,k} + R_{ijik,t} = 0.$$

Since covariant derivative commutes with contractions, the previous relation yields

$$R_{ijkt,i} = R_{jt,k} - R_{jk,t}, \tag{1.35}$$

whence, contracting again this time with respect to j and k, we get

$$R_{ikkt,i} = R_{kt,k} - R_{kk,t}$$

that is,

$$2R_{kt,\,k} = S_t. \tag{1.36}$$

Now because of (1.33) and (1.34) we have

$$R_{kt} = \frac{S}{m}\delta_{kt},$$

and using again the fact that the metric tensor is parallel we deduce that

$$R_{kt,l} = \frac{1}{m}S_l\delta_{kt}.$$

Now tracing with respect to k and l, we get

$$R_{kt,k} = \frac{1}{m}S_t; \tag{1.37}$$

substituting in (1.36), we obtain

$$\left(\frac{2}{m} - 1\right)S_t = 0,$$

and we conclude that if $m \geq 3$ and M is connected, then the scalar curvature, and therefore λ, are constant. We have thus proved the well-known result of Schur:

Theorem 1.6. *Let $(M, \langle\,,\,\rangle)$ be a connected Riemannian manifold of dimension $m \geq 3$. If*

$$\mathrm{Ric} = \lambda\langle\,,\,\rangle$$

for some $\lambda \in C^\infty(M)$, then M is Einstein.

Note that, if $m \geq 3$, then $(M, \langle , \rangle)$ is Einstein if and only if the *traceless Ricci tensor*

$$T = \mathrm{Ric} - \frac{S}{m} \langle , \rangle \tag{1.38}$$

is identically null. Observe also that, in our o.n. coframe,

$$T_{ij} = R_{ij} - \frac{S}{m} \delta_{ij}. \tag{1.39}$$

Using (1.37) (sometimes called *Schur's identities*) we shall obtain a remarkable formula, an infinitesimal version of the *Kazdan-Warner obstruction*, that will be discussed in detail below (see Section 2.1.2).

1.2   Comparison theorems

1.2.1   Ricci identities

We now want to recall that the curvature tensor can be interpreted as an obstruction to the validity of Schwarz's theorem for mixed derivatives of third order and higher. Since this will be useful in the sequel, we consider the case of a function $u : M \to \mathbb{R}$ which we assume to be at least $C^3(M)$. If

$$du = u_i \theta^i \tag{1.40}$$

for some smooth coefficients u_i, the *Hessian* of u is defined as the 2-covariant tensor $\mathrm{Hess}(u) = \nabla du$ of components u_{ij} given by

$$u_{ij}\theta^j = du_i - u_k \theta_i^k, \tag{1.41}$$

that is,

$$\mathrm{Hess}(u) = u_{ij}\theta^j \otimes \theta^i. \tag{1.42}$$

The *Laplacian* of u is the trace of the Hessian, that is,

$$\Delta u = \mathrm{tr}\,(\mathrm{Hess}(u)) = u_{ii}. \tag{1.43}$$

One can verify that the operator Δ defined in (1.43) is the Laplace-Beltrami operator associated to the metric $\langle , \rangle$ (see for instance [Pet06b]). Since second derivatives commute we expect $\mathrm{Hess}(u)$ to be a symmetric tensor. This can be verified as follows: we differentiate equation (1.40) and use the structure equations to get

$$0 = du_i \wedge \theta^i + u_i d\theta^i = (u_{ij}\theta^j + u_k \theta_i^k) \wedge \theta^i - u_i \theta_k^i \wedge \theta^k$$
$$= u_{ij}\theta^j \wedge \theta^i$$
$$= \frac{1}{2}(u_{ij} - u_{ji})\theta^j \wedge \theta^i,$$

thus

$$0 = \sum_{1 \leq j < i \leq m} (u_{ij} - u_{ji})\theta^j \wedge \theta^i;$$

since $\{\theta^j \wedge \theta^i\}$ $(1 \leq j < i \leq m)$ is a basis for the 2-forms we deduce

$$u_{ij} = u_{ji}, \tag{1.44}$$

as expected. The *third derivatives* of u are defined by the rule

$$u_{ijk}\theta^k = du_{ij} - u_{kj}\theta_i^k - u_{ik}\theta_j^k. \tag{1.45}$$

Note that according to (1.44), the right-hand side is symmetric with respect to i and j and therefore

$$u_{ijk} = u_{jik}. \tag{1.46}$$

We now differentiate (1.41) and use the structure equations to get

$$du_{ik} \wedge \theta^k - u_{ij}\theta_k^j \wedge \theta^k = -du_t \wedge \theta_i^t + u_k\theta_t^k \wedge \theta_i^t - u_k\Theta_i^k$$
$$= -(u_{tk}\theta^k + u_k\theta_t^k) \wedge \theta_i^t + u_k\theta_t^k \wedge \theta_i^t - \frac{1}{2}u_k R_{ijt}^k \theta^j \wedge \theta^t,$$

thus

$$(du_{ik} - u_{tk}\theta_i^t - u_{it}\theta_k^t) \wedge \theta^k = -\frac{1}{2}u_t R_{ijk}^t \theta^j \wedge \theta^k,$$

and, by (1.45),

$$u_{ikj}\theta^j \wedge \theta^k = -\frac{1}{2}u_t R_{ijk}^t \theta^j \wedge \theta^k.$$

Skew-symmetrizing we get

$$\frac{1}{2}(u_{ikj} - u_{ijk})\theta^j \wedge \theta^k = -\frac{1}{2}u_t R_{ijk}^t \theta^j \wedge \theta^k,$$

thus

$$u_{ijk} = u_{ikj} + u_t R_{ijk}^t = u_{ikj} + u_t R_{tijk}. \tag{1.47}$$

Identities like (1.47) are generally called *Ricci identities*.

As a first application of (1.47) we derive the well-known *Bochner-Weitzenböck formula*. We denote with $|\nabla u|^2$ the squared norm of the vector field ∇u and with $|\text{Hess}(u)|^2$ the squared norm of $\text{Hess}(u)$ with respect to the natural fiber metric induced by $\langle \, , \, \rangle$ on the tensor bundle $T_0^2(M)$ (see e.g. [Lee97], [Pet06a]). Note that, in components, $|\nabla u|^2 = u_i u_i$ and $|\text{Hess}(u)|^2 = u_{ij} u_{ij}$.

Lemma 1.7. *Let $u \in C^3(M)$; then*

$$\frac{1}{2}\Delta|\nabla u|^2 = |\text{Hess}(u)|^2 + \text{Ric}\,(\nabla u, \nabla u) + \langle \nabla \Delta u, \nabla u \rangle. \tag{1.48}$$

Proof. We set $v = \sum_{k=1}^{m} (u_k)^2 = u_k u_k$. To compute Δv we need to trace the second covariant derivative of v. Using the Leibniz rule we have

$$v_i = 2u_k u_{ki}$$

and differentiating once more

$$v_{it} = 2u_{kt} u_{ki} + 2u_k u_{kit}.$$

Tracing with respect to i and t we obtain

$$\frac{1}{2}\Delta|\nabla u|^2 = u_{kt} u_{kt} + u_k u_{ktt} = |\text{Hess}(u)|^2 + u_k u_{ktt}; \qquad (1.49)$$

however, from (1.47) we deduce

$$u_{ktt} = u_{ttk} + u_s R_{stkt} = u_{ttk} + u_s R_{sk},$$

from which

$$u_k u_{ktt} = \langle \nabla \Delta u, \nabla u \rangle + \text{Ric}\,(\nabla u, \nabla u). \qquad (1.50)$$

Putting together (1.49) and (1.50) we obtain the desired identity. $\qquad\square$

We end this section with another useful commutation relation, not so easily available in the literature. First we define the "(1,1) versions" of the Hessian (of a sufficiently smooth function u on M) and of the Ricci tensor, i.e., the two tensor field of type $(1,1)$, respectively denoted by $\text{hess}(u)$ and ric, such that, for all $X, Y \in \mathfrak{X}(M)$,

$$\langle \text{hess}(u)(X), Y \rangle = \text{Hess}(u)(X, Y) \qquad (1.51)$$

and

$$\langle \text{ric}\,(X), Y \rangle = \text{Ric}\,(X, Y). \qquad (1.52)$$

Note that, with the notation of section 1.1.2, we can write

$$\text{hess}(u)(X) = (\text{Hess}(u)(X, \cdot))^\sharp$$

and

$$\text{ric}\,(X) = (\text{Ric}\,(X, \cdot))^\sharp.$$

We are now ready to prove the following

Lemma 1.8. *Let $u \in C^\infty(M)$. Then*

$$u_{tkkt} - u_{kktt} = \frac{1}{2}\langle \nabla S, \nabla u \rangle + \text{tr}\,(\text{hess}(u) \circ \text{ric}). \qquad (1.53)$$

Proof. We start from the commutation relations (1.47). By taking covariant derivative we deduce

$$u_{ijkt} - u_{ikjt} = u_{st} R_{sijk} + u_s R_{sijk,t}. \tag{1.54}$$

Differentiating both sides of (1.45), using the structure equations and (1.45) itself, we arrive at

$$u_{ijkl}\theta^l \wedge \theta^k = -\frac{1}{2}(u_{tj} R_{tilk} + u_{it} R_{tjlk})\theta^l \wedge \theta^k,$$

from which, interchanging k and l and adding, we deduce

$$u_{ijkl} - u_{ijlk} = u_{tj} R_{tikl} + u_{it} R_{tjkl}. \tag{1.55}$$

Now, (1.53) follows immediately from (1.54), (1.55), (1.36) and tracing. $\square$

1.2.2 Cut locus and regularity of the distance function

We recall a few facts on the cut locus and the Riemannian distance function that will be repeatedly used in the sequel, referring to Chavel's book [Cha06] for proofs and further details.

Let o be a point in the complete manifolds $(M, \langle\,,\,\rangle)$, and let γ be a geodesic issuing from o. It is known that γ is locally minimizing. A point q in the image of γ is said to be a *cut point* for o along γ if γ minimizes the distance from o to q, but ceases to be minimizing after beyond q. The set of cut points of o along geodesic issuing emanating from o is the *cut locus* of o, and is denoted by cut(o). It turns out that cut(o) is a closed set of measure zero with respect to the Riemannian measure, and that the set $D_o = M \setminus \text{cut}(o)$ is an open starshaped domain, which is in fact the maximal domain of the normal geodesic coordinates centered at o. At the tangent space level, we say that v is in the tangent cut locus of o, Cut(o), if the geodesic γ_v with initial velocity v minimizes distances for $t \in [0, 1]$ and does not minimize distances for $t > 1$. Thus cut(o) is the image of Cut(o) under the exponential map exp_o, and the set $E_o = \{tv \in T_o M : v \in \text{Cut}(o), 0 \le t < 1\}$, is the maximal starshaped domain with respect to 0 on which $\exp_o$ is a diffeomorphism, and $D_o = \exp_o(E_o)$. Moreover if $r(x)$ denotes the Riemannian distance function from o, namely, $r(x) = d(x, o) = |\exp^{-1}(x)|$, then $r(x)$ is smooth on $D_o \setminus \{o\}$.

Let $B_R(o)$ denote the geodesic ball centered at o with radius R, and let $\partial B_R(o)$ be its its boundary.

Following R.L. Bishop, [Bis77] we say that q is an *ordinary* cut point for o if there are two or more minimizing geodesics joining o and q. Cut points which are not ordinary are said to be *singular*.

Bishop proves that ordinary cut points are dense in cut(o) ([Bis77], Main Theorem). Since it is easy to verify that the distance function $r(x)$ is not C^1 at ordinary cut points (see [Bis77], Proposition), we deduce that if $r(x)$ is smooth on the punctured ball $B_R(o) \setminus \{o\}$, then $B_R(o) \cap \text{cut}(o) = \emptyset$.

1.2.3   The Laplacian comparison theorem

Now we show how (1.47) is the starting point to derive the classical Laplacian comparison theorem without using Jacobi fields. Fix a reference point o in $(M, \langle\,,\,\rangle)$, and let γ be a minimizing geodesic parameterized by arclength issuing from o; we adopt the standard notation $\dot{\gamma}$ to denote the tangent vector of γ. Note that, since γ is a geodesic, we have $\nabla_{\dot{\gamma}}\dot{\gamma} = 0$. We define a unit vector field $Y \perp \dot{\gamma}$ along γ by parallel translation (see e.g. [Lee97]); note that $\gamma(t)$ is an integral curve of ∇r, that is, $\dot{\gamma}(t) = (\nabla r)(\gamma(t))$. To perform calculations we let $\{\theta^i\}$ be a local orthonormal coframe and $\{e_i\}$ its dual frame. Then

$$dr = r_i \theta^i \quad \text{and} \quad Y = Y^j e_j.$$

By Gauss' lemma (see for instance [dC92]) $r_i r_i \equiv 1$ and since γ is a geodesic we obtain

$$r_i r_{ij} = 0, \quad j = 1, \dots m. \tag{1.56}$$

Therefore

$$Y^j r_{ij} r_i = 0.$$

Differentiating the latter equation and using the fact that Y is parallel yields

$$(r_i Y^j r_{ijk} + r_{ij} r_{ik} Y^j)\theta^k = 0;$$

hence, if $\dot{\gamma} = \dot{\gamma}^k e_k$, since $r_{ijk} = r_{jik}$,

$$r_{ist} Y^i \dot{\gamma}^s Y^t = -r_{ij} r_{ik} Y^j Y^k. \tag{1.57}$$

Now in formula (1.47) we take $u(x) = r(x)$ to deduce

$$r_{ijk} \dot{\gamma}^k Y^j Y^i - r_{ist} Y^i \dot{\gamma}^s Y^t = -R_{ijkt} Y^i \dot{\gamma}^j Y^k \dot{\gamma}^t. \tag{1.58}$$

Thus, inserting (1.57) into (1.58), we get

$$r_{ijk} \dot{\gamma}^k Y^j Y^i + r_{ij} r_{ik} Y^j Y^k = -R_{ijkt} Y^i \dot{\gamma}^j Y^k \dot{\gamma}^t. \tag{1.59}$$

Recalling the definition (1.51) of hess we let

$$\mathrm{hess}(r)(Y) = \nabla_Y \nabla r,$$

so that

$$\mathrm{Hess}(r)(Y, X) = \langle \mathrm{hess}(r)(Y), X \rangle.$$

Having set

$$\mathrm{hess}^2(r)(Y) = \mathrm{hess}(r)(\mathrm{hess}(r)(Y)),$$

we define

$$\mathrm{Hess}^2(r)(Y, X) = \langle \mathrm{hess}^2(r)(Y), X \rangle.$$

Then, since $\nabla_{\dot\gamma}Y = 0$, (1.59) can be reinterpreted in the form

$$\frac{d}{dt}(\mathrm{Hess}(r)(\gamma)(Y,Y)) + \mathrm{Hess}^2(r)(\gamma)(Y,Y) = -\mathrm{Sect}_\gamma(Y \wedge \dot\gamma). \qquad (1.60)$$

Note that (1.56) rewrites as

$$\mathrm{hess}(r)(\nabla r) \equiv 0. \qquad (1.61)$$

We sum (1.60) over an orthonormal basis $\{Y^i\}$ $(i = 2,\ldots,m)$ of $\dot\gamma^\perp$, and use (1.61) to get

$$\frac{d}{dt}(\Delta r)(\gamma) + |\mathrm{Hess}(r)|^2(\gamma) = -\mathrm{Ric}(\nabla r, \nabla r)(\gamma). \qquad (1.62)$$

Thus, using Newton's inequality

$$|\mathrm{Hess}(r)|^2 \geq \frac{(\Delta r)^2}{m-1},$$

we obtain

$$\frac{d}{dt}(\Delta r \circ \gamma) + \frac{(\Delta r \circ \gamma)^2}{m-1} \leq -\mathrm{Ric}(\nabla r \circ \gamma, \nabla r \circ \gamma). \qquad (1.63)$$

We recall that, in the literature, $\mathrm{Ric}(\nabla r, \nabla r)$ is called *radial Ricci curvature*. It follows that, if we assume

$$\mathrm{Ric}(\nabla r, \nabla r) \geq -(m-1)G(r) \qquad (1.64)$$

for some function $G \in C^0([0,+\infty))$, then

$$\frac{d}{dt}(\Delta r \circ \gamma) + \frac{(\Delta r \circ \gamma)^2}{m-1} \leq (m-1)G(t). \qquad (1.65)$$

Now we recall that $\Delta r = (\sqrt{g}(r,u))^{-1}\partial\sqrt{g}/\partial r$ (see for instance [Cha06]), where $\sqrt{g}$ is the square root of the determinant of the metric in polar geodesic coordinates (r,u) centered at o. Also, $\sqrt{g} = \det\mathcal{A}(r,u)$ where $\mathcal{A}(r,u)$ is the matrix solution of the differential equation in $u^\perp \subset T_oM$,

$$\mathcal{A}''(r,u) + \mathcal{R}(r,u)\mathcal{A}(r,u) = 0,$$

satisfying the initial conditions $\mathcal{A}(0,u) = 0$, $\mathcal{A}'(0,u) = Id$, and $\mathcal{R}(r,u)$ is the composition of the curvature operator at $\exp_o(ru)$ with parallel translation along the geodesic $\gamma_u(t) = \exp_o(tu)$ (see again [Cha06], p. 114). Thus

$$\mathcal{A}(r,u) = r\,\mathrm{Id} + O(r^2) \quad \text{and} \quad \mathcal{A}'(r,u) = \mathrm{Id} + O(r)$$

and we conclude that

$$\Delta r = \log(\det\mathcal{A})' = \mathrm{tr}(\mathcal{A}'\mathcal{A}^{-1}) = \frac{m-1}{r} + O(r) \qquad (1.66)$$

(see [Pet06a], page 136, for a different derivation).

Hence, having set $\varphi(t) = \Delta r \circ \gamma$, using (1.65) and (1.66) and again the fact that γ is parameterized by arclength we deduce that, under assumption (1.64),

$$\begin{cases} \varphi'(t) + \dfrac{\varphi(t)^2}{m-1} \le (m-1)G(t), \\[2mm] \varphi(t) = \dfrac{m-1}{t} + o(1) \quad \text{as } t \to 0^+. \end{cases} \tag{1.67}$$

Of course, in order to make sense from the analytical point of view, (1.67) has to be interpreted with the image of γ inside of the domain D_o of the normal geodesic coordinates centered at o, or, in other words, outside the cut locus of o. To analyze (1.67) we now need two simple calculus lemmas.

Lemma 1.9. *Let* $G \in C^0([0,+\infty))$ *and let* $\varphi, \psi \in C^2((0,+\infty)) \cap C^1([0,+\infty))$ *be solutions of the problems:*

$$\text{i)} \begin{cases} \varphi'' - G\varphi \le 0, \\ \varphi(0) = 0; \end{cases} \qquad \text{ii)} \begin{cases} \psi'' - G\psi \ge 0, \\ \psi(0) = 0, \ \psi'(0) > 0. \end{cases} \tag{1.68}$$

If $\varphi(r) > 0$ *for* $r \in (0,T)$ *and* $\psi'(0) \ge \varphi'(0)$, *then* $\psi(r) > 0$ *in* $(0,T)$ *and*

$$\frac{\varphi'}{\varphi} \le \frac{\psi'}{\psi}, \quad \psi \ge \varphi \text{ on } (0,T). \tag{1.69}$$

Proof. Since $\psi'(0) > 0$, $\psi > 0$ in a neighborhood of 0. We observe in passing that if G is assumed to be nonnegative, then, integrating (1.68) ii), we have

$$\psi'(r) \ge \psi'(0) + \int_0^r G(s)\psi(s)\,ds,$$

so that ψ' is positive in the interval where $\psi \ge 0$, and we conclude that, in fact, $\psi > 0$ on $(0,+\infty)$. In the general case, where no assumption is made on the sign of G, we let

$$\beta = \sup\{t : \psi > 0 \text{ in } (0,t)\};$$

$$\tau = \min\{\beta, T\}.$$

The function $\psi'\varphi - \psi\varphi' \in C^0([0,+\infty))$ vanishes in $r = 0$, and it satisfies

$$(\psi'\varphi - \psi\varphi')' = \psi''\varphi - \psi\varphi'' \ge 0$$

in $(0,\tau)$. Thus, $\psi'\varphi - \psi\varphi' \ge 0$ on $[0,\tau)$, and, dividing through by $\varphi\psi$, we deduce that

$$\frac{\psi'}{\psi} \ge \frac{\varphi'}{\varphi} \quad \text{in } (0,\tau).$$

Integrating between ε and r, $0 < \varepsilon < r < \tau$, yields

$$\varphi(r) \le \frac{\varphi(\varepsilon)}{\psi(\varepsilon)}\psi(r)$$

and, since

$$\lim_{\varepsilon \to 0^+} \frac{\varphi(\varepsilon)}{\psi(\varepsilon)} = \frac{\varphi'(0)}{\psi'(0)} \le 1,$$

we conclude that in fact

$$\varphi(r) \le \psi(r) \text{ in } [0, \tau).$$

Since $\varphi > 0$ in $(0, T)$ by assumption, this in turn forces $\tau = T$, for, otherwise, $\tau = \beta < T$ and we would have $\varphi(\beta) > 0$, while, by continuity, $\psi(\beta) = 0$, a contradiction. $\qquad\square$

Lemma 1.10. *Let $G \in C^0([0, +\infty))$ and let $g_i \in C^1((0, T_i))$, $i = 1, 2$ be solutions of the Riccati differential inequalities*

$$\text{i) } g_1' + \frac{g_1^2}{\alpha} - \alpha G \le 0; \quad \text{ii) } g_2' + \frac{g_2^2}{\alpha} - \alpha G \ge 0 \tag{1.70}$$

satisfying the condition

$$g_i(t) = \frac{\alpha}{t} + O(1) \text{ as } t \to 0^+ \tag{1.71}$$

for some $\alpha > 0$. Then $T_1 \le T_2$ and $g_1(t) \le g_2(t)$ in $(0, T_1)$.

Proof. Since $\tilde{g}_i = \alpha^{-1}g_i$ satisfy the conditions in the statement with $\alpha = 1$, without loss of generality we assume $\alpha = 1$. Observe that the functions $g_i(s) - \frac{1}{s}$ are bounded and integrable in a neighborhood of $s = 0$, thus we define $\varphi_i \in C^2((0, T_i)) \cap C^1([0, T_i))$ on $[0, T_i)$, by setting

$$\varphi_i(t) = te^{\int_0^t (g_i(s) - \frac{1}{s})\, ds}.$$

Then $\varphi_i(0) = 0$, $\varphi_i > 0$ on $(0, T_i)$ and straightforward computations show that

$$\varphi_i'(t) = g_i(t)\varphi_i(t), \quad \varphi_i'(0) = 1$$

and

$$\varphi_1'' \le G\varphi_1 \quad \text{on } (0, T_1);$$
$$\varphi_2'' \ge G\varphi_2 \quad \text{on } (0, T_2).$$

An application of Lemma 1.9 shows that $T_1 \le T_2$ and $g_1 = \frac{\varphi_1'}{\varphi_1} \le \frac{\varphi_2'}{\varphi_2} = g_2$ on $(0, T_1)$, as required. $\qquad\square$

We are now ready to prove the next *Laplacian comparison theorem*, which is a simplified (but sufficient for our purposes) version of that appearing in [PRS08]:

Theorem 1.11. *Let $(M, \langle\,,\,\rangle)$ be a complete manifold of dimension $m \geq 2$. Having fixed a reference point $o \in M$, let $r(x) = \mathrm{dist}_M(x, o)$. Assume that the radial Ricci curvature $\mathrm{Ric}(\nabla r, \nabla r)$ of M satisfies*

$$\mathrm{Ric}(\nabla r, \nabla r) \geq -(m-1)G(r) \tag{1.72}$$

for some nonnegative function $G \in C^0([0, +\infty))$. Let $h \in C^2([0, +\infty))$ be a solution of the problem

$$\begin{cases} h'' - Gh \geq 0, \\ h(0) = 0, \quad h'(0) = 1. \end{cases} \tag{1.73}$$

Then the inequality

$$\Delta r(x) \leq (m-1)\frac{h'(r(x))}{h(r(x))} \tag{1.74}$$

holds pointwise on $M \backslash (\{o\} \cup \mathrm{cut}(o))$ and weakly on all of M.

Proof. Fix any $x \in M \backslash (\{o\} \cup \mathrm{cut}(o))$ and let $\gamma : [0, l] \to M$ be a minimizing geodesic from o to x parameterized by arclength. We then arrive to (1.67), where the differential inequality is in $(0, l]$. Since $g = (m-1)\frac{h'}{h}$ satisfies

$$g'(t) + \frac{g(t)^2}{m-1} \geq (m-1)G(t) \quad \text{on } (0, +\infty) \tag{1.75}$$

and (1.71) with $\alpha = m-1$, an application of Lemma 1.10 to (1.67) and (1.75) gives

$$\varphi(t) \leq (m-1)\frac{h'(t)}{h(t)} \quad \text{in } (0, l].$$

Thus, in particular, since $\gamma(l) = x$ and $r(x) = l$,

$$\Delta r(x) \leq (m-1)\frac{h'(r(x))}{h(r(x))},$$

showing the validity of (1.74) pointwise within the cut locus. It remains to show the validity of (1.74) weakly in all of M, which is guaranteed by the following lemma. □

Lemma 1.12. *Set $D_o = M \backslash \mathrm{cut}(o)$ and suppose that*

$$\Delta r \leq \alpha(r) \quad \text{pointwise on } D_o \backslash \{o\}, \tag{1.76}$$

for some $\alpha \in C^0((0, +\infty))$. Let $v \in C^2(\mathbb{R})$ be nonnegative and set $u(x) = v(r(x))$ on M. Suppose either

$$\text{i) } v' \leq 0 \qquad \text{or} \qquad \text{ii) } v' \geq 0. \tag{1.77}$$

Then we respectively have

$$\text{i) } \Delta u \geq v''(r) + \alpha(r)v'(r); \qquad \text{ii) } \Delta u \leq v''(r) + \alpha(r)v'(r) \tag{1.78}$$

weakly on M.

Proof. Let E_o be the maximal star-shaped domain in T_oM on which $\exp_o$ is a diffeomorphism onto its image D_o, so that we have $\mathrm{cut}(o) = \partial(\exp_o(E_o))$. Since E_o is a star-shaped domain, we can exhaust E_o by a family $\{E_o^n\}$ of relatively compact, star-shaped domains with smooth boundary. We set $D_o^n = \exp_o(E_o^n)$ so that

$$\overline{D}_o^n \subset D_o^{n+1} \quad \text{and} \quad \bigcup_n D_o^n = D_o.$$

The fact that each E_o^n is star-shaped implies

$$\frac{\partial r}{\partial \nu_n} = \langle \nabla r, \nu_n \rangle > 0 \quad \text{on } \partial D_o^n, \tag{1.79}$$

where ν_n denotes the outward unit normal to ∂D_o^n. Now we assume the validity of (1.77) i). Since $r \in C^\infty(D_o^n \setminus \{o\})$, computing we get

$$\Delta u \geq v'' + \alpha(r)v' \quad \text{pointwise on } D_o^n \setminus \{o\}. \tag{1.80}$$

Let $0 \leq \varphi \in C_0^\infty(M)$, where $C_0^\infty(M)$ denotes the set of smooth functions with compact support on M. We claim that, $\forall n$,

$$\int_{D_o^n} u\Delta\varphi \geq \int_{D_o^n} (v'' + \alpha(r)v')\varphi + \varepsilon_n,$$

where $\varepsilon_n \to 0$ as $n \to +\infty$. Since $M = D_o \cup \mathrm{cut}(o)$ and $\mathrm{cut}(o)$ has measure 0, inequality (1.78) i) will follow by letting $n \to +\infty$. To prove the claim we fix $\delta > 0$ small and we apply the second Green formula (see e.g. [GT01]) on $\overline{D}_o^n \setminus B_\delta(o)$ to obtain

$$\int_{D_o^n \setminus B_\delta(o)} u\Delta\varphi = \int_{D_o^n \setminus B_\delta(o)} \varphi\Delta u - \int_{\partial D_o^n \setminus \partial B_\delta(o)} \left(\varphi \frac{\partial u}{\partial \nu_n} - u \frac{\partial \varphi}{\partial \nu_n} \right), \tag{1.81}$$

where ν_n is the outward unit normal to $\partial D_o^n \setminus \partial B_\delta(o)$. We note that, according to (1.77) i) and (1.79),

$$\frac{\partial u}{\partial \nu_n} = v'(r) \frac{\partial r}{\partial \nu_n} \leq 0 \quad \text{on } \partial D_o^n.$$

Using this, (1.79) and (1.81) we obtain

$$\int_{D_o^n} u\Delta\varphi \geq \int_{D_o^n} (v'' + \alpha(r)v')\varphi + \varepsilon_n + I_\delta,$$

with

$$\varepsilon_n = \int_{\partial D_o^n} u \frac{\partial \varphi}{\partial \nu_n},$$

$$I_\delta = \int_{B_\delta(o)} [u\Delta\varphi - (v'' + \alpha(r)v')\varphi] - \int_{\partial B_\delta(o)} \left[u \frac{\partial \varphi}{\partial r} - \varphi \frac{\partial u}{\partial r} \right],$$

where the notation $\frac{\partial \cdot}{\partial r}$ means $\langle \nabla r, \nabla \cdot \rangle$. Clearly, $I_\delta \to 0$ as $\delta \downarrow 0^+$; on the other hand, since $\varphi \in C_0^\infty(M)$ and cut(o) has measure 0, using the divergence and Lebesgue theorems we see that, as $n \to +\infty$,

$$\varepsilon_n = \int_{D_o^n} \mathrm{div}(u\nabla\varphi) \to \int_{D_o} \mathrm{div}(u\nabla\varphi) = \int_M \mathrm{div}(u\nabla\varphi) = 0.$$

This proves the claim and the validity of (1.78) i). The case (1.77) ii) and (1.78) ii) can be dealt with in a similar way.   □

Remark. We note that, for the above proofs to work, it is not necessary that (1.72) holds on the entire M: instead, for instance, if (1.72) is valid on $B_R(o)$, then (1.74) holds on $B_R(o)\backslash(\{o\} \cup \mathrm{cut}(o))$ and weakly on $B_R(o)$.

1.2.4   The Bishop-Gromov comparison theorem

We now show how to get from the previous results a (somewhat generalized) version of what is known in the literature as the *Bishop-Gromov comparison theorem* (see also [PRS08]). We denote with $\mathrm{vol}\, B_R(o)$ and $\mathrm{vol}\, \partial B_R(o)$ the volume of the geodesic ball $B_R(o)$ and of its boundary $\partial B_R(o)$, respectively.

Theorem 1.13. *Let $(M, \langle\,,\,\rangle)$ be a complete, m-dimensional Riemannian manifold satisfying*

$$\mathrm{Ric}\,(\nabla r, \nabla r) \geq -(m-1)G(r) \quad on\ M \tag{1.82}$$

for some $G \in C^0([0,+\infty))$, $G \geq 0$, where $r(x) = \mathrm{dist}(x,o)$. Let $h \in C^2([0,+\infty))$ be the nonnegative solution of the problem

$$\begin{cases} h'' - G(t)h = 0, \\ h(0) = 0,\ h'(0) = 1. \end{cases} \tag{1.83}$$

Then, for almost every $R > 0$, the function

$$R \mapsto \frac{\mathrm{vol}\,\partial B_R(o)}{h(R)^{m-1}} \tag{1.84}$$

is nonincreasing, and

$$\mathrm{vol}\,\partial B_R(o) \leq \omega_m h(R)^{m-1}, \tag{1.85}$$

where ω_m is the volume of the unit sphere in $\mathbb{R}^m$. Moreover,

$$R \mapsto \frac{\mathrm{vol}\,B_R(o)}{\int_0^R h(t)^{m-1}\,dt} \tag{1.86}$$

is a nonincreasing function on $(0,+\infty)$.

Since it will be used in the proof of Theorem 1.13, and also in the next chapters, we first recall the useful *co-area formula*.

We denote with $W^{1,1}(M)$ the Sobolev space consisting of functions in $L^1(M)$ with (weak) gradient also in $L^1(M)$. We also denote with A_t^f the t-level set ($t \in \mathbb{R}$) of a function f on M, i.e., $A_t^f = \{x \in M | f(x) = t\}$. Following Schoen and Yau (see [SY94], p.89) we can state the following

Proposition 1.14. *Let M be a compact Riemannian manifold with boundary and $f \in W^{1,1}(M)$. For any nonnegative measurable function g on M the following formula holds:*

$$\int_M g = \int_{-\infty}^{+\infty} \left(\int_{A_t^f} \frac{g}{|\nabla f|} \right) dt. \tag{1.87}$$

For a proof see the classical [Fed69]. Note that, in particular, if $f(x) = r(x) = \text{dist}_M\, x, o$, equation (1.87) becomes

$$\int_M g = \int_0^D \left(\int_{\partial B_t(o)} g \right) dt, \tag{1.88}$$

where $D = \sup_M r(x)$.

Proof of Theorem 1.13. In case o is a pole of M (see [GW79] and also the proof of Theorem 6.8 in Chapter 6) one integrates the divergence of the radial vector field

$$X = h(r(x))^{-m+1} \nabla r$$

on concentric balls $B_R(o)$, and uses the divergence and Laplacian comparison theorems. However, in general, objects are nonsmooth and inequalities are intended in the sense of distributions. Therefore, we have to take some extra care. The Laplacian comparison theorem asserts that

$$\Delta r(x) \le (m-1) \frac{h'(r(x))}{h(r(x))} \tag{1.89}$$

pointwise on the open, star-shaped, full measured set $M \setminus \text{cut}(o)$ and weakly on all of M. Thus, $\forall\ 0 \le \varphi \in \text{Lip}_0(M)$,

$$-\int \langle \nabla r, \nabla \varphi \rangle \le (m-1) \int \frac{h'(r(x))}{h(r(x))} \varphi. \tag{1.90}$$

$\forall\, \varepsilon > 0$, consider the radial cut-off function

$$\varphi_\varepsilon(x) = \rho_\varepsilon(r(x)) h(r(x))^{-m+1}, \tag{1.91}$$

where ρ_ε is the piecewise linear function

$$\rho_\varepsilon(t) = \begin{cases} 0, & \text{if } t \in [0, r) \\ \frac{t-r}{\varepsilon}, & \text{if } t \in [r, r+\varepsilon) \\ 1, & \text{if } t \in [r+\varepsilon, R-\varepsilon) \\ \frac{R-t}{\varepsilon}, & \text{if } t \in [R-\varepsilon, R) \\ 0, & \text{if } t \in [R, +\infty). \end{cases} \tag{1.92}$$

Note that

$$\nabla\varphi_\varepsilon = \left\{ -\frac{\chi_{R-\varepsilon, R}}{\varepsilon} + \frac{\chi_{r, r+\varepsilon}}{\varepsilon} - (m-1)\frac{h'(r(x))}{h(r(x))}\rho_\varepsilon \right\} h(r(x))^{-m+1}\nabla r,$$

for almost all $x \in M$, where $\chi_{s,t}$ is the characteristic function of the annulus $B_t(o)\backslash B_s(o)$. Therefore, using φ_ε into (1.90) and simplifying, we get

$$\frac{1}{\varepsilon}\int_{B_R(o)\backslash B_{R-\varepsilon}(o)} h(r(x))^{-m+1} \le \frac{1}{\varepsilon}\int_{B_{r+\varepsilon}(o)\backslash B_r(o)} h(r(x))^{-m+1}.$$

Using the co-area formula (1.87)) we deduce that

$$\frac{1}{\varepsilon}\int_{R-\varepsilon}^{R} \mathrm{vol}(\partial B_t(o))\, h(t)^{-m+1} \le \frac{1}{\varepsilon}\int_{r}^{r+\varepsilon} \mathrm{vol}(\partial B_t(o))\, h(t)^{-m+1}$$

and, letting $\varepsilon \downarrow 0$,

$$\frac{\mathrm{vol}(\partial B_R(o))}{h(R)^{m-1}} \le \frac{\mathrm{vol}(\partial B_r(o))}{h(r)^{m-1}} \tag{1.93}$$

for almost all $0 < r < R$. Letting $r \to 0$ and recalling that $h(r) \sim r$ and $\mathrm{vol}(\partial B_r) \sim \omega_m r^{m-1}$ as $r \to 0$ (which can be deduced, for instance, integrating equation (1.66) on a geodesic ball and using the divergence theorem and Gauss' lemma), we conclude that, for almost any $R > 0$,

$$\mathrm{vol}\,\partial B_R(o) \le \omega_m h(R)^{m-1}.$$

To prove the second statement we note that, as observed in [CGT82], for general real valued functions $f(t) \ge 0$, $g(t) > 0$, if $t \mapsto \frac{f(t)}{g(t)}$ is decreasing, then $t \mapsto \frac{\int_0^t f}{\int_0^t g}$ is

decreasing. Indeed, since f/g is decreasing, if $0 < r < R$,

$$\int_0^r f \int_r^R g = \int_0^r g\frac{f}{g}\int_r^R g \ge \frac{f(r)}{g(r)}\int_0^r g \int_r^R g \ge \int_0^r g \int_r^R g\frac{f}{g} = \int_0^r g \int_r^R f$$

whence

$$\int_0^r f \int_0^R g = \int_0^r f \int_0^r g + \int_0^r f \int_r^R g \ge \int_0^r f \int_0^r g + \int_0^r g \int_r^R f = \int_0^r g \int_0^R f.$$

In particular, applying this observation to (1.93) and using the co-area formula (1.87) we deduce that

$$r \mapsto \frac{\operatorname{vol} B_r(o)}{\int_0^r h(t)^{m-1}\,dt}$$

is decreasing, concluding the proof. $\qquad\Box$

We conclude with the following analytical result whose proof can be found in [PRS08]:

Proposition 1.15. *Assume h is a solution of*

$$\begin{cases} h'' - H^2(1+r^2)^{\delta/2}h = 0, \\ h(0) = 0, \quad h'(0) = 1 \end{cases}$$

where $H > 0$ and $\delta \geq -2$. Set

$$H' = \begin{cases} H, & \text{if } \delta > -2 \\ \frac{1}{2}(1 + \sqrt{1 + 4H^2}), & \text{if } \delta = -2. \end{cases}$$

Then,

$$\frac{h'}{h}(r) \leq H'r^{\delta/2}(1 + o(1)) \quad \text{as } r \to +\infty.$$

Moreover, there exists a constant $C > 0$ such that for $r > 1$,

$$h(r) \leq C \begin{cases} \exp\left(\frac{2H'}{2+\delta}(1+r)^{1+\delta/2}\right), & \text{if } \delta \geq 0 \\ r^{-\delta/4}\exp\left(\frac{2H'}{2+\delta}r^{1+\delta/2}\right), & \text{if } -2 < \delta < 0 \\ r^{H'}, & \text{if } \delta = -2. \end{cases}$$

1.2.5 The Hessian comparison theorem

For the sake of completeness we recall here the following *Hessian comparison theorem*; however, since it will only be used a very few times in the sequel, we only give its statement (the interested reader can find the proof in [PRS08]). Recall that the *radial sectional curvature* K_{rad} of a manifold is the sectional curvature of a 2-plane containing ∇r.

Theorem 1.16. *Let $(M, \langle\,,\,\rangle)$ be a complete manifold of dimension m. Having fixed a reference point $o \in M$, let $r(x) = \operatorname{dist}_M(x, o)$, and let $D_o = M \setminus \operatorname{cut}(o)$ be the domain of the normal geodesic coordinates centered at o. Given a smooth even function G on $\mathbb{R}$, let h be the solution of the Cauchy problem*

$$\begin{cases} h'' - Gh = 0, \\ h(0) = 0, \quad h'(0) = 1, \end{cases} \tag{1.94}$$

and let $I = [0, r_0) \subseteq [0, +\infty)$ be the maximal interval where h is positive. If the radial sectional curvature of M satisfies

$$K_{rad} \geq -G(r(x)) \quad on \ B_{r_0}(o), \tag{1.95}$$

then

$$\mathrm{Hess}(r) \leq \frac{h'(r(x))}{h(r(x))}\{\langle \ , \ \rangle - dr \otimes dr\} \tag{1.96}$$

on $D_o \setminus \{o\} \cup B_{r_0}(o)$ in the sense of quadratic forms. On the other hand, if

$$K_{rad} \leq -G(r(x)) \quad on \ B_{r_0}(o), \tag{1.97}$$

then

$$\mathrm{Hess}(r) \geq \frac{h'(r(x))}{h(r(x))}\{\langle \ , \ \rangle - dr \otimes dr\}. \tag{1.98}$$

Remark. By taking traces in Theorem 1.16 we immediately obtain the corresponding estimates for Δr. However, as we have seen in Theorem 1.11, the estimate from above for the Laplacian of the distance function holds under the weaker assumption that the radial Ricci curvature (and not the full radial sectional curvature) is bounded from below by $-(m-1)G(r(x))$. Furthermore the estimate in this latter case can be extended, in weak form, to the entire manifold. This is not the case for the above estimates on $\mathrm{Hess}(r)$.

1.3   Some formulas for immersed submanifolds

Let $(N, \langle \ , \ \rangle)$ and M be respectively a Riemannian manifold and a manifold of dimensions n and m, with $m \leq n$. Let $f : M \to N$ be an immersion and let $f^* \langle \ , \ \rangle$ be the metric induced on M by f, where f^* denotes the pull-back (note that $f^* \langle \ , \ \rangle$ is indeed a metric since f is an immersion). If g is a given metric on M and $f : M \to N$ is an immersion we will say that f is an *isometric immersion* if $g = f^* \langle \ , \ \rangle$. To simplify notation we use the symbol $\langle \ , \ \rangle$ on M to denote the induced metric; more generally, from now on we shall omit the pull-back notation, being clear from the context where forms or tensors are considered.

We fix the following indexes convention:

$$1 \leq i, j, k, \ldots \leq m, \quad m+1 \leq \alpha, \beta, \gamma, \ldots \leq n, \quad 1 \leq a, b, c, \ldots \leq n.$$

Definition 1.17. *Given the isometric immersion $f : M \to (N, \langle \ , \ \rangle)$, a Darboux coframe along f is a local o.n. coframe $\{\theta^i\}$ on N such that*

$$\theta^\alpha = 0 \quad on \ M. \tag{1.99}$$

In particular, for a Darboux coframe along f we have

$$\langle \ , \ \rangle = \sum_{i=1}^{m} \left(\theta^i\right)^2 \quad on \ M. \tag{1.100}$$

The dual $\{e_a\}$ of a Darboux coframe is called a *Darboux frame along f* and condition (1.99) is equivalent to say that the vectors $\{e_i\}$ (locally) span (the image of) TM in TN, while $\{e_\alpha\}$ are orthogonal to TM (and span in fact $TM^\perp$, the normal bundle that will be defined afterward).

The existence of Darboux coframes along f can, of course, be proved analytically but the above geometric meaning is "evidence" of their existence. We let $\{\theta^a_b\}$ be the Levi-Civita connection forms relative to $\{\theta^a\}$ on N. Then $\{\theta^i_j\}$ on M are the Levi-Civita connection forms of the induced metric. Indeed, obviously,

$$\theta^i_j + \theta^j_i = 0 \quad \text{on } M. \tag{1.101}$$

Furthermore, from the first structure equations on N and by (1.99),

$$d\theta^i = -\theta^i_j \wedge \theta^j. \tag{1.102}$$

We now recall the following elementary but useful

Lemma 1.18 (Cartan's lemma). *Let $U \subset M$ be an open set of the Riemannian manifold $(M, \langle\,,\,\rangle)$. Let $\{\theta^i\}$ be a local basis of T^*U, and assume that a set of 1-forms $\{\omega^i_J\}$ on U, where J is any set of indices, satisfies $\sum_i \omega^i_J \wedge \theta^i = 0$. Then, there exist smooth functions $b^i_{J,k}$ on U such that*

$$\omega^i_J = b^i_{J,k}\theta^k \quad \text{and} \quad b^i_{J,k} = b^k_{J,i},$$

that is, the matrix $B = (b^i_{J,k})^i_k$ is an $n \times n$ symmetric matrix of 1-forms.

Proof. We can write ω^i_J as $\omega^i_J = b^i_{J,k}\theta^k$ for some smooth functions $b^i_{J,k}$ on U. Then from $\sum_i \omega^i_J \wedge \theta^i = 0$ we deduce

$$0 = \sum_{i,k} b^i_{J,k}\theta^k \wedge \theta^i = \sum_{i<k}(b^i_{J,k} - b^k_{J,i})\theta^k \wedge \theta^i,$$

which easily implies the thesis. $\square$

To obtain further information we differentiate equations (1.99), use (1.102) and (1.99) again to obtain

$$0 = d\theta^\alpha = -\theta^\alpha_i \wedge \theta^i - \theta^\alpha_\beta \wedge \theta^\beta = -\theta^\alpha_i \wedge \theta^i.$$

Hence, by Cartan's lemma 1.18, there exist (locally defined) smooth functions h^α_{ij} such that

$$\theta^\alpha_i = h^\alpha_{ij}\theta^j, \tag{1.103}$$

with

$$h^\alpha_{ij} = h^\alpha_{ji}. \tag{1.104}$$

The h^α_{ij} are the coefficients of the *second fundamental tensor* II (a tensor along f) of the immersion, which in the present setting is defined by

$$\text{II} = h^\alpha_{ij}\theta^i \otimes \theta^j \otimes e_\alpha. \tag{1.105}$$

One can verify that II is globally defined and symmetric. The *mean curvature vector field* is

$$H = \frac{1}{m}\mathrm{II}(e_i, e_i) = \frac{1}{m}h_{ii}^{\alpha}e_{\alpha}.$$

If ν is a globally defined unit normal vector field, we define the *mean curvature in the direction of ν* as

$$h^{\nu} = \langle H, \nu \rangle .$$

If $m + 1 = n$ and both the hypersurface M and N are orientable, we can choose Darboux frames along f *preserving orientations*, that is such that $\theta^1 \wedge \cdots \wedge \theta^{m+1}$ and $\theta^1 \wedge \cdots \wedge \theta^m$ give the correct orientations respectively of N and M. In this case, the vector field e_{m+1} dual to θ^{m+1} on N is a global normal vector field on M. The mean curvature in the direction of e_{m+1} is called the *mean curvature of the immersed hypersurface*. Moreover, we have the following definitions:

- if $\mathrm{II} \equiv 0$ on M, then the immersion is said to be *totally geodesic*;

- if $\mathrm{II} - H \langle \, , \, \rangle \equiv 0$ on M, then the immersion is said to be *totally umbilical*, and an *umbilical point* p is a point of M where $\mathrm{II}_p - H_p \langle \, , \, \rangle_p = 0$;

- if $H \equiv 0$ on M, then the immersion is said to be *minimal* (this last definition comes from the variational principle of the area functional).

On M we can consider the second structure equations

$$d\theta_j^i = -\theta_k^i \wedge \theta_j^k + \Omega_j^i \tag{1.106}$$

with Ω_j^i the curvature forms

$$\Omega_j^i = \frac{1}{2}\,^M\! R_{jkl}^i \theta^k \wedge \theta^l. \tag{1.107}$$

Our aim is to relate the curvature of M with that of N. We let

$$\Theta_b^a = \frac{1}{2}\,^N\! R_{bcd}^a \theta^c \wedge \theta^d$$

be the curvature forms on N. Pulling back the second structure equations of N and using (1.107), (1.103) we obtain

$$d\theta_j^i = -\theta_k^i \wedge \theta_j^k - \theta_{\alpha}^i \wedge \theta_j^{\alpha} + \Theta_j^i$$

$$= -\theta_k^i \wedge \theta_j^k + h_{ik}^{\alpha}h_{jl}^{\alpha}\theta^k \wedge \theta^l + \frac{1}{2}\,^N\! R_{jkl}^i \theta^k \wedge \theta^l.$$

Therefore, skew-symmetrizing in k and l,

$$\Omega_j^i = d\theta_j^i + \theta_k^i \wedge \theta_j^k = \frac{1}{2}\left(h_{ik}^{\alpha}h_{jl}^{\alpha} - h_{il}^{\alpha}h_{jk}^{\alpha} + \,^N\! R_{jkl}^i\right)\theta^k \wedge \theta^l.$$

From the above we thus obtain the *Gauss equations*

$$^M R^i_{jkl} = {}^N R^i_{jkl} + h^\alpha_{ik} h^\alpha_{jl} - h^\alpha_{il} h^\alpha_{jk}. \tag{1.108}$$

We now differentiate (1.103) and use (1.102) and the structure equations to obtain a classical set of commutation relations:

$$
\begin{aligned}
0 = d\theta^\alpha_i - d\big(h^\alpha_{ij}\theta^j\big) &= -\theta^\alpha_j \wedge \theta^j_i - \theta^\alpha_\gamma \wedge \theta^\gamma_i + \Theta^\alpha_i \\
&\quad - dh^\alpha_{ij} \wedge \theta^j + h^\alpha_{ij}\theta^j_k \wedge \theta^k + h^\alpha_{ij}\theta^j_\gamma \wedge \theta^\gamma \\
&= h^\alpha_{jk}\theta^j_i \wedge \theta^k - h^\gamma_{ik}\theta^\alpha_\gamma \wedge \theta^k - dh^\alpha_{ij} \wedge \theta^j + h^\alpha_{ij}\theta^j_k \wedge \theta^k \\
&\quad + \frac{1}{2}{}^N R^\alpha_{ijk}\theta^j \wedge \theta^k.
\end{aligned}
$$

Setting

$$h^\alpha_{ijk}\theta^k = dh^\alpha_{ij} - h^\alpha_{kj}\theta^k_i - h^\alpha_{ik}\theta^k_j + h^\beta_{ij}\theta^\alpha_\beta \tag{1.109}$$

the above rewrites as

$$\left(h^\alpha_{ijk}\theta^k + \frac{1}{2}{}^N R^\alpha_{ijk}\theta^k \right) \wedge \theta^j = 0.$$

Hence, by Cartan's lemma 1.18 there exist l^α_{ijk}, with

$$l^\alpha_{ijk} = l^\alpha_{ikj},$$

such that

$$h^\alpha_{ijk}\theta^k + \frac{1}{2}{}^N R^\alpha_{ijk}\theta^k = l^\alpha_{ijk}.$$

However,

$$h^\alpha_{ikj}\theta^k + \frac{1}{2}{}^N R^\alpha_{ikj}\theta^k = l^\alpha_{ikj};$$

thus, subtracting and recalling the symmetries of the curvature tensor,

$$h^\alpha_{ijk} - h^\alpha_{ikj} = -{}^N R^\alpha_{ijk}. \tag{1.110}$$

These commutation rules are known as *Codazzi equations*.

Note that the coefficients h^α_{ijk} defined in (1.109) are the coefficients of the covariant derivative of II.

Although for the sequel we will not use the Van der Waerden-Bortolotti covariant derivation, we briefly describe it in a few words in the above formalism. Given the immersion $f : M \to (N, \langle\,,\,\rangle)$, on M we have a well-defined bundle, the *normal bundle* $TM^\perp$ that pointwise is the orthogonal complement of $T_p M$ in $T_p N$. Given a Darboux coframe along f we locally define a covariant derivative by setting

$$D e_\alpha = \theta^\beta_\alpha \otimes e_\beta.$$

$\left\{\theta^{\alpha}_{\beta}\right\}$ are called the *connection forms* (one verifies that this definition is meaningful globally). We let the *curvature forms* Φ^{α}_{β} be defined *via* the second structure equations as follows:

$$d\theta^{\alpha}_{\beta} = -\theta^{\alpha}_{\gamma} \wedge \theta^{\gamma}_{\beta} + \Phi^{\alpha}_{\beta}, \tag{1.111}$$

and setting

$$\Phi^{\alpha}_{\beta} = \frac{1}{2}{}^{\perp}R^{\alpha}_{\beta ij}\theta^{i} \wedge \theta^{j};$$

the ${}^{\perp}R^{\alpha}_{\beta ij}$ are called the components of the *normal curvature tensor*. Comparing (1.111) with the pull-back of the second structure equations of N, that is,

$$d\theta^{\alpha}_{\beta} = -\theta^{\alpha}_{\gamma} \wedge \theta^{\gamma}_{\beta} - \theta^{\alpha}_{i} \wedge \theta^{i}_{\beta} + \Theta^{\alpha}_{\beta}$$

we deduce

$$\Phi^{\alpha}_{\beta} = \theta^{\alpha}_{i} \wedge \theta^{\beta}_{i} + \Theta^{\alpha}_{\beta}.$$

A simple computation as those presented above shows that

$$^{\perp}R^{\alpha}_{\beta ij} = h^{\alpha}_{ki}h^{\beta}_{kj} - h^{\alpha}_{kj}h^{\beta}_{ki} + {}^{N}R^{\alpha}_{\beta ij}. \tag{1.112}$$

These equations are often called the *Ricci equations.*

Chapter 2

Pointwise conformal metrics

At the beginning of this chapter we introduce the basic formalism and the derivation of the geometric Yamabe equation. Then, we concentrate on the case where M is compact to illustrate the interplay between geometry and analysis, with a few illuminating examples such as the Kazdan-Warner obstruction, a result of Obata on Einstein manifolds, the far-reaching "generalization" of Bidaut-Véron and Véron and a result of Escobar. Along the way we give a detailed proof, which inspires to P. Petersen's treatise [Pet06a], of a famous rigidity result of Obata. In this way, we hope to provide some geometrical feeling on the subject of this monograph that will enable us to proceed with the noncompact case: the case of the rest of our investigation.

2.1 The Yamabe equation

2.1.1 The derivation of the Yamabe equation

Let $(M, \langle\,,\,\rangle)$ be a Riemannian manifold and consider a pointwise conformal deformation of the metric $\langle\,,\,\rangle$, that is, a new metric on M of the form

$$\widetilde{\langle\,,\,\rangle} = \varphi^2 \langle\,,\,\rangle, \tag{2.1}$$

with φ a strictly positive smooth function. Denoting with $\widetilde{R}$ the curvature tensor of $\widetilde{\langle\,,\,\rangle}$, we want to determine the relationship between $\widetilde{R}$ and R. Let $\{\theta^i\}$, $i = 1, \ldots, m = \dim M$, be a local orthonormal coframe on $(M, \langle\,,\,\rangle)$ with corresponding Levi-Civita connection forms $\{\theta^i_j\}$. Then, in the new metric $\widetilde{\langle\,,\,\rangle}$,

$$\widetilde{\theta}^i = \varphi\theta^i, \quad i = 1, \ldots, m, \tag{2.2}$$

is a local orthonormal coframe on $\left(M, \widetilde{\langle\,,\,\rangle}\right)$. To determine the corresponding connection forms we could use the general theory developed in Chapter 1, but it is easy to deduce that, if $d\varphi = \varphi_t \theta^t$, the 1-forms

$$\widetilde{\theta}_j^i = \theta_j^i + \frac{\varphi_j}{\varphi}\theta^i - \frac{\varphi_i}{\varphi}\theta^j \tag{2.3}$$

are skew-symmetric and satisfy the first structure equation. Thus, they are the desired connection forms relative to the coframe defined in (2.2). In order to determine the curvature forms, we use the structure equations and the expression for the components of the Hessian (see equation (1.41)) to compute

$$\widetilde{\Theta}_j^i = d\widetilde{\theta}_j^i + \widetilde{\theta}_k^i \wedge \widetilde{\theta}_j^k$$

$$= d\theta_j^i + d\left(\frac{\varphi_j}{\varphi}\right)\wedge\theta^i + \frac{\varphi_j}{\varphi}d\theta^i - d\left(\frac{\varphi_i}{\varphi}\right)\wedge\theta^j - \frac{\varphi_i}{\varphi}d\theta^j + \widetilde{\theta}_k^i \wedge \widetilde{\theta}_j^k$$

$$= -\theta_k^i \wedge \theta_j^k + \Theta_j^i + (-\varphi^{-2}\varphi_k\varphi_j\theta^k + \varphi^{-1}d\varphi_j)\wedge\theta^i - \varphi^{-1}\varphi_j\theta_k^i\wedge\theta^k$$

$$\quad - (-\varphi^{-2}\varphi_k\varphi_i\theta^k + \varphi^{-1}d\varphi_i)\wedge\theta^j + \varphi^{-1}\varphi_i\theta_k^j\wedge\theta^k$$

$$\quad + \left(\theta_k^i + \frac{\varphi_k}{\varphi}\theta^i - \frac{\varphi_i}{\varphi}\theta^k\right)\wedge\left(\theta_j^k + \frac{\varphi_j}{\varphi}\theta^k - \frac{\varphi_k}{\varphi}\theta^j\right)$$

$$= \Theta_j^i + \left(\frac{\varphi_{jk}}{\varphi} - 2\frac{\varphi_j\varphi_k}{\varphi^2}\right)\theta^k\wedge\theta^i - \left(\frac{\varphi_{ik}}{\varphi} - 2\frac{\varphi_i\varphi_k}{\varphi^2}\right)\theta^k\wedge\theta^j - \frac{\varphi_k\varphi_k}{\varphi^2}\theta^i\wedge\theta^j,$$

that is,

$$\widetilde{\Theta}_j^i = \Theta_j^i + \left(\frac{\varphi_{jk}}{\varphi} - 2\frac{\varphi_k\varphi_j}{\varphi^2}\right)\delta_t^i\theta^k\wedge\theta^t - \left(\frac{\varphi_{ik}}{\varphi} - 2\frac{\varphi_i\varphi_k}{\varphi^2}\right)\delta_t^j\theta^k\wedge\theta^t$$

$$\quad - \frac{\varphi_l\varphi_l}{\varphi^2}\delta_k^i\delta_t^j\theta^k\wedge\theta^t.$$

Hence, anti-symmetrizing the coefficients of the wedge products on the right-hand side, and recalling the definition of the curvature tensor, we obtain

$$\varphi^2\widetilde{R}_{jkt}^i = R_{jkt}^i + \left(\frac{\varphi_{jk}}{\varphi} - 2\frac{\varphi_k\varphi_j}{\varphi^2}\right)\delta_t^i - \left(\frac{\varphi_{jt}}{\varphi} - 2\frac{\varphi_t\varphi_j}{\varphi^2}\right)\delta_k^i \tag{2.4}$$

$$\quad - \left(\frac{\varphi_{ik}}{\varphi} - 2\frac{\varphi_i\varphi_k}{\varphi^2}\right)\delta_t^j + \left(\frac{\varphi_{it}}{\varphi} - 2\frac{\varphi_i\varphi_t}{\varphi^2}\right)\delta_k^j$$

$$\quad - \frac{\varphi_l\varphi_l}{\varphi^2}\left(\delta_k^i\delta_t^j - \delta_t^i\delta_k^j\right).$$

Taking traces with respect to i and k we have

$$\varphi^2\widetilde{R}_{jt} = R_{jt} - (m-2)\frac{\varphi_{jt}}{\varphi} + 2(m-2)\frac{\varphi_j\varphi_t}{\varphi^2} - (m-3)\frac{\varphi_l\varphi_l}{\varphi^2}\delta_t^j - \frac{\varphi_{kk}}{\varphi}\delta_t^j. \tag{2.5}$$

Thus, denoting with $|\nabla\varphi|$, Hess(φ) and $\Delta\varphi$ respectively the length of the gradient, the Hessian and the Laplacian of φ in the metric $\langle\,,\,\rangle$, and recalling that R_{jt} (resp.

$\widetilde{R}_{jt})$ are the components of the Ricci tensor Ric with respect to the orthonormal basis θ^i (resp. of $\widetilde{\text{Ric}}$ with respect to $\widetilde{\theta}^i$), we have

$$\widetilde{\text{Ric}} = \text{Ric} - (m-2)\frac{1}{\varphi}\,\text{Hess}(\varphi) + 2(m-2)\frac{1}{\varphi^2}\,d\varphi \otimes d\varphi \qquad (2.6)$$

$$- (m-3)\frac{|\nabla\varphi|^2}{\varphi^2}\langle\,,\,\rangle - \frac{\Delta\varphi}{\varphi}\langle\,,\,\rangle.$$

A further tracing of (2.5) with respect to j and t yields

$$\varphi^2\widetilde{S} = S - 2(m-1)\frac{\Delta\varphi}{\varphi} - (m-1)(m-4)\frac{|\nabla\varphi|^2}{\varphi^2}. \qquad (2.7)$$

In case $m = \dim M \geq 3$, we set

$$\varphi = u^{\frac{2}{m-2}} \quad \text{so that} \quad \widetilde{\langle\,,\,\rangle} = u^{\frac{4}{m-2}}\langle\,,\,\rangle. \qquad (2.8)$$

In this case, (2.7) immediately gives

$$c_m\Delta u - Su + \widetilde{S}u^{\frac{m+2}{m-2}} = 0, \qquad (2.9)$$

with $c_m = 4\frac{m-1}{m-2}$. In case $m = 2$ we set

$$\varphi = e^u \quad \text{so that} \quad \widetilde{\langle\,,\,\rangle} = e^{2u}\langle\,,\,\rangle. \qquad (2.10)$$

In this case (2.7) gives

$$2\Delta u - S + \widetilde{S}e^{2u} = 0. \qquad (2.11)$$

Equations (2.9) and (2.11) are the classical *Yamabe equations*.

We conclude this section with an immediate application of (2.7) in the compact case, improving on a result of Obata, [Oba62b]. First we need the next simple

Lemma 2.1. *Let* $(M,\langle\,,\,\rangle)$ *be a compact Riemannian manifold. Then, every homothetic diffeomorphism is an isometry.*

Proof. Let $\varphi : M \to M$ be a diffeomorphism such that $\varphi^*\langle\,,\,\rangle = c^2\langle\,,\,\rangle$ for some constant $c > 0$. By contradiction, suppose $c \neq 1$. Without loss of generality, we can assume $c > 1$ (indeed, in case $c < 1$ it suffices to consider φ^{-1}). Now, for every $n \in \mathbb{N}$, let $\varphi^{(n)} : M \to M$ be the n-th iterate of φ. Then

$$\left(\varphi^{(n)}\right)^*\langle\,,\,\rangle = c^{2n}\langle\,,\,\rangle,$$

proving that, for any fixed $p \neq q \in M$,

$$c^{2n}d(p,q) = d\left(\varphi^{(n)}(p), \varphi^{(n)}(q)\right) \leq \text{diam}(M) < +\infty.$$

Since $c > 1$, taking the limit as $n \to +\infty$, we obtain the desired contradiction. $\square$

We are now ready to prove the following

Theorem 2.2. *Let* $(M, \langle\, ,\, \rangle)$ *be a compact manifold of dimension* $m \geq 3$ *(for the ease of exposition) with scalar curvature* $s(x) \leq 0$. *Let* $\psi : M \to M$ *be a conformal diffeomorphism with scalar curvature* $\widetilde{S}(x)$. *Thus* ψ *is an isometry if and only if* $\widetilde{S}(x) = kS(x)$ *for some* $k \in (0, +\infty)$.

Proof. If ψ is an isometry, clearly $k = 1$. *Vice versa*, to simplify notation let $c_m = 4\frac{m-1}{m-2}$, $\sigma = \frac{m+2}{m-2}$ and $k = a^{1-\sigma}$ for some $a \in (0, +\infty)$. Since ψ is conformal and $m \geq 3$, from equation (2.9) we deduce

$$c_m \Delta u = S(x)\left[u - a^{1-\sigma} u^{\sigma}\right] \tag{2.12}$$

with $u > 0$, $u \in C^{\infty}(M)$ such that

$$\psi^* \langle\, ,\, \rangle = u^{\frac{4}{m-2}} \langle\, ,\, \rangle.$$

Define the vector field

$$W = c_m \left[\left(\frac{u}{a}\right)^{\sigma-1} - 1\right] \nabla u;$$

computing its divergence, using (2.12) and the divergence theorem we have

$$\int_M \left[\left(\frac{u}{a}\right)^{\sigma-1} - 1\right]^2 S(x) u = c_m(\sigma - 1) \int_M \frac{1}{u}\left(\frac{u}{a}\right)^{\sigma-1} |\nabla u|^2.$$

Since $S(x) \leq 0$ it follows that $|\nabla u| \equiv 0$ and $u \equiv a$, that is $\psi : M \to M$ is a homothety. The result now follows from Lemma 2.1. $\qquad\square$

We shall come back to this kind of problems again in Corollary 2.9 below and later in the complete noncompact case.

2.1.2   The Kazdan-Warner obstruction

Let now T denote the traceless Ricci tensor, that is (see equation (1.38))

$$T = \mathrm{Ric} - \frac{S}{m} \langle\, ,\, \rangle.$$

T enables us to immediately find an obstruction to the existence of a conformally deformed metric as in (2.8) or (2.10) with assigned scalar curvature $\widetilde{S}(x)$. We shall consider the case $m \geq 3$ so that we shall provide a necessary condition for the existence of a positive solution on M of equation (2.9). Indeed, with respect to a local orthonormal coframe $\{\theta^i\}$ we have

$$T_{ij} = R_{ij} - \frac{S}{m}\delta_{ij}.$$

On the other hand, from (1.36),

$$R_{li,i} = \frac{1}{2}S_l.$$

Thus, tracing the covariant derivative

$$T_{ij,k} = R_{ij,k} - \frac{S_k}{m}\delta_{ij}$$

with respect to j and k yields

$$T_{ik,k} = R_{ik,k} - \frac{S_k}{m}\delta_{ik} = \frac{m-2}{2m}S_i. \tag{2.13}$$

Now, let X be a vector field on M and consider the vector field W associated to the 1-form $T(X,\cdot)$ using the duality induced by the metric, namely,

$$W = T(X,\cdot)^\sharp = X^i T_i^j e_j, \quad T_i^j = T_{ij}, \tag{2.14}$$

where $\{e_j\}$ is the orthonormal frame dual to the coframe $\{\theta^j\}$. Since the covariant derivative satisfies the Leibniz rule, the divergence of W is given by

$$\operatorname{div} W = W_k^k = X_k^i T_{ik} + X^i T_{ik,k}. \tag{2.15}$$

On the other hand, using the symmetry of T_{ik} and (1.20),

$$T_{ik}X_k^i = \frac{1}{2}\left(X_k^i + X_i^k\right)T_{ik} = \frac{1}{2}(\mathcal{L}_X\langle\,,\,\rangle)_{ik}T_{ik}.$$

It follows that

$$\operatorname{div} W = \frac{1}{2}(\mathcal{L}_X\langle\,,\,\rangle)_{ik}T_{ik} + \frac{m-2}{2m}X(S). \tag{2.16}$$

We now recall that a vector field X is said to be *conformal* if it generates a local 1-parameter group φ_t of local conformal diffeomorphisms, that is

$$(\varphi_t^*\langle\,,\,\rangle)_p = \rho_t(p)\langle\,,\,\rangle_p, \quad p \in M$$

for some positive function ρ_t. Using the formula which expresses the Lie derivative $\mathcal{L}_X$ as a derivative along the flow generated by X, namely,

$$\mathcal{L}_X\langle\,,\,\rangle = \lim_{t\to 0}\frac{\varphi_t^*\langle\,,\,\rangle - \langle\,,\,\rangle}{t},$$

(see e.g. [Lee03], p. 472 ff.) one proves that X is conformal if and only if

$$\mathcal{L}_X\langle\,,\,\rangle = \lambda\langle\,,\,\rangle$$

for some function $\lambda = \lambda(p)$, which, using (1.19), is easily seen to be equal to $\frac{2}{m}(\operatorname{div} X)$. Substituting into (2.16) and using the fact that T is traceless, we deduce that

$$\operatorname{div} W = \frac{m-2}{2m}X(S) \tag{2.17}$$

whenever X is a conformal vector field on M. Thus we have the next result which is known as the *Kazdan-Warner obstruction* (see [KW74a] and [BE87]):

Theorem 2.3. *Let* $(M, \langle\,,\,\rangle)$ *be a compact manifold of dimension* $m \geq 3$ *and* $\widetilde{\langle\,,\,\rangle}$ *a metric on* M *conformally related to* $\langle\,,\,\rangle$ *with scalar curvature* $\widetilde{S}$. *Then, for each conformal vector field* X *on* $(M, \langle\,,\,\rangle)$ *we have*

$$\int_M X(\widetilde{S})\, \widetilde{d\mathrm{vol}} = 0. \tag{2.18}$$

Proof. Since X is conformal with respect to the metric $\langle\,,\,\rangle$, it generates a flow of local diffeomorphisms which are conformal transformations for the metric $\langle\,,\,\rangle$ and therefore also for the conformally related metric $\widetilde{\langle\,,\,\rangle}$. It follows that X is conformal also in the metric $\widetilde{\langle\,,\,\rangle}$. Considering the analogous (2.17) in the metric $\widetilde{\langle\,,\,\rangle}$ and integrating with the aid of the divergence theorem we obtain (2.18). □

Remark. As a further application of the method of a moving frame, we give here an explicit formula for $\mathcal{L}_X\widetilde{\langle\,,\,\rangle}$ in terms of $\mathcal{L}_X\langle\,,\,\rangle$ and $\langle\,,\,\rangle$, from which it can be again deduced that a vector field X conformal w.r.t. the metric $\langle\,,\,\rangle$ is conformal also in the metric $\widetilde{\langle\,,\,\rangle}$. Indeed we have the following

Lemma 2.4. *Let* $X \in \mathfrak{X}(M)$ *be a vector field on the Riemannian manifold* $(M, \langle\,,\,\rangle)$, *and let* $\widetilde{\langle\,,\,\rangle} = \varphi^2\langle\,,\,\rangle$, $\varphi > 0$, *be a conformally deformed metric. Then*

$$\mathcal{L}_X\widetilde{\langle\,,\,\rangle} = \varphi^2 \mathcal{L}_X\langle\,,\,\rangle + 2\varphi\langle X, \nabla\varphi\rangle \langle\,,\,\rangle. \tag{2.19}$$

Proof. We choose a local o.n. coframe $\{\theta^i\}$ with associated frame $\{e_i\}$. First we observe that (2.2) implies that $\widetilde{e}_i = \varphi^{-1} e_i$; then, with the notation of section 1.1.2 we have

$$X = X^i e_i = \widetilde{X}^i \widetilde{e}_i \tag{2.20}$$

and

$$\nabla X = X^i_k \theta^k \otimes e_i, \quad \widetilde{\nabla} X = \widetilde{X}^i_k \widetilde{\theta}^k \otimes \widetilde{e}_i \tag{2.21}$$

with $X^i_k \theta^k = (dX^i + X^j \theta^i_j)$ and $\widetilde{X}^i_k \widetilde{\theta}^k = (d\widetilde{X}^i + \widetilde{X}^j \widetilde{\theta}^i_j)$. From (2.20) we deduce

$$\widetilde{X}^i = \varphi X^i, \tag{2.22}$$

while a computation using (2.21) and (2.3) shows that

$$\widetilde{X}^i_k = X^i_k + \varphi^{-1}\left(X^i \varphi_k + X^j \varphi_j \delta_{ik} - \varphi_i X^k\right), \tag{2.23}$$

which implies

$$\widetilde{X}^i_k + \widetilde{X}^i_k = X^i_k + X^i_k + 2\varphi^{-1} X^j \varphi_j \delta_{ik}. \tag{2.24}$$

Equation (2.19) now follows easily from (2.24) and (1.20). □

Example. The m-dimensional standard sphere $\mathbb{S}^m$, $m \geq 3$, can be realized, outside the North pole N and the South pole S, as the model manifold (see also Chapter 4, section 4.3)

$$\left((0,\pi) \times \mathbb{S}^{m-1}, dr \otimes dr + (\sin r)^2 d\theta^2 \right),$$

where $d\theta^2$ denotes the standard metric on $\mathbb{S}^{m-1}$. Using this isometric representation, a conformal vector field X on $\mathbb{S}^m$ is given by

$$X = \sin r \nabla r, \quad X_N = X_S = 0.$$

Now, consider any smooth, radial function $\tilde{s}(r(x))$ on $\mathbb{S}^m$. Then, by the co-area formula (1.87),

$$
\begin{aligned}
\int_{\mathbb{S}^m} X(\tilde{s}) \, \widetilde{dvol} &= \int_{\mathbb{S}^m} \left\langle X, \frac{d\tilde{s}}{dr} \nabla r \right\rangle \widetilde{dvol} \\
&= \int_0^\pi \left(\int_{\partial B_r} \frac{d\tilde{s}}{dr} \sin r \, \widetilde{dvol} \right) dr \\
&= \int_0^\pi \frac{d\tilde{s}}{dr} \sin r \, \widetilde{vol}(\partial B_r).
\end{aligned}
$$

Accordingly, if $\tilde{s}(r)$ is monotonic and nonconstant, we deduce

$$\int_{\mathbb{S}^m} X(\tilde{s}) \, \widetilde{dvol} \neq 0.$$

It follows from Theorem 2.3 that any nonconstant monotonic function $\tilde{s}(r(x))$ cannot be the scalar curvature of a pointwise conformal deformation of the canonical metric of $\mathbb{S}^m$.

2.1.3   The Weyl and Cotton tensors

We now establish a relation between

$$\tilde{T} = \widetilde{\mathrm{Ric}} - \frac{\tilde{S}}{m} \widetilde{\langle,\rangle} \quad \text{and} \quad T = \mathrm{Ric} - \frac{S}{m} \langle,\rangle, \tag{2.25}$$

that will be used below in the proof of Theorem 2.8. Suppose, once again, that (2.1) holds, that is,

$$\widetilde{\langle,\rangle} = \varphi^2 \langle,\rangle.$$

Using (2.6) and (2.7), after some computations we obtain

$$\widetilde{T} = \widetilde{\mathrm{Ric}} - \frac{\widetilde{S}}{m}\langle\,,\,\rangle$$

$$= T - (m-2)\left\{\frac{1}{\varphi}\,\mathrm{Hess}(\varphi) - 2\frac{1}{\varphi^2}\,d\varphi \otimes d\varphi\right\}$$

$$+ \frac{m-2}{m}\left\{\frac{\Delta\varphi}{\varphi} - 2\frac{|\nabla\varphi|^2}{\varphi^2}\right\}\langle\,,\,\rangle \qquad (2.26)$$

$$= T + (m-2)\varphi\left\{\mathrm{Hess}(\varphi^{-1}) - \frac{\Delta(\varphi^{-1})}{m}\langle\,,\,\rangle\right\}.$$

Next, we observe that using (2.4) and (2.5) we are able to detect a part of the curvature tensor which is naturally invariant with respect to a conformal change of the metric. Indeed, from (2.5) we have

$$(m-2)\left\{\frac{\varphi_{jt}}{\varphi} - 2\frac{\varphi_t\varphi_j}{\varphi^2}\right\} = R_{jt} - \varphi^2\widetilde{R}_{jt} - \left[(m-3)\frac{\varphi_l\varphi_l}{\varphi^2} + \frac{\varphi_{ll}}{\varphi}\right]\delta_t^j,$$

and inserting into (2.4) gives

$$\varphi^2\left[\widetilde{R}_{jkt}^i - \frac{1}{m-2}\left(\widetilde{R}_{ik}\delta_t^j - \widetilde{R}_{jk}\delta_t^i + \widetilde{R}_{jt}\delta_k^i - \widetilde{R}_{it}\delta_k^j\right)\right]$$

$$= R_{jkt}^i - \frac{1}{m-2}\left(R_{ik}\delta_t^j - R_{jk}\delta_t^i + R_{jt}\delta_k^i - R_{it}\delta_k^j\right)$$

$$+ \frac{1}{m-2}\left\{2\frac{\Delta\varphi}{\varphi} + (m-4)\frac{|\nabla\varphi|^2}{\varphi^2}\right\}\left(\delta_k^i\delta_t^j - \delta_t^i\delta_k^j\right).$$

On the other hand, by (2.7)

$$2\frac{\Delta\varphi}{\varphi} + (m-4)\frac{|\nabla\varphi|^2}{\varphi^2} = -\frac{1}{(m-1)}\left(\varphi^2\widetilde{S} - S\right)$$

and we obtain

$$\varphi^2\left[\widetilde{R}_{jkt}^i - \frac{1}{m-2}\left(\widetilde{R}_{ik}\delta_t^j - \widetilde{R}_{jk}\delta_t^i + \widetilde{R}_{jt}\delta_k^i - \widetilde{R}_{it}\delta_k^j\right)\right.$$

$$\left. + \frac{\widetilde{S}}{(m-1)(m-2)}\left(\delta_k^i\delta_t^j - \delta_t^i\delta_k^j\right)\right]$$

$$= R_{jkt}^i - \frac{1}{m-2}\left(R_{ik}\delta_t^j - R_{jk}\delta_t^i + R_{jt}\delta_k^i - R_{it}\delta_k^j\right)$$

$$+ \frac{S}{(m-1)(m-2)}\left(\delta_k^i\delta_t^j - \delta_t^i\delta_k^j\right).$$

It follows that the $(1,3)$-tensor, called the *Weyl tensor*, defined as

$$W = \mathrm{Riem} - \left[\frac{1}{m-2}\left(R_{ik}\delta_t^j - R_{jk}\delta_t^i + R_{jt}\delta_k^i - R_{it}\delta_k^j\right)\right. \qquad (2.27)$$

$$\left. - \frac{S}{(m-1)(m-2)}\left(\delta_k^i\delta_t^j - \delta_t^i\delta_k^j\right)\right]\theta^k \otimes \theta^t \otimes \theta^j \otimes e_i$$

is invariant under a conformal change of the metric.

Remark. It is worth noting that the corresponding $(0,4)$-version of W is *not* conformally invariant.

Taking covariant derivatives we obtain

$$W^i_{jks,t} = R^i_{jks,t} - \frac{1}{m-2}(R_{ik,t}\delta_{js} - R_{is,t}\delta_{jk} + R_{js,t}\delta_{ik} - R_{jk,t}\delta_{is})$$
$$+ \frac{S_t}{(m-1)(m-2)}(\delta_{ik}\delta_{js} - \delta_{is}\delta_{jk}),$$

so that taking the divergence with respect to the first index, that is, $W^t_{jks,t}$, using (1.35) and (1.36) we get

$$W^t_{jks,t} = R^t_{jks,t} - \frac{1}{m-2}R_{tk,t}\delta_{js} + \frac{1}{m-2}R_{ts,t}\delta_{jk} - \frac{1}{m-2}R_{js,k} + \frac{1}{m-2}R_{jk,s}$$
$$+ \frac{S_k}{(m-1)(m-2)}\delta_{sj} - \frac{S_s}{(m-1)(m-2)}\delta_{jk}$$
$$= -R_{jk,s} + R_{js,k} + \frac{1}{m-2}R_{jk,s} - \frac{1}{m-2}R_{js,k}$$
$$+ \frac{1}{m-2}\left(\frac{1}{m-1} - \frac{1}{2}\right)S_k\delta_{js} - \frac{1}{m-2}\left(\frac{1}{m-1} - \frac{1}{2}\right)S_s\delta_{jk}$$
$$= \frac{1-m+2}{m-2}R_{jk,s} + \frac{-1+m-2}{m-2}R_{js,k}$$
$$+ \frac{1}{m-2}\frac{3-m}{2(m-1)}S_k\delta_{js} - \frac{1}{m-2}\frac{3-m}{2(m-1)}S_s\delta_{jk}$$
$$= \frac{m-3}{m-2}C_{jsk},$$

where C_{jsk} are the components of the *Cotton tensor* C, i.e.,

$$C_{jsk} = R_{js,k} - R_{jk,s} + \frac{1}{2(m-1)}(S_s\delta_{jk} - S_k\delta_{js}).$$

Note that, if $m = 3$, then div W $\equiv 0$; moreover, it is possible to prove that if $m = 3$, then W $\equiv 0$ (see for instance [Eis49]).

The Cotton tensor can also be interpreted in the following way. Let

$$A = \text{Ric} - \frac{S}{2(m-1)}\langle \, , \, \rangle \qquad (2.28)$$

be the *Schouten tensor* of components

$$A_{ij} = R_{ij} - \frac{S}{2(m-1)}\delta_{ij}.$$

Clearly A is symmetric, hence taking covariant derivatives

$$A_{ij,k} = A_{ji,k},$$

but for the last two indices one immediately verifies that

$$A_{ij,k} - A_{ik,j} = C_{ijk}.$$

Hence we can think of the Cotton tensor as the obstruction for the Schouten tensor to be Codazzi; in other words, A is a Codazzi tensor if and only if the Cotton tensor C is identically null. Although we shall not make use of A in what follows, we recall that it enables to write, for $m \geq 3$, the decomposition of the curvature tensor: denoting again with W the $(0,4)$ version of the Weyl tensor we have

$$R = \mathrm{W} + \frac{1}{m-2} A \oslash \langle\,,\,\rangle,$$

where $\oslash$ is the *Kulkarni-Nomizu product*; for the definition of this latter and more information on the Schouten tensor we refer to A. Besse's treatise [Bes08].

The importance of the Weyl and the Cotton tensors will be emphasized in Theorem 2.6 below, which is a classical result due to Weyl (case $m \geq 4$) and to Schouten (case $m = 3$).

We first recall the following

Definition 2.5. *A Riemannian manifold $(M, \langle\,,\,\rangle)$ of dimension $m \geq 2$ is said to be* locally conformally flat *if, for every $p \in M$, there exist an open set $U \ni p$ and a function $\varphi \in C^\infty(U)$, $\varphi > 0$, such that the manifold $(U, \varphi^2 \langle\,,\,\rangle)$ is flat.*

Remark. Recall that, by the classical *Riemann theorem* (see for instance [Spi79]), the flat manifold $(U, \varphi^2 \langle\,,\,\rangle)$ is locally isometric to $\mathbb{R}^m$. It follows that M is locally conformally flat if each point $x \in M$ has a coordinate chart (U, ξ) such that $\xi^* \langle\,,\,\rangle_{\mathbb{R}^m}$ is pointwise conformally related to $\langle\,,\,\rangle$. Namely, $\xi : U \to \mathbb{R}^m$ is a conformal imbedding.

Obviuously, flat spaces are locally conformally flat. Here are some more interesting examples of both conformally flat and nonconformally flat manifolds.

Example. Every 2-dimensional Riemannian manifold is locally conformally flat. This fact is known as "the existence of isothermic coordinates" on any smooth surface and was established by Korn and Lichtenstein, [Kor14], [Lic16], assuming that the Riemannian metric at hand has Hölder continuous coefficients. For a conceptually easier proof we refer the reader to the paper [Che55] by Chern. The case of real analytic Riemannian metrics goes back to Gauss. Some regularity condition on the coefficients of the metric is needed as shown by Hartman and Wintner, [HW53]. Now, since every smooth manifold supports a smooth Riemannian metric, using isothermic coordinates one concludes that any orientable smooth surface possesses an underlying complex structure.

Example. The standard sphere $\mathbb{S}^m$ and the standard hyperbolic space $\mathbb{H}^m$ are locally conformally flat. As for the hyperbolic space, we are identifying $\mathbb{H}^m$ with either its Poincaré or its half-space model.

Example. An important class of m-dimensional, locally conformally flat manifolds are those admitting a conformal immersion into the standard sphere $\mathbb{S}^m$. In general, such an immersion does not exist. For instance, consider the flat (hence conformally flat) torus $\mathbb{T}^m$. By standard topological arguments, if $\mathbb{T}^m$ were immersed into $\mathbb{S}^m$, the immersion would be a covering map. Therefore, the fundamental group of $\mathbb{T}^m$ would inject into the fundamental group of $\mathbb{S}^m$, and this is clearly impossible. A general obstruction to the existence of conformal immersions into spheres is represented by the value of the *Yamabe invariant* of the manifold. Let $(M, \langle \, , \, \rangle)$ be a Riemannian manifold of dimension $m \geq 3$ with scalar curvature $S(x)$. The Yamabe invariant of M is the real constant

$$Y(M) = \inf_{v \in C_0^\infty(M) \backslash \{0\}} \frac{\int_M |\nabla v|^2 + \frac{m-2}{4(m-1)} S(x) v^2}{\left(\int_M v^{\frac{2m}{m-2}}\right)^{\frac{m-2}{m}}}.$$

Note that, in case $S(x) \equiv 0$, $Y(M)$ reduces to the ordinary Sobolev constant. By a result of Schoen and Yau, if M has a conformal immersion into $\mathbb{S}^m$, then $Y(M) = Y(\mathbb{S}^m) > 0$; see Chapter 6 in [SY94]. By way of example, it is a simple matter to verify that the Riemannian product $M = \mathbb{T}^{m-1} \times \mathbb{R}$ satisfies $Y(M) = 0$. Indeed, it is a complete, flat manifold with sub-linear volume growth.

Example. Suppose that $m - k \geq 2$. Then, it can be shown that the Riemannian product $\mathbb{S}^{m-k-1} \times \mathbb{H}^{k+1}$ is conformally diffeomorphic to $\mathbb{S}^m \backslash \mathbb{S}^k$, where $\mathbb{S}^k \subset \mathbb{S}^m$ is an equatorial k-sphere. In particular, $\mathbb{S}^{m-k-1} \times \mathbb{H}^{k+1}$ is locally conformally flat.

Example. If M has constant sectional curvature, then the Riemannian products $M \times \mathbb{R}$ and $M \times \mathbb{S}^1$ are locally conformally flat.

Example. In general, the Riemannian product of locally conformally flat manifolds is not locally conformally flat. For instance, the product of standard spheres $\mathbb{S}^m \times \mathbb{S}^m$, $m \geq 2$, is an example of a compact, simply connected, non-locally conformally flat manifold. A direct verification is possible. However, note that this follows directly from Kuiper's Theorem 2.7 below. Indeed, if $\mathbb{S}^m \times \mathbb{S}^m$ were locally conformally flat, by Theorem 2.7 it would be conformally diffeomorphic to the standard sphere $\mathbb{S}^{2m}$. In particular, the m-th de Rham cohomology groups would satisfy

$$H_{dR}^m(\mathbb{S}^m \times \mathbb{S}^m) \simeq H_{dR}^m(\mathbb{S}^{2m}),$$

but this is impossible. Indeed, for instance, a Mayer-Vietoris argument (see for example [Lee11]) shows that $H_{dR}^m(\mathbb{S}^{2m}) = 0$. On the other hand, by the Künneth formula, (see [GHV72])

$$H_{dR}^m(\mathbb{S}^m \times \mathbb{S}^m) \simeq \oplus_{k=0}^m H_{dR}^k(\mathbb{S}^m) \otimes H_{dR}^{m-k}(\mathbb{S}^m) \neq 0,$$

because, by Poincaré duality, $H_{dR}^m(\mathbb{S}^m) = H_{dR}^0(\mathbb{S}^m) = \mathbb{R}$.

One of the most important results in Riemann surfaces theory is the Riemann-Köbe uniformization theorem (see for instance [For91]), according to which
every simply connected Riemann surface is bi-holomorphic either to the complex
plane $\mathbb{C}$, to the open unit disk $\mathbb{B}^m \subset \mathbb{C}$ or to the Riemann sphere $\mathbb{S}^2$. Note that a
bi-holomorphism is a (orientation preserving) conformal diffeomorphism. There-
fore, recalling the existence theorem for isothermic coordinates alluded to above,
we conclude e.g. that every 2-dimensional, compact, simply connected (hence orientable) Riemannian manifold is conformally diffeomorphic to $\mathbb{S}^2$. In general, Riemannian manifolds of dimension $m \geq 3$ do not enjoy any such uniformization
property. An obstruction is given in the following (see [Eis49], [Bes08], [SY94])

Theorem 2.6. *Let $(M, \langle \, , \, \rangle)$ be a Riemannian manifold, $\dim M = m \geq 3$. A necessary and sufficient condition for $(M, \langle \, , \, \rangle)$ to be locally conformally flat is that*

$$\begin{cases} C \equiv 0 & \text{if } m = 3, \\ W \equiv 0 & \text{if } m \geq 4. \end{cases}$$

A complete classification of locally conformally flat manifolds is unknown. In
general, in higher dimensions, uniformization does require curvature restrictions.
However, in the compact setting, we have the following seminal result due to N.
Kuiper, [Kui49].

Theorem 2.7. *Let $(M, \langle \, , \, \rangle)$ be a compact, simply connected, locally conformally
flat manifold of dimension $m \geq 3$. Then, M is conformally diffeomorphic to the
standard sphere $\mathbb{S}^m$.*

Proof. The idea of the proof is as follows. Details can be found in Chapter 6 of
Schoen-Yau's book [SY94]. Since M is locally conformally flat, every point $x \in M$
has a neighborhood U_α conformally imbedded into $\mathbb{R}^m$, hence into $\mathbb{S}^m$. Let ξ_α be
such a conformal imbedding. Consider the couple (U_α, ξ_α). If (U_β, ξ_β) is a second
local conformal imbedding, the transition function $\xi_\alpha \circ \xi_\beta^{-1}$ is a local conformal
automorphism of $\mathbb{S}^m$. The classical Liouville theorem then shows that there exists
a global conformal transformation $\psi_{(\alpha,\beta)} \in \mathrm{Conf}\,(\mathbb{S}^m)$ such that

$$\psi_{(\alpha,\beta)} = \xi_\alpha \circ \xi_\beta^{-1}, \text{ on } \xi_\beta\,(U_\alpha \cap U_\beta).$$

Note that, with the obvious meaning of the symbols, the following cycle conditions
are satisfied:

$$\psi_{(\alpha,\beta)} \circ \psi_{(\beta,\alpha)} = id, \tag{2.29}$$
$$\psi_{(\alpha,\beta)} \circ \psi_{(\beta,\gamma)} \circ \psi_{(\gamma,\alpha)} = id.$$

We are now in the position to define a conformal immersion $\Phi : M \to \mathbb{S}^m$. Indeed,
let $x_0 \in M$ and let $(U_{\alpha_0}, \xi_{\alpha_0})$ be a fixed conformal imbedding with $x_0 \in U_{\alpha_0}$. For
every $x \in M$ let γ be a chosen path from $\gamma\,(0) = x_0$ to $\gamma\,(1) = x$ which is covered

by a finite chain of elements $(U_{\alpha_0}, \xi_{\alpha_0}), \ldots, (U_{\alpha_n}, \xi_{\alpha_n})$ such that $U_{\alpha_j} \cap U_{\alpha_{j+1}} \neq \emptyset$. We define

$$\Phi(x) = \xi_{\alpha_0}(x), \text{ on } U_{\alpha_0},$$

and

$$\Phi(x) = \psi_{(a_0, \alpha_1)} \circ \cdots \circ \psi_{(\alpha_{n-1}, \alpha_n)} \circ \xi_n(x), \text{ on } U_{\alpha_n}.$$

Since M is simply connected, using the cycle conditions (2.29) together with a monodromy argument shows that Φ is well defined and, by construction, Φ is a conformal immersion. It remains to show that Φ is a diffeomorphism. This follows from standard topological arguments. Indeed, since M is compact and $\mathbb{S}^m$ is connected, Φ is a covering map. But $\mathbb{S}^m$ is simply connected, hence Φ is a bijection. $\qquad\square$

2.2 Some applications in the compact case

2.2.1 A rigidity result of Obata

We now prove a rigidity result for compact Einstein manifolds due to M. Obata, [Oba62a]. Its proof relies on the transformation laws (2.26) and on a rigidity result for complete Riemannian manifolds supporting nontrivial solutions of certain differential inequalities; see Theorem 2.10 below.

Theorem 2.8. *Let $(M, \langle\,,\,\rangle)$ be a compact, Einstein manifold of dimension $m \geq 3$ and let $\widetilde{\langle\,,\,\rangle}$ be a pointwise conformal deformation of $\langle\,,\,\rangle$. If $(M, \widetilde{\langle\,,\,\rangle})$ has constant scalar curvature $\widetilde{S}$, then $(M, \widetilde{\langle\,,\,\rangle})$ is Einstein. Furthermore, if $(M, \widetilde{\langle\,,\,\rangle})$ is not conformally diffeomorphic to the standard sphere $\mathbb{S}^m$, then $\widetilde{\langle\,,\,\rangle} = c^2 \langle\,,\,\rangle$, for some constant $c > 0$.*

Remark. This result has been generalized by J. Escobar, [Esc90], to the case where M has a nonempty boundary $\partial M \neq \emptyset$. In this situation, one also requires that the inclusion $\imath : \partial M \hookrightarrow (M, \langle\,,\,\rangle)$ is totally geodesic and that $\imath : \partial M \hookrightarrow (M, \widetilde{\langle\,,\,\rangle})$ is minimal. Thus, in the conclusion of the theorem, the standard sphere $\mathbb{S}^m$ is replaced by the standard hemisphere $\mathbb{S}^m_+$. Escobar's proof follows closely Obata's original argument. We shall limit ourselves to proving Theorem 2.8, since in any case Escobar's theorem will result as a consequence of Theorem 2.16 of section 2.2.3.

Before proving the theorem we point out the following simple, interesting, consequence, to be compared with Theorem 2.2 . Later on, in the setting of complete, non-Einstein manifolds, we shall give a version of the next result assuming that the conformal diffeomorphism at hand preserves the scalar curvature (see Theorem 5.9).

Corollary 2.9. *Let $(M, \langle\,,\,\rangle)$ be a compact Einstein manifold of dimension $m \geq 3$ which is not conformally diffeomorphic to $\mathbb{S}^m$. Let $\varphi \in \mathrm{Conf}(M)$ be a conformal* *diffeomorphism such that* $\varphi^* \langle \, , \, \rangle$ *has constant scalar curvature. Then* $\varphi \in \mathrm{Iso}\,(M)$, *that is, it is a Riemannian isometry.*

Proof. By Theorem 2.8, $\varphi^* \langle \, , \, \rangle = c^2 \langle \, , \, \rangle$, for some constant $c > 0$. Therefore, the result follows from Lemma 2.1. $\qquad\qquad\qquad\qquad\qquad\qquad\qquad\qquad\qquad\qquad$ $\square$

Now we give the

Proof of Theorem 2.8. For the sake of convenience, we set $\widetilde{\langle \, , \, \rangle} = \varphi^{-2} \langle \, , \, \rangle$ and denote with $\sim$ quantities that refer to the metric $\widetilde{\langle \, , \, \rangle}$. According to (2.26),

$$T = \widetilde{T} + (m-2)\,\varphi \left\{ \widetilde{\mathrm{Hess}}(\varphi^{-1}) - \frac{\widetilde{\Delta}\,(\varphi^{-1})}{m} \widetilde{\langle \, , \, \rangle} \right\}.$$

Since $\langle \, , \, \rangle$ is Einstein we have $T \equiv 0$, and we deduce that, in components with respect to a local orthonormal coframe $\left\{ \widetilde{\theta}^j \right\}$ for $\widetilde{\langle \, , \, \rangle}$,

$$-\widetilde{T}_{ij} = (m-2)\,\varphi \left\{ \left(\varphi^{-1}\right)_{ij} - \frac{\widetilde{\Delta}\,(\varphi^{-1})}{m}\delta_{ij} \right\}. \tag{2.30}$$

Since $\widetilde{T}$ is a traceless tensor,

$$\varphi^{-1} \left|\widetilde{T}\right|^2_{\widetilde{\langle \, , \, \rangle}} = -\,(m-2)\left(\varphi^{-1}\right)_{ij}\widetilde{T}_{ij}.$$

On the other hand, according to equation (2.13),

$$\widetilde{T}_{ik,k} = \frac{m-2}{2m}\widetilde{S}_i$$

and since $\widetilde{S}$ is constant, we obtain

$$\widetilde{T}_{ik,k} = 0. \tag{2.31}$$

Thus, if we let $\left\{ \widetilde{e}_j \right\}$ be the dual frame of $\left\{ \widetilde{\theta}^j \right\}$, and define the vector field $\widetilde{W}$ by the formula

$$\widetilde{W} = \left(\varphi^{-1}\right)_i \widetilde{T}_{ij}\widetilde{e}_j, \tag{2.32}$$

where $d\left(\varphi^{-1}\right) = \left(\varphi^{-1}\right)_i \widetilde{\theta}^i$, a computation that uses (2.31) shows that

$$\widetilde{\mathrm{div}}\widetilde{W} = \left(\varphi^{-1}\right)_{ij}\widetilde{T}_{ij} + \left(\varphi^{-1}\right)_i \widetilde{T}_{ik,k} = \left(\varphi^{-1}\right)_{ij}\widetilde{T}_{ij},$$

and we conclude that

$$\varphi^{-1} \left|\widetilde{T}\right|^2_{\widetilde{\langle \, , \, \rangle}} = -\,(m-2)\,\widetilde{\mathrm{div}}\widetilde{W}.$$

Integrating on $\left(M, \widetilde{\langle\, , \rangle}\right)$ and using the divergence theorem we deduce that

$$\int_M \varphi^{-1} \left|\widetilde{T}\right|^2_{\widetilde{\langle\, , \rangle}} \widetilde{dvol} = 0.$$

Accordingly, $\widetilde{T} \equiv 0$ and $\left(M, \widetilde{\langle\, , \rangle}\right)$ is an Einstein manifold.

Suppose now that the m-dimensional manifold $(M, \langle\, , \rangle)$ is not conformally diffeomorphic to $\mathbb{S}^m$, $m \geq 3$. Having set as usual $\widetilde{\langle\, , \rangle} = u^{\frac{4}{m-2}} \langle\, , \rangle$, we shall prove that u is identically equal to a positive constant c^2. Let S be the the scalar curvature of $\langle\, , \rangle$, which is constant since $(M, \langle\, , \rangle)$ is Einstein. We first claim that either $S = \widetilde{S} = 0$ or $S \cdot \widetilde{S} > 0$. To see this, we recall that S and $\widetilde{S}$ are related by the Yamabe equation

$$c_m \Delta u - Su + \widetilde{S} u^{\frac{m+2}{m-2}} = 0, \quad c_m = 4\frac{m-1}{m-2}.$$

Suppose that $S \leq 0$ and $\widetilde{S} \geq 0$. Then,

$$\Delta u \geq 0,$$

and since $(M, \langle\, , \rangle)$ is compact we deduce that u is a positive constant. Therefore, inserting this information into the Yamabe equation, we conclude that

$$0 \geq S = \widetilde{S} u^{\frac{4}{m-2}} \geq 0,$$

proving that $S = \widetilde{S} = 0$, and the claim follows. The same conclusion holds if we assume instead that $S \geq 0$ and $\widetilde{S} \leq 0$. Thus, we need to consider three possible cases.

First case: $S = \widetilde{S} = 0$. Using the Yamabe equation once more gives $\Delta u = 0$ and, therefore, u is a positive constant.

Second case: $S < 0$ and $\widetilde{S} < 0$. Up to rescaling $\widetilde{\langle\, , \rangle}$ by a positive constant, we can assume that $S = \widetilde{S} < 0$. Let

$$u(x_0) = \max_M u, \quad u(x_1) = \min_M u.$$

Then, by the usual maximum principle, $\Delta u(x_0) \leq 0$ which implies, according to the Yamabe equation,

$$Su(x_0) \left(1 - u^{\frac{4}{m-2}}(x_0)\right) \leq 0.$$

As a consequence,

$$u(x) \leq u(x_0) \leq 1.$$

Similarly, since u achieves its minimum at x_1 we have $\Delta u(x_1) \geq 0$. This latter, in turn, implies

$$1 \leq u(x_1) \leq u(x),$$

and, therefore, $u \equiv 1$.

Third case: $S > 0$ and $\tilde{S} > 0$. Let us observe that, since $\widetilde{\langle \, , \, \rangle} = u^{\frac{4}{m-2}} \langle \, , \, \rangle$ and both the metrics are Einstein, setting

$$v = u^{-\frac{2}{m-2}},$$

by (2.26) we have

$$\mathrm{Hess}(v) - \frac{\Delta v}{m} \langle \, , \, \rangle = 0 \quad \text{on } M. \tag{2.33}$$

We define the vector field $X = \nabla v$ so that $\mathrm{div}\, X = \Delta v$ and

$$\frac{1}{2}\mathcal{L}_X \langle \, , \, \rangle = \mathrm{Hess}(v).$$

From the above, we deduce

$$\frac{1}{2}\mathcal{L}_X \langle \, , \, \rangle = \frac{\mathrm{div}\, X}{m} \langle \, , \, \rangle,$$

that is, X is a conformal vector field. We now show, using again the moving frame formalism, that since M is Einstein, $f = \mathrm{div}\, X = \Delta v$ satisfies the equation

$$\mathrm{Hess}(f) + \frac{S}{m\,(m-1)}f \langle \, , \, \rangle = 0, \tag{2.34}$$

where, we recall, $s > 0$. First we observe that the previous equation can be rewritten as

$$\mathrm{Hess}(\Delta v) + \Delta v \frac{S}{m\,(m-1)} \langle \, , \, \rangle = 0,$$

or, in components,

$$(\Delta v)_{kt} = v_{iikt} = -\Delta v \frac{S}{m\,(m-1)}\delta_{kt};$$

then we note that, by (2.33), we have

$$v_{ij} = \frac{v_{tt}}{m}\delta_{ij}, \tag{2.35}$$

from which we deduce

$$v_{ijk} = \frac{v_{ttk}}{m}\delta_{ij} \tag{2.36}$$

and

$$v_{ijkl} = \frac{v_{ttkl}}{m}\delta_{ij}. \tag{2.37}$$

By (1.54) applied to v we deduce

$$v_{iikl} = (\Delta v)_{kl} = v_{ikil} + v_{sl}R_{siik} + v_s R_{siik,l}$$
$$= v_{kiil} - v_{sl}R_{sk} - v_s R_{sk,l} = v_{kiil} - v_{sl}R_{sk},$$

since $R_{sk,l} = 0$. Then, using (2.35), (2.36) and (2.37) we obtain

$$(\Delta v)_{kl} = v_{kiil} - \frac{v_{ii}}{m}\delta_{sl}R_{sk} = \frac{v_{ssil}\delta_{ki}}{m} - \frac{v_{ii}}{m}R_{kl},$$

which easily implies (2.34).

In case f is constant, we get $f = \Delta v = 0$ and, since M is compact, v itself must be constant. This implies that also u is constant, as desired. Finally, in case f is nonconstant, we can apply the next result to conclude that $(M, \langle\, ,\, \rangle)$ is isometric to the sphere $\mathbb{S}_{k^2}^m$ of constant curvature $k^2 = S/m\,(m-1)$. But this implies that $(M, \langle\, ,\, \rangle)$ is conformally diffeomorphic to the standard sphere $\mathbb{S}^m$, against our initial assumption. □

The following result is again due to Obata, [Oba62a] (for the analogous statement for $\mathbb{S}_+^m$, the upper hemisphere, see Theorem 2.14 below); see also [PR].

Theorem 2.10. *Let $(M, \langle\, ,\, \rangle)$ be a complete Riemannian manifold of dimension m. Suppose that there exists a nonconstant, smooth function $f : M \to \mathbb{R}$ such that, for some constant $k > 0$,*

$$\mathrm{Hess}(f) + k^2 f \langle\, ,\, \rangle = 0. \tag{2.38}$$

Then $(M, \langle\, ,\, \rangle)$ is isometric to the sphere $S_{k^2}^m$ of constant sectional curvature k^2.

Proof. First, we show that f has a critical point. Indeed, by contradiction, suppose that $\nabla f \neq 0$ on M. Consider the smooth vector field $X = \nabla f/|\nabla f|$ on M. Since $|X| \in L^\infty(M)$ and $(M, \langle\, ,\, \rangle)$ is geodesically complete, then X is complete (see e.g. [Lee03], Chapter 12). Let $\gamma : \mathbb{R} \to M$ be an integral curve of X, namely $X_{\gamma(s)} = \dot{\gamma}(s)$, for every $s \in \mathbb{R}$. A direct computation that uses (2.38) shows that, for every vector field Y,

$$\langle D_{\dot{\gamma}}\dot{\gamma}, Y \rangle = \frac{1}{|\nabla f|}\mathrm{Hess}\,(f)\,(\dot{\gamma}, Y) - \frac{1}{|\nabla f|}\mathrm{Hess}\,(f)\,(\dot{\gamma}, \dot{\gamma})\,\langle \dot{\gamma}, Y \rangle = 0.$$

Therefore, γ is a unit speed geodesic of M. Evaluating (2.38) along γ we deduce that the function $y(s) = f \circ \gamma(s)$ satisfies

$$y'' = -k^2 y$$

which is oscillatory. Let $s_0 \in \mathbb{R}$ be such that $y'(s_0) = 0$. Then, recalling that γ is an integral curve of X, we conclude

$$0 = y'(s_0) = \langle \nabla f(\gamma(s_0)), \dot{\gamma}(s_0) \rangle = |\nabla f(\gamma(s_0))| \neq 0,$$

a contradiction.

Let $o \in M$ be a critical point of f and set $r(x) = d(x, o)$. Note that $f(o) \neq 0$ for otherwise, having fixed any unit speed geodesic γ issuing from o, we would have that $y(s) = f \circ \gamma(s)$ solves the Cauchy problem

$$\begin{cases} y'' = -k^2 y, \\ y(0) = 0, \\ y'(0) = 0, \end{cases}$$

and, hence, $y(s) \equiv 0$. Since this would be true for every geodesic γ we should conclude that $f \equiv 0$, a contradiction. Thus, without loss of generality, we assume that $f(o) = 1$. We claim that, for every $x \in M$, it holds that

$$f(x) = \cos(kr(x)). \tag{2.39}$$

Indeed, consider a unit speed, minimizing geodesic $\gamma : [0, r(x)] \to M$ from $\gamma(0) = o$ to $\gamma(r(x)) = x$. As noted above, the smooth function $y(s) = f \circ \gamma(s)$ is the solution of the Cauchy problem

$$\begin{cases} y'' = -k^2 y, \\ y(0) = 1, \\ y'(0) = 0. \end{cases}$$

Therefore, $y(s) = \cos(ks)$. Evaluating at $s = r(x)$ we conclude the validity of (2.39). Now, observe that the function $\cos(ks)$ is strictly decreasing on $(0, \pi/k)$. It follows from (2.39) that

$$r(x) = k^{-1} \arccos(f(x))$$

is smooth on the geodesic ball $B_{\pi/k}(o) \setminus \{o\}$. Applying the Bishop density result of Chapter 1 we therefore conclude that

$$cut(o) \cap B_{\pi/k}(o) = \emptyset. \tag{2.40}$$

Therefore, the exponential map $\exp_o : \mathbb{B}^m_{\pi/k}(0) \subset T_o M \to B_{\pi/k}(o)$ is a diffeomorphism. Let us introduce geodesic polar coordinates (r, θ) on $T_o M$. Furthermore, let us consider a local orthonormal coframe $\{\theta^\alpha\}$ on $\mathbb{S}^{m-1}$ with dual frame $\{E_\alpha\}$. Thus, the standard metric of $\mathbb{S}^{m-1}$ is written as $d\theta^2 = \sum \theta^\alpha \otimes \theta^\alpha$. We extend both $\{\theta^\alpha\}$ and $\{E_\alpha\}$ radially. Then, by Gauss' lemma,

$$\langle \, , \, \rangle = dr \otimes dr + \sigma_{\alpha\beta}(r, \theta) \, \theta^\alpha \otimes \theta^\beta.$$

Furthermore, since $\langle \, , \, \rangle$ is infinitesimally Euclidean and the standard metric of $\mathbb{R}^m \approx T_o M$ is written as

$$\langle \, , \, \rangle_{\mathbb{R}^m} = dr \otimes dr + r^2 \delta_{\alpha\beta} \theta^\alpha \otimes \theta^\beta,$$

we have the further condition

$$\sigma_{\alpha\beta}(r, \theta) = \delta_{\alpha\beta} r^2 + o(r^2), \text{ as } r \searrow 0. \tag{2.41}$$

Now we use the fact that

$$\mathcal{L}_{\nabla r} \langle \, , \, \rangle = 2\mathrm{Hess}\,(r)\,, \text{ on } B_{\pi/k}\,(o) \setminus \{o\}$$

where $\mathcal{L}_{\nabla r}$ is the Lie derivative in the radial direction ∇r. Thus, by the definition of Lie derivative we deduce that

$$2\,\mathrm{Hess}(r)(E_\alpha, E_\beta) = \mathcal{L}_{\nabla r} \langle \, , \, \rangle (E_\alpha, E_\beta) = \nabla r(\sigma_{\alpha\beta}). \tag{2.42}$$

Since

$$\nabla r = -\frac{\nabla f}{|\nabla f|},$$

the Hessian of the distance function r can be expressed as

$$\begin{aligned}
\mathrm{Hess}\,(r)\,(E_\alpha, E_\beta) &= -\left\langle D_{E_\alpha} \frac{\nabla f}{|\nabla f|}, E_\beta \right\rangle \\
&= -\frac{1}{|\nabla f|}\mathrm{Hess}\,(f)\,(E_\alpha, E_\beta) \\
&= \frac{k^2 \cos\,(kr)}{k \sin\,(kr)}\,\langle E_\alpha, E_\beta \rangle \\
&= k \cot\,(kr)\,\sigma_{\alpha\beta}.
\end{aligned}$$

Inserting into (2.42), and recalling (2.41) we obtain that $\sigma_{\alpha\beta}$ are the (unique) solutions of the asymptotic Cauchy problems

$$\begin{cases} \partial_r \sigma_{\alpha\beta} = 2k \cot\,(kr)\,\sigma_{\alpha\beta}, \text{ on } B_{\pi/k}\,(o) \setminus \{o\}\,, \\ \sigma_{\alpha\beta}\,(r, \theta) = \delta_{\alpha\beta} r^2 + o\,(r^2)\,, \text{ as } r \searrow 0. \end{cases}$$

Integrating, finally gives

$$\sigma_{\alpha\beta} = k^{-2} \sin^2\,(kr)\,\delta_{\alpha\beta}.$$

We have thus established that, in polar coordinates of $\mathbb{B}^m_{\pi/k}\,(0) \setminus \{o\} = (0, \pi/k) \times \mathbb{S}^{m-1} \subset T_o M$, it holds that

$$\langle \, , \, \rangle = dr \otimes dr + k^{-2} \sin^2\,(kr)\,d\theta^2.$$

Since

$$\left((0, \pi/k) \times \mathbb{S}^{m-1}, dr \otimes dr + k^{-2} \sin^2\,(kr)\,d\theta^2\right) \tag{2.43}$$

is isometric to the m-dimensional 2-punctured sphere $\mathbb{S}^m_{k^2} \setminus \{2 \text{ points}\}$ of constant curvature k^2, we conclude that the geodesic ball $B_{\pi/k}\,(o) \subset M$ is isometric to $\mathbb{S}^m_{k^2} \setminus \{\text{point}\}$. To complete the proof it suffices to show that

$$B_{\pi/k}\,(o) = M \setminus \{\text{point}\}. \tag{2.44}$$

Indeed, suppose we have already proved this fact. Then, by Seifert-Van Kampen's theorem, (see [Lee11]) M is simply connected. Moreover, $M \setminus \{\text{point}\}$ has constant curvature k^2 because it is isometric to $S_{k^2}^m \setminus \{\text{point}\}$. By continuity, M itself has constant curvature k^2. Therefore, the Hopf classification theorem (see [Pet06b]) tells us that M must be isometric to $S_{k^2}^m$, as desired. The proof of (2.44) combines Morse theoretic and cut-locus arguments. First of all, we observe that $\partial B_{\pi/k}(o)$ is made up entirely by nondegenerate critical points of f. Therefore, by Morse's lemma, (see [Mil63], [Pet06b]) $\partial B_{\pi/k}(o)$ is a discrete, compact (hence finite) set. Indeed, for every $x \in \partial B_{\pi/k}(o)$, by continuity we have

$$|\nabla f|(x) = -k \sin(\pi) = 0,$$

and, by assumption,

$$\text{Hess}(f)(x) = -k^2 f(x) \langle, \rangle_x,$$

where, $f(x) = \cos(\pi) = -1$. In particular, Hess(f) is (strictly) negative definite on $\partial B_{\pi/k}(o)$, as claimed. To conclude, we note that $\partial B_{\pi/k}(o)$ is connected. In fact,

$$\partial B_{\pi/k}(o) = \exp_o\left(\partial \mathbb{B}_{\pi/k}^m(0)\right). \tag{2.45}$$

To see this, we recall that

$$\exp_o\left(\partial \mathbb{B}_{\pi/k}^m(0) \cap T_o M \setminus c(o)\right) = \partial B_{\pi/k}(o) \cap M \setminus cut(o), \tag{2.46}$$

where $c(o) \subset T_o M$ is the tangential cut-locus of $o \in M$. Let $\bar{v} \in \partial \mathbb{B}_{\pi/k}^m(0) \cap c(o)$. By definition, $\bar{x} = \exp_o \bar{v} \in cut(o)$. Since, by (2.40),

$$B_{\pi/k}(o) \subset M \setminus cut(o),$$

and, on a generic complete Riemannian manifold,

$$\exp_o\left(\overline{\mathbb{B}_{\pi/k}^m(0)}\right) \subseteq \overline{B_{\pi/k}(o)},$$

we must conclude that $\bar{x} \notin B_{\pi/k}(o)$, that is,

$$\exp_o\left(\partial \mathbb{B}_{\pi/k}^m(0) \cap c(o)\right) \subseteq \partial B_{\pi/k}(o) \cap cut(o).$$

Conversely, let $x \in \partial B_{\pi/k}(o) \cap cut(o)$. By definition, there is a unit speed geodesic $\gamma(t) = \exp_o(vt) : [0, +\infty) \to M$ from $\gamma(0) = o$ to $\gamma(\pi/k) = x$ that does not minimize distances past π/k. Then, $v\pi/k \in \partial \mathbb{B}_{\pi/k}^m(0) \cap c(o)$ and $\exp_o(v\pi/k) = x$. Summarizing,

$$\exp_o\left(\partial \mathbb{B}_{\pi/k}^m(0) \cap c(o)\right) = \partial B_{\pi/k}(o) \cap cut(o). \tag{2.47}$$

From (2.46) and (2.47) we conclude the validity of (2.45). $\qquad\qquad\square$

2.2.2 A result by M. F. Bidaut-Véron and L. Véron

The vector field $\widetilde{W}$ defined in (2.32), with $\widetilde{T}_{ij}$ given in (2.30), will suggest to us how to proceed to provide a proof of Theorem 2.12 below. First we give an intrinsic definition: by direct manipulation of the quantities involved we have

$$
\begin{aligned}
\widetilde{W} &= (\varphi^{-1})_i\left\{-(\varphi^{-1})_{ij} + \frac{(\varphi^{-1})_{kk}}{m}\delta_{ij}\right\}(m-2)\varphi\tilde{e}_j \\
&= \frac{m-2}{m}\varphi(\varphi^{-1})_{kk}(\varphi^{-1})_i\tilde{e}_i - (m-2)\varphi(\varphi^{-1})_{ij}(\varphi^{-1})_i\tilde{e}_j \\
&= \frac{m-2}{m}\varphi(\widetilde{\Delta}\varphi^{-1})\widetilde{\nabla}\varphi^{-1} - (m-2)\varphi\widetilde{\mathrm{Hess}}(\varphi^{-1})\left(\widetilde{\nabla}\varphi^{-1},\cdot\right)^\sharp \\
&= (m-2)\varphi\left\{\frac{\widetilde{\Delta}\varphi^{-1}}{m}\widetilde{\nabla}\varphi^{-1} - \widetilde{\mathrm{Hess}}(\varphi^{-1})\left(\widetilde{\nabla}\varphi^{-1},\right)^\sharp\right\} \\
&= (m-2)\varphi\left\{\frac{\widetilde{\Delta}\varphi^{-1}}{m}\widetilde{\nabla}\varphi^{-1} - \frac{1}{2}\widetilde{\nabla}\left|\widetilde{\nabla}\varphi^{-1}\right|^2_{\widetilde{(\,,\,)}}\right\}.
\end{aligned}
$$

We set the following general

Definition 2.11. *Given $u \in C^2(M)$, $u > 0$, the vector field*

$$
Z = u^\alpha\left\{\frac{1}{2}\nabla|\nabla u|^2 - \frac{\Delta u}{m}\nabla u\right\}, \qquad \alpha \in \mathbb{R}, \tag{2.48}
$$

is called an Obata type vector field.

Thus $\widetilde{W}$ above is an Obata vector field (modulo the multiplicative constant $2-m$) with respect to the metric $\langle\,,\,\rangle$. Furthermore, the function involving φ satisfies equation (2.7) with the roles of $\langle\,,\,\rangle$ and $\widetilde{\langle\,,\,\rangle}$ interchanged, that is

$$
\varphi^2 S = \widetilde{S} - 2(m-1)\frac{\widetilde{\Delta}\varphi}{\varphi} - (m-1)(m-4)\frac{\left|\widetilde{\nabla}\varphi\right|^2_{\widetilde{(\,,\,)}}}{\varphi^2}.
$$

This, in turn, implies

$$
\begin{aligned}
\widetilde{\Delta}\varphi^{-1} &= -\frac{1}{\varphi^2}\widetilde{\Delta}\varphi + \frac{2}{\varphi^3}\left|\widetilde{\nabla}\varphi\right|^2_{\widetilde{(\,,\,)}} \\
&= \frac{S}{2(m-1)}\varphi - \frac{\widetilde{S}}{2(m-1)}\varphi^{-1} + \frac{m}{2}\frac{\left|\widetilde{\nabla}\varphi\right|^2_{\widetilde{(\,,\,)}}}{\varphi^3} \\
&= -\frac{\widetilde{S}}{2(m-1)}\varphi^{-1} + \frac{S}{2(m-1)}(\varphi^{-1})^{-1} + \frac{m}{2}\frac{\left|\widetilde{\nabla}\varphi^{-1}\right|^2_{\widetilde{(\,,\,)}}}{\varphi^{-1}}.
\end{aligned}
$$

Therefore, $u = \varphi^{-1}$ is a positive solution of the differential equation

$$\widetilde{\Delta} u = \frac{m}{2} \frac{\left|\widetilde{\nabla} u\right|^2_{\widetilde{\langle , \rangle}}}{u} + \frac{S}{2(m-1)} \left(u^{-1} - \frac{\widetilde{S}}{S} u \right).$$

We have thus obtained the further suggestion to consider the vector field Z defined in (2.48), with u a positive solution of a differential equation of the type

$$\Delta u = (\beta+1) \frac{|\nabla u|^2}{u} + \frac{1}{\beta} \left(u^{1+\beta(1-\sigma)} - \lambda u \right), \quad \beta \in \mathbb{R} \setminus \{0\}, \, \sigma > 1, \, \lambda \in \mathbb{R}. \quad (2.49)$$

Note that, if

$$\Delta u = -f(u, |\nabla u|), \quad f = f(u,t) \in C^1(\mathbb{R}^2),$$

a straightforward computation which exploits the Bochner-Weitzenböck formula (1.48) gives

$$\operatorname{div} Z = u^\alpha \left\{ |\operatorname{Hess}(u)|^2 - \frac{(\Delta u)^2}{m} \right\}$$
$$+ u^{\alpha-1} \left\{ \frac{\alpha}{m} f(u, |\nabla u|) + \frac{1-m}{m} u \frac{\partial f}{\partial u}(u, |\nabla u|) \right\} |\nabla u|^2$$
$$+ u^\alpha \operatorname{Ric}(\nabla u, \nabla u) + \alpha u^{\alpha-1} \operatorname{Hess}(u)(\nabla u, \nabla u)$$
$$+ \frac{1-m}{m} u^\alpha \frac{\partial f}{\partial t}(u, |\nabla u|) |\nabla u|^{-1} \operatorname{Hess}(u)(\nabla u, \nabla u).$$

Hence, if f has the special form

$$f(u, |\nabla u|) = \delta \frac{|\nabla u|^2}{u} + g(u), \quad \delta \in \mathbb{R},$$

the previous formula becomes

$$\operatorname{div} Z = u^\alpha \left\{ |\operatorname{Hess}(u)|^2 - \frac{(\Delta u)^2}{m} \right\}$$
$$+ u^{\alpha-1} \left\{ \frac{\alpha}{m} g(u) + \frac{1}{m}(\alpha+m-1)\delta \frac{|\nabla u|^2}{u} + \frac{1-m}{m} u g'(u) \right\} |\nabla u|^2$$
$$+ u^\alpha \operatorname{Ric}(\nabla u, \nabla u)$$
$$+ u^{\alpha-1} \left\{ \alpha + 2\delta \frac{1-m}{m} \right\} \operatorname{Hess}(u)(\nabla u, \nabla u).$$

To get rid of the last term containing $\operatorname{Hess}(u)(\nabla u, \nabla u)$ we observe that, given $\gamma \in \mathbb{R}$,

$$\operatorname{div}\left(\gamma u^\beta |\nabla u|^2 \nabla u \right) = 2\gamma u^\beta \operatorname{Hess}(u)(\nabla u, \nabla u) + \gamma u^\beta |\nabla u|^2 \Delta u + \beta\gamma u^{\beta-1} |\nabla u|^4.$$

We choose $\beta = \alpha - 1$, $\gamma = \frac{(m-1)\delta}{m} - \frac{\alpha}{2}$ so that, for the vector field

$$V = Z + \left(\frac{m-1}{m}\delta - \frac{\alpha}{2}\right)u^{\alpha-1}|\nabla u|^2\nabla u, \tag{2.50}$$

we have

$$\text{div}\, V = u^\alpha\left\{|\text{Hess}(u)|^2 - \frac{(\Delta u)^2}{m}\right\}$$

$$+ u^{\alpha-1}\left\{\frac{\alpha(m+2) - 2(m-1)\delta}{2m}g(u) - \frac{m-1}{m}ug'(u)\right\}|\nabla u|^2$$

$$+ \frac{u^{\alpha-2}}{2m}\left\{3m\alpha\delta - m\alpha(\alpha-1) - 2\delta^2(m-1)\right\}|\nabla u|^4$$

$$+ u^\alpha\,\text{Ric}(\nabla u, \nabla u).$$

Now, if $\delta = -(\beta + 1)$ and $g(u) = \frac{1}{\beta}u - \frac{1}{\beta}u^{1+\beta(1-\sigma)}$ from the previous relation we obtain

$$\text{div}\, V = u^\alpha\left\{|\text{Hess}(u)|^2 - \frac{(\Delta u)^2}{m}\right\} \tag{2.51}$$

$$- \frac{1}{2m\beta}[2(m-1)\beta\sigma + (m+2)\alpha]u^{\alpha+\beta(1-\sigma)}|\nabla u|^2$$

$$- \frac{1}{2m}\left[m\alpha^2 + (2+3\beta)m\alpha + 2(m-1)(\beta+1)^2\right]u^{\alpha-2}|\nabla u|^4$$

$$+ \frac{1}{2m}\left\{2m\,\text{Ric}(\nabla u, \nabla u) + \frac{\lambda}{\beta}[\alpha(m+2) + 2(m-1)\beta]|\nabla u|^2\right\}u^\alpha.$$

We are now ready to prove the following beautiful result first due to Bidaut-Véron and Véron, [BVV91].

Theorem 2.12. *Let $(M, \langle\,,\,\rangle)$ be a compact manifold of dimension $m \geq 2$ and Ricci curvature satisfying*

$$\text{Ric} \geq k > 0.$$

Let φ be a positive solution of

$$\Delta\varphi - \lambda\varphi + \varphi^\sigma = 0 \quad on\ M \tag{2.52}$$

for some constant $\sigma > 1$, $\lambda > 0$. Then φ is constant provided

(i) *$m = 2$ and $\lambda \leq 2k/(\sigma - 1)$;*

(ii) *$m \geq 3$, $\lambda \leq mk/(m-1)(\sigma - 1)$, $\sigma \leq (m+2)/(m-2)$ and*

    (A) *either at least one of the last two inequalities is strict or*

    (B) *$(M, \langle\,,\,\rangle)$ has constant scalar curvature S and it is not isometric to $S^m_{k^2}$, the sphere of constant curvature $k^2 = s/(m-1)m$.*

Proof. We set $u = \varphi^{-1/\beta}$, $\beta \neq 0$. Then u is positive and satisfies equation (2.49), hence (2.51) holds. With the aid of the divergence theorem we obtain

$$0 = \int_M 2mu^\alpha \left\{ |\mathrm{Hess}(u)|^2 - \frac{1}{m}(\Delta u)^2 \right\} - A \int_M u^{\alpha-2}|\nabla u|^4 \tag{2.53}$$
$$+ \int_M u^\alpha \left\{ 2m\,\mathrm{Ric}\,(\nabla u, \nabla u) + D|\nabla u|^2 \right\} - B \int_M u^{\alpha-\beta(\sigma-1)}|\nabla u|^2,$$

where, for ease of notation, we have set

$$A = m\alpha^2 + (3\beta + 2)m\alpha + 2(m-1)(1+\beta)^2,$$
$$B = \frac{1}{\beta}[2(m-1)\beta\sigma + (m+2)\alpha], \quad D = \frac{\lambda}{\beta}[(m+2)\alpha + 2(m-1)\beta].$$

Next we observe that, by Newton's inequality

$$|\mathrm{Hess}(u)|^2 \geq \frac{1}{m}(\Delta u)^2,$$

the first integral on the right-hand side of the above is nonnegative. The idea of the proof is to find, under the conditions listed in (i) and (ii), $\beta \neq 0$ and $\alpha \in \mathbb{R}$, such that

$$A \leq 0, \quad B \leq 0, \quad \text{and } \mathrm{Ric} + \frac{D}{2m} \geq 0 \tag{2.54}$$

and at least one of the above inequalities is strict. Once this is achieved, then (2.53) implies that $\nabla u \equiv 0$; thus u and therefore φ are constant. Let $y = 1 + 1/\beta$, $\delta = -\alpha/\beta$, so that $y, \delta \in \mathbb{R}$, $y \neq 1$. Rewriting A, B and D in terms of y and δ, the inequalities to be established become

$$\begin{array}{ll} \text{(a)} & 2\frac{m-1}{m}y^2 - 2\delta y + \delta^2 - \delta \leq 0, \\ \text{(b)} & 2\sigma\frac{m-1}{m+2} \leq \delta, \\ \text{(c)} & 2\frac{m}{m+2}\mathrm{Ric} \geq \lambda\left(\delta - 2\frac{m-1}{m+2}\right)\langle\,,\,\rangle, \end{array} \tag{2.55}$$

with at least one strict inequality. If either (i) or (ii) holds, then, for $m \geq 2$,

$$2\frac{m}{m+2}\mathrm{Ric} \geq \lambda\left(2\frac{m-1}{m+2}\sigma - 2\frac{m-1}{m+2}\right)\langle\,,\,\rangle \tag{2.56}$$

and setting

$$\delta = 2\sigma\frac{m-1}{m+2} > 0,$$

inequalities (2.55) (b) and (c) are satisfied. In order to find a value $y \neq 1$ which satisfies (2.55) (a) with strict inequality, it suffices that the quadratic polynomial in y on the left-hand side has two distinct solutions, which, taking into account our choice of σ, in turn amounts to the validity of the inequality $(m+1) - (m-2)\sigma > 0$.

This inequality being always trivially satisfied if $m = 2$, it remains to analyze the case where $m > 2$, $\sigma = (m + 2) / (m - 2)$. We assume that $(M, \langle \, , \, \rangle)$ has constant scalar curvature S and show that if φ is a positive nonconstant solution of (2.49), then $(M, \langle \, , \, \rangle)$ is isometric to the sphere $\mathbb{S}^m_{k^2}$ of constant sectional curvature $k^2 = S/(m-1)m$. Our assumption that φ is not constant implies that u is positive, and nonconstant; by the condition on σ, equality holds in (2.55) (b) and (a) with $y = \frac{m}{m-2}$, so that $A = B = 0$. Since the third and the fourth summands in (2.53) vanish, while the integrands in the first and third integral are nonnegative, they must vanish identically. In particular we must have

$$|\text{Hess}(u)|^2 = \frac{(\Delta u)^2}{m}, \text{ on } M$$

and therefore, by the equality case in the Cauchy-Schwarz inequality,

$$\text{Hess}(u) = \frac{\Delta u}{m} \langle \, , \, \rangle, \text{ on } M.$$

Now we proceed as in the proof of Theorem 2.8. We set $X = \nabla u$ and we observe that $f = \text{div} \, X$ satisfies

$$\text{Hess}(f) + \frac{S}{m(m-1)} f \langle , \rangle = 0.$$

Note that f is not constant, for otherwise the above would imply $Sf \equiv 0$ and since $S \geq mk > 0$ by the assumption on the Ricci curvature, we would conclude that

$$f = \Delta u = 0$$

so that u is constant on M. Obata's Theorem 2.10 implies the conclusion. $\square$

Remark. 1. Equation (2.52) is, of course, a normalized version of the more general case

$$\Delta \varphi - \lambda \varphi + \mu \varphi^\sigma = 0 \quad \text{on } M, \tag{2.57}$$

with $\lambda, \mu > 0$ and $\varphi > 0$. In this situation the required condition on λ depends on μ.

2. Similarly to what happened in the proof of Theorem 2.8, if we consider equation (2.57) with $\lambda \leq 0$, $\mu \geq 0$, then $\Delta \varphi \leq 0$ on the compact manifold $(M, \langle , \rangle)$, and hence φ is a positive constant by the maximum principle. For $\lambda \geq 0$, $\mu \leq 0$, then $\Delta \varphi - \lambda \varphi \geq 0$ and we obtain the same conclusion. Finally, for $\lambda \leq 0$, $\mu \leq 0$, $\Delta \varphi + (\mu \varphi^{\sigma-1}) \varphi \leq 0$ and again φ is constant.

3. The above observation also shows that Theorem 2.8 is a consequence of Theorem 2.12 for $m \geq 3$. Indeed, if S is the scalar curvature of $(M, \langle , \rangle)$ and $\widetilde{\langle , \rangle} = \varphi^{\frac{4}{m-2}} \langle \, , \, \rangle$, $\varphi > 0$, then $c_m \Delta \varphi - S\varphi + \widetilde{S}\varphi = 0$, with $\widetilde{S}$ the constant scalar curvature of $\widetilde{\langle , \rangle}$. However, the Einstein condition is not required for $(M, \langle , \rangle)$ in Theorem 2.12.

2.2.3   A version of Theorem 2.12 on manifolds with boundary

We shall now give a version of Theorem 2.12 in case M is compact with nonempty boundary ∂M. For simplicity we consider the case that M is oriented. In the next result we explicate the boundary term in applying the divergence theorem to (2.51).

Lemma 2.13. *Let $(M, \langle\,,\,\rangle)$ be an m-dimensional oriented compact Riemannian manifold with boundary ∂M. Let* II *and h denote respectively the second fundamental tensor and the mean curvature of the embedding $\partial M \hookrightarrow M$ in the direction of the outward unit normal vector field ν. Let $u \in C^3(M)$ and let $\widetilde{u} = u|_{\partial M}$. If V is the vector field defined in (2.50), that is*

$$V = u^\alpha \left\{ \frac{1}{2} \nabla |\nabla u|^2 - \frac{\Delta u}{m} \nabla u \right\} - \left(\frac{m-1}{m}(\beta + 1) + \frac{\alpha}{2} \right) u^{\alpha-1} |\nabla u|^2 \nabla u,$$

we have

$$
\begin{aligned}
2m \int_M \operatorname{div} V = {}& \int_{\partial M} 2 \widetilde{u}^\alpha [(m-1)\operatorname{Hess}(u)(\nu,\nu) - (m+1)\Delta \widetilde{u}] \frac{\partial u}{\partial \nu} \\
& - \int_{\partial M} [2(m-1)(\beta+1) + 3m\alpha] \widetilde{u}^{\alpha-1} |\nabla \widetilde{u}|^2 \frac{\partial u}{\partial \nu} \\
& - \int_{\partial M} [2(m-1)(\beta+1) + m\alpha] \widetilde{u}^{\alpha-1} \left(\frac{\partial u}{\partial \nu} \right)^3 \quad\quad (2.58) \\
& + \int_{\partial M} 2(m-1) h \widetilde{u}^\alpha \left(\frac{\partial u}{\partial \nu} \right)^2 + 2m \widetilde{u}^\alpha \langle \operatorname{II}(\nabla \widetilde{u}, \nabla \widetilde{u}), \nu \rangle.
\end{aligned}
$$

Proof. Let $\{e_1, \ldots, e_{m-1}, e_m = \nu\}$ be a Darboux frame along $\partial M \hookrightarrow M$. Set

$$h_{ij} = \langle \operatorname{II}(e_i, e_j), \nu \rangle,$$

so that

$$h = \frac{1}{m-1} h_{kk},$$

where $1 \le i, j, k \le m-1$. Then, with obvious meaning of the notation we have

$$\widetilde{u}_i = u_i, \quad u_m = \frac{\partial u}{\partial \nu} = \langle \nabla u, \nu \rangle.$$

It therefore follows that

$$
\begin{cases}
\widetilde{u}_{ij} = u_{ij} + u_m h_{ij}, \\
\left(\dfrac{\partial u}{\partial \nu} \right)_i = u_{mi} - \widetilde{u}_j h_{ij}.
\end{cases}
$$

Hence, using the identity $\left\langle \nabla |\nabla u|^2, X \right\rangle = 2 \operatorname{Hess}(u)(X, \nabla u)$, we get

$$2m \int_{\partial M} \langle V, \nu \rangle = \int_{\partial M} -2\tilde{u}^\alpha \Delta u \frac{\partial u}{\partial \nu} + 2m\tilde{u}^\alpha \operatorname{Hess}(u)(\nabla u, \nu)$$

$$- \int_{\partial M} [2(m-1)(\beta+1) + m\alpha] \tilde{u}^{\alpha-1} |\nabla u|^2 \frac{\partial u}{\partial \nu}$$

$$= \int_{\partial M} 2\tilde{u}^\alpha [(m-1) \operatorname{Hess}(u)(\nu, \nu) - \Delta \tilde{u}] \frac{\partial u}{\partial \nu}$$

$$+ \int_{\partial M} 2(m-1) h \tilde{u}^\alpha \left(\frac{\partial u}{\partial \nu} \right)^2$$

$$- \int_{\partial M} [2(m-1)(\beta+1) + m\alpha] \tilde{u}^{\alpha-1} \left[|\nabla \tilde{u}|^2 + \left(\frac{\partial u}{\partial \nu} \right)^2 \right] \frac{\partial u}{\partial \nu}$$

$$+ \int_{\partial M} 2m\tilde{u}^\alpha \langle \mathrm{II}(\nabla \tilde{u}, \nabla \tilde{u}), \nu \rangle + 2m\tilde{u}^\alpha \left\langle \nabla \widetilde{\frac{\partial u}{\partial \nu}}, \nabla \tilde{u} \right\rangle.$$

We use the first Green formula (see e.g. [GT01]) to deal with the last integrand of the above formula, that is,

$$\int_{\partial M} 2m\tilde{u}^\alpha \left\langle \nabla \widetilde{\frac{\partial u}{\partial \nu}}, \nabla \tilde{u} \right\rangle = - \left(\int_{\partial M} 2m\tilde{u}^\alpha \Delta \tilde{u} \frac{\partial u}{\partial \nu} + 2m\alpha \tilde{u}^{\alpha-1} |\nabla \tilde{u}|^2 \frac{\partial u}{\partial \nu} \right).$$

Formula (2.58) now follows at once. □

In the next result we shall make use of the following extension of Obata's theorem 2.10 due to Escobar [Esc90]. Its proof is similar to that of Theorem 2.10 and it will therefore be omitted.

Theorem 2.14. *Let* $(M, \langle \, , \, \rangle)$ *be an m-dimensional Riemannian compact manifold with boundary* ∂M. *Assume that there exists a nonconstant function* f *on* M *satisfying*

$$\begin{cases} \operatorname{Hess}(f) + kf \langle, \rangle = 0 & \text{on } M, \\ \frac{\partial f}{\partial \nu} \equiv 0 & \text{on } \partial M \end{cases} \tag{2.59}$$

with k *a positive constant. Then* $(M, \langle \, , \, \rangle)$ *is isometric to* $\mathbb{S}^m_+ \left(\sqrt{k} \right)$, *the upper hemisphere of radius* $k^{-1/2}$ *with the induced metric from* $\mathbb{R}^{m+1}$.

We are now ready for the next

Lemma 2.15. *Let* $(M, \langle \, , \, \rangle)$ *be an m-dimensional,* $m \geq 3$, *compact manifold with boundary* ∂M *such that the embedding* $\partial M \hookrightarrow M$ *is totally geodesic. Assume that* $(M, \langle \, , \, \rangle)$ *is Einstein but not Ricci flat. Let* v *be a nonconstant function satisfying*

$$\begin{cases} \operatorname{Hess}(v) - \frac{\Delta v}{m} \langle, \rangle = 0 & \text{on } M, \\ \frac{\partial v}{\partial \nu} \equiv 0 & \text{on } \partial M \end{cases} \tag{2.60}$$

with ν *the outward unit normal to* ∂M. *Then the scalar curvature* S *is positive and* $(M, \langle, \rangle)$ *is isometric to* $\mathbb{S}^m_+ \left(\sqrt{\frac{m(m-1)}{S}} \right)$.

Proof. Let $X = \nabla v$. Then, because of the first equation of (2.60) X is a conformal vector field and hence, since S is constant, $f = \text{div} X = \Delta v$ satisfies (2.34), that is

$$\text{Hess}(f) + \frac{S}{m(m-1)} f \langle,\rangle \quad \text{on } M. \tag{2.61}$$

We shall now show that

$$\frac{\partial f}{\partial \nu} \equiv 0 \quad \text{on M.} \tag{2.62}$$

Towards this end we observe that, with respect to a Darboux frame $\{e_i, e_m = \nu\}$, with $1 \leq i \leq m - 1$ along the embedding $\partial M \hookrightarrow M$, from conformality of X we have

$$X_{im} + X_{mi} = \frac{2}{m}(\text{div } X)\delta_{im} = 0. \tag{2.63}$$

Thus differentiating in the e_k direction, $1 \leq k \leq m - 1$,

$$X_{imk} + X_{mik} = 0. \tag{2.64}$$

Now $X_m = \langle X, \nu \rangle = \frac{\partial v}{\partial \nu} \equiv 0$ on ∂M because of (2.60), thus $X_{mik} = 0$ on ∂M for $1 \leq i, k \leq m - 1$. From (2.64) we then deduce

$$X_{imk} \equiv 0 \quad \text{on } \partial M. \tag{2.65}$$

Now, since X is conformal

$$X_{ab} + X_{ba} = \frac{2}{m}(\text{div} X)\delta_{ab}, \quad 1 \leq a, b \leq m$$

and for $1 \leq k \leq m - 1$

$$X_{kk} = \frac{\text{div} X}{m} \quad \text{(no sum over } k\text{)}.$$

Hence

$$X_{kkm} = \frac{1}{m}(\text{div } X)_m \quad \text{(no sum over } k\text{)}. \tag{2.66}$$

From the Ricci identities (1.47),

$$X_{kkm} - X_{kmk} = X_s R_{skkm} \quad \text{(no sum over } k\text{)}.$$

Thus using (2.65)

$$\sum_{k=1}^{m-1} X_{kkm} = -X_t R_{tm} \quad \text{on } \partial M.$$

But $(M, \langle\,,\,\rangle)$ is Einstein and therefore $R_{sm} = 0$ for $1 \leq s \leq m - 1$. Furthermore $X_m = 0$, hence

$$\sum_{k=1}^{m-1} X_{kkm} = 0. \tag{2.67}$$

Using (2.66) and (2.67) we find

$$\frac{m}{m-1}(\operatorname{div}X)_m = \sum_{k=1}^{m-1} X_{kkm} = 0 \quad \text{on } \partial M$$

that is, the validity of (2.62). Thus (2.60) is satisfied by $f = \Delta v$ with $k = \frac{S}{m(m-1)}$ and the conclusion follows at once from Theorem 2.14 once we prove that f is nonconstant and $S > 0$. We reason by contradiction. If f were constant, by (2.61) since $S \neq 0$ we would have $f \equiv 0$ so that $X = \nabla v$ would be a Killing field and $\Delta v \equiv 0$. Since $\operatorname{div}(v\nabla v) = v\Delta v + |\nabla v|^2 = |\nabla v|^2$ and $\frac{\partial v}{\partial \nu} \equiv 0$ the divergence theorem yields

$$\int_M |\nabla v|^2 \equiv 0,$$

that is, v is constant contradicting the assumption of the lemma. Thus f is non-constant. Next, to show that $S > 0$ we trace (2.61) to obtain

$$\begin{cases} \Delta f + \frac{S}{m(m-1)}f = 0 & \text{on } M, \\ \frac{\partial f}{\partial \nu} \equiv 0 & \text{on } \partial M. \end{cases}$$

From the divergence theorem

$$0 = \int_{\partial M} f\frac{\partial f}{\partial \nu} = \int_M \operatorname{div}(f\nabla f) = \int_M f\Delta f + |\nabla f|^2 = -\int_M f^2 \frac{S}{m(m-1)} + |\nabla f|^2$$

and thus

$$S = m(m-1)\frac{\int_M |\nabla f|^2}{\int_M f^2} > 0. \qquad \square$$

We are now ready to prove

Theorem 2.16. *Let $(M, \langle\,,\,\rangle)$ be a compact manifold of dimension $m \geq 2$ with boundary ∂M. Assume that the embedding $\partial M \hookrightarrow M$ is totally geodesic and that*

$$\operatorname{Ric} \geq k > 0. \tag{2.68}$$

Let $\sigma > 1$, $\lambda > 0$ and φ a positive solution of

$$\begin{cases} \Delta\varphi - \lambda\varphi + \varphi^\sigma = 0 & \text{on } M, \\ \frac{\partial\varphi}{\partial\nu} \equiv 0 & \text{on } \partial M \end{cases} \tag{2.69}$$

with ν the outward unit normal to ∂M. Then φ is constant provided

(i) *$m = 2$ and $\lambda \leq \frac{2k}{\sigma-1}$;*

(ii) *$m \geq 3$, $\lambda \leq \frac{m}{m-1}\frac{k}{\sigma-1}$, $\sigma \leq \frac{m+2}{m-2}$ and*

   (A) *either at least one of the last two inequalities is strict or*

(B) $(M, \langle\,,\,\rangle)$ is Einstein and it is not isometric to an upper hemisphere of constant curvature.

Proof. We set $u = \varphi^{-1/\beta}$ with $\beta \neq 0$. Then u satisfies (2.49) and $\frac{\partial u}{\partial \nu} \equiv 0$ on ∂M. Using (2.51), (2.58), $\frac{\partial u}{\partial \nu} \equiv 0$ and the fact that $\partial M \hookrightarrow M$ is totally geodesic we get again (2.53). Now the proof proceeds exactly as in Theorem 2.12 up to

$$\mathrm{Hess}(u) = \frac{\Delta u}{m} \langle\,,\,\rangle \quad \text{on } M.$$

Since $\frac{\partial u}{\partial \nu} \equiv 0$ on ∂M we can now apply Lemma 2.15 and we reach a contradiction. $\qquad\square$

Remark. As it is apparent from the proof of Lemma 2.15, we may substitute the assumption that $(M, \langle\,,\,\rangle)$ is Einstein in (B) with $(M, \langle\,,\,\rangle)$ has constant scalar curvature and $\mathrm{Ric}(\nu, X) = 0$ for every $X \in T\partial M^\perp \subseteq TM$ on ∂M.

Remark.   1. An observation similar to the Remark on page 61 holds also here. Point 1. applies *verbatim* to the more general equation

$$\Delta \varphi - \lambda \varphi + \mu \varphi^\sigma = 0 \quad \text{on } M \tag{2.70}$$

with $\lambda, \mu > 0$ and $\varphi > 0$.

2. If λ and μ have different sign, that is, $\lambda \leq 0$ and $\mu \geq 0$ or $\lambda \geq 0$ and $\mu \leq 0$, $\varphi > 0$, then from (2.69)

$$0 = \int_{\partial M} \frac{\partial \varphi}{\partial \nu} = \int_M \Delta \varphi = \int_M \lambda \varphi - \mu \varphi^\sigma,$$

so that

$$\varphi\left(\lambda - \mu \varphi^{\sigma-1}\right) \equiv 0.$$

From this latter we easily deduce $\lambda = \mu = 0$ and φ is constant from (2.69) and compactness of M, without requiring any further assumption.

3. Finally, if $\lambda, \mu < 0$ we have

$$\Delta \varphi - \lambda \varphi = -\mu \varphi^\sigma \geq 0.$$

Suppose first that φ assumes its maximum at a point $x_0 \in \partial M$. Then, by the boundary point Lemma (see [PW67] for more details) one has $\frac{\partial \varphi}{\partial \nu}(x_0) > 0$, contradicting $\frac{\partial \varphi}{\partial \nu}(x_0) \leq 0$ on ∂M. Hence φ has to assume its maximum at $x_0 \in M \setminus \partial M$. Similarly φ attains its minimum at $x_1 \in M \setminus \partial M$. Then

$$0 \geq \Delta \varphi(x_0) = \lambda \varphi(x_0) - \mu \varphi^\sigma(x_0),$$

and therefore

$$\frac{\mu}{\lambda} \varphi^{\sigma-1}(x_0) \leq 1.$$

Analogously, since $\Delta\varphi(x_1) \geq 0$ one gets

$$\frac{\mu}{\lambda}\varphi^{\sigma-1}(x_1) \geq 1.$$

Since $\frac{\mu}{\lambda} > 0$ and φ is positive we conclude that $\varphi(x_0) = \varphi(x_1)$ and therefore φ is constant.

Remark. In Theorem 2.16 we can substitute the assumption that the embedding of the boundary in M is totally geodesic with that of a convex boundary (see for instance [PRS03a]). It seems also worth mentioning that using this result Ilias obtained some sharp Sobolev constants on M (see [Ili96] and [PRS03a], Theorem 3.5).

With the above observation and with the aid of Lemma 2.19, from Theorem 2.16 we deduce the following result of Escobar (see [Esc90], Theorem 4.1):

Theorem 2.17. *Let* $(M, \langle\,,\,\rangle)$ *be a compact Einstein manifold of dimension* $m \geq 3$ *and with boundary* ∂M *totally geodesic in* $(M, \langle\,,\,\rangle)$. *Let* $\widetilde{\langle\,,\,\rangle} = \varphi^{\frac{4}{m-2}}$ *be a conformal change of metric with constant scalar curvature* $\widetilde{S}$ *and with* $\partial M \hookrightarrow \left(M, \widetilde{\langle\,,\,\rangle}\right)$ *minimal. Then* $\widetilde{\langle\,,\,\rangle}$ *is Einstein and if* $(M, \langle\,,\,\rangle)$ *is not isometric to an upper hemisphere of constant curvature,* φ *is constant, that is, the conformal change is an homothety.*

2.2.4 A rigidity result of Escobar

In this section we prove a nice rigidity result due to Escobar, [Esc90]. Towards this aim we need some preliminary facts contained in the next two lemmas.

Lemma 2.18. *Let* $(M, \langle\,,\,\rangle)$ *be a manifold with boundary, constant scalar curvature* S *and trace-free Ricci tensor* T. *Let* ν *be the outward unit normal to* $i : \partial M \hookrightarrow M$ *and let* X *be a conformal vector field on* M *such that, for some* $\psi \in C^\infty(\partial M)$, $X|_{\partial M} = \psi\nu$. *Then*

$$\int_{\partial M} \psi T(\nu, nu) = 0. \tag{2.71}$$

Proof. We let $W = T(X,\,)^\sharp$ as in (2.14) so that, according to (2.15),

$$\operatorname{div} W = X_{i,k}T_{ik} + X_i T_{ik,k}. \tag{2.72}$$

Since X is conformal,

$$\mathcal{L}_X \langle\,,\,\rangle = \frac{2}{m} \operatorname{div} X \langle\,,\,\rangle,$$

or, in other words,

$$X_{ik} + X_{ki} = \left(\frac{2}{m} \operatorname{div} X\right)\delta_{ik}.$$

Hence

$$X_{ik}T_{ik} = \frac{1}{2}(X_{ik} + X_{ki})T_{ik} = \left(\frac{1}{m}\,\mathrm{div}\,X\right)T_{kk} = 0,$$

since T is trace-free. Thus using (2.72) and the divergence theorem,

$$\int_{\partial M} T(X,\nu) = \int_M (\mathrm{div}\,T)(X).$$

However, since the scalar curvature S is constant, from (2.13) $\mathrm{div}\,T \equiv 0$ and from the above we immediately deduce (2.71). $\qquad\qquad\qquad\square$

Remark. For $m = \dim M = 2$ there is no need to assume S constant, see equation (2.13).

Lemma 2.19. *Let $(M, \langle\,,\,\rangle)$ be a manifold with boundary ∂M and $\widetilde{\langle\,,\,\rangle} = \varphi^2 \langle\,,\,\rangle$ for some $\varphi > 0$, $\varphi \in C^\infty(M)$. Let ν be the unit outward normal of $\partial M \to (M, \langle\,,\,\rangle)$ and h_{ij} the coefficient of the second fundamental form in the direction of ν. Set $\tilde{h}_{ij}$ for the analogous quantities for $\partial M \to \left(M, \widetilde{\langle\,,\,\rangle}\right)$ with respect to $\tilde{\nu} = \varphi^{-1}\nu$. Then, for $1 \le i,j \le m-1$,*

$$\tilde{h}_{ij} = \varphi^{-1}h_{ij} + \varphi^{-2}\frac{\partial\varphi}{\partial\nu}\delta_{ij}, \qquad (2.73)$$

and for the mean curvature

$$\tilde{h} = \varphi^{-1}\left(h + \frac{\partial\log\varphi}{\partial\nu}\right). \qquad (2.74)$$

In particular, if $\partial M \to (M, \langle\,,\,\rangle)$ and $\partial M \to \left(M, \widetilde{\langle\,,\,\rangle}\right)$ are both minimal, then $\frac{\partial\varphi}{\partial\nu} \equiv 0$ on ∂M.

Proof. A simple computation. Indeed, with the notation of Chapter 1 and section 2.1.1 we let $\{\theta^i\}$ be a Darboux frame along the inclusion map $\imath : \partial M \to (M, \langle\,,\,\rangle)$. Thus $\theta^m = 0$ on ∂M and h_{ij} are defined by the requirement

$$\theta_i^m = h_{ij}\theta^j, \quad 1 \le i,j,\dots \le m-1.$$

Similarly, $\tilde{\theta}^i = \varphi\theta^i$ is a Darboux frame along the inclusion $\tilde{\imath} : \partial M \to \left(M, \widetilde{\langle\,,\,\rangle}\right)$. Thus $\tilde{\theta}^m = 0$ and

$$\tilde{\theta}_i^m = \tilde{h}_{ij}\tilde{\theta}^j, \quad 1 \le i,j,\dots \le m-1.$$

To relate h_{ij} with $\tilde{h}_{ij}$ we recall from (2.3) that

$$\tilde{\theta}_i^m = \theta_i^m + \frac{\varphi_i}{\varphi}\theta^m - \frac{\varphi_m}{\varphi}\theta^i, \quad d\varphi = \varphi_j\theta^j + \varphi_m\theta^m.$$

Hence

$$\varphi \widetilde{h}_{ij}\theta^j = h_{ij}\theta^j - \frac{\varphi_m}{\varphi}\delta^i_j\theta^j$$

and it follows that

$$\widetilde{h}_{ij} = \varphi^{-1}h_{ij} - \varphi^{-2}\varphi_m\delta_{ij},$$

from which (2.73), and thus (2.74), follow immediately. □

Lemma 2.20. *Let (M, g_0) be a compact Einstein manifold such that $\partial M \hookrightarrow (M, g_0)$ is totally geodesic and let $g = \varphi^{-2}g_0$, $\varphi > 0$, $\varphi \in C^\infty(M)$ be a conformally related metric with constant scalar curvature S and such that $\partial M \hookrightarrow (M, g)$ has constant mean curvature h with respect to the outward unit (with respect to g) normal ν. Let T be the trace-free Ricci tensor of g. Then*

$$\int_M \varphi^{-1}|T|^2 = -(m-2)\int_{\partial M} h\varphi^{-1}T(\nu,\nu) \tag{2.75}$$

(where of course we are integrating with respect to the volume element of g on M and ∂M).

Proof. We set $\widetilde{g}$ for g_0 so that $\widetilde{g} = \varphi^2 g$. From (2.26)

$$\widetilde{T}_{ij} = T_{ij} + (m-2)\varphi\left\{(\varphi^{-1})_{ij} - \frac{\Delta(\varphi^{-1})}{m}\delta_{ij}\right\},$$

and since $\widetilde{g}$ by assumption is Einstein we get

$$-T_{ij} = (m-2)\varphi\left\{(\varphi^{-1})_{ij} - \frac{\Delta(\varphi^{-1})}{m}\delta_{ij}\right\}.$$

The fact that T is traceless yields

$$-|T|^2 = (m-2)\varphi(\varphi^{-1})_{ij}T_{ij}.$$

Integrating the above on (M, g) we obtain

$$-\int_M \varphi^{-1}|T|^2 = (m-2)\int_M (\varphi^{-1})_{ij}T_{ij}. \tag{2.76}$$

Again from (2.13), since (M, g) has constant scalar curvature

$$T_{ik,k} = 0. \tag{2.77}$$

Similarly to what we did in the proof of Lemma 2.18 we consider the vector field

$$W = T(\nabla\varphi^{-1})^\sharp.$$

Taking its divergence in the metric g we have (see (2.72))

$$\text{div } W = \left(\varphi^{-1}\right)_{ij} T_{ij} + \left(\varphi^{-1}\right)_i T_{ik,k}.$$

Hence, using (2.77) and the divergence theorem,

$$\int_{\partial M} T\left(\nabla\left(\varphi^{-1}\right), \nu\right) = \int_M \left(\varphi^{-1}\right)_{ij} T_{ij}. \tag{2.78}$$

We need now to write the left-hand side of (2.78) appropriately. First we show that if $Y \in T\partial M$, then $T(Y,\nu) = 0$. Since $T = \text{Ric} - \frac{S}{m}\langle\,,\,\rangle$, this is clearly equivalent to showing that

$$\text{Ric}_{(M,g)}(Y,\nu) = 0 \tag{2.79}$$

(and this is what we expect if we want to show that g is Einstein, see Theorem 2.21). By assumption $\partial M \hookrightarrow (M, \widetilde{g})$ is totally geodesic and therefore it follows from Lemma 2.19 that, since g and $\widetilde{g}$ are conformally related, $\partial M \hookrightarrow (M, \widetilde{g})$ is totally umbilical. Moreover, from (2.74)

$$0 = h + \varphi^{-1}\frac{\partial\varphi}{\partial\nu}, \tag{2.80}$$

with h the constant mean curvature of $\partial M \hookrightarrow (M, g)$. In other words

$$\varphi h = -\frac{\partial\varphi}{\partial\nu}.$$

In this case we can rewrite (2.73) of Lemma 2.19, since $\partial M \hookrightarrow (M, \widetilde{g})$ is totally geodesic, as

$$h_{ij} = h\delta_{ij}$$

for $1 \le i, j, \ldots, \le m - 1$. Since h is constant, from the above we deduce

$$h_{ijk} = 0.$$

On the other hand, from the Codazzi equations,

$$h_{ijk} - h_{ikj} = -R_{ijk}^m.$$

Tracing with respect to i and k we deduce

$$R_{mj} = 0,$$

where R_{ij} are the components of the Ricci tensor of (M, g). This proves (2.79). It follows that (2.78) becomes

$$\int_{\partial M} \frac{\partial\left(\varphi^{-1}\right)}{\partial\nu} T(\nu,\nu) = \int_M \left(\varphi^{-1}\right)_{ij} T_{ij}.$$

On the other hand, we can rewrite (2.80) as

$$\frac{\partial(\varphi^{-1})}{\partial\nu} = \varphi^{-1}h.$$

Substituting into the above we get

$$\int_{\partial M} \varphi^{-1}hT(\nu,\nu) = \int_M \left(\varphi^{-1}\right)_{ij}T_{ij}. \tag{2.81}$$

Hence (2.75) follows immediately from (2.81) and (2.76). □

We are now ready to prove Escobar's result ([Esc90]).

Theorem 2.21. *Let $\mathbb{B}^m$ be the (open) unit ball in $\mathbb{R}^m$ and $\overline{\mathbb{B}^m}$ the closed unit ball of $\mathbb{R}^m$; let g be a metric on $\overline{\mathbb{B}^m}$ conformally related to the Euclidean metric $\langle\,,\,\rangle$. Assume that S, the scalar curvature of $\left(\overline{\mathbb{B}^m},g\right)$, is constant and that $\partial\mathbb{B}^m \hookrightarrow \left(\overline{\mathbb{B}^m},g\right)$ has constant mean curvature h with respect to the outward unit (w.r.t. g) normal ν. Then $\left(\overline{\mathbb{B}^m},g\right)$ has constant sectional curvature and it is therefore isometric to a geodesic ball in a space form with an appropriate radius depending on h.*

Proof. We shall prove that $\left(\overline{\mathbb{B}^m},g\right)$ has constant sectional curvature: the remaining part of the conclusion is a well-known result of É. Cartan (see e.g. [Car88]). We introduce on $\overline{\mathbb{B}^m}$ the auxiliary metric

$$g_0 = \frac{4}{\left(1+|x|^2\right)^2}\langle\,,\,\rangle.$$

Note that g_0 is obtained from the upper hemisphere $\left(\mathbb{S}^m_+,\bar{g}\right)$, $\bar{g}$ denoting the standard metric induced on $\mathbb{S}^m$ by the inclusion $\mathbb{S}^m \hookrightarrow \mathbb{R}^{m+1}$, by stereographic projection π from the South pole $\pi : \overline{\mathbb{B}^m} \to \mathbb{S}^m_+$

$$\pi : x \mapsto \left(\frac{2x}{1+|x|^2}, \frac{1-|x|^2}{1+|x|^2}\right).$$

It follows that $\left(\overline{\mathbb{B}^m},g_0\right)$ is Einstein and $\partial\mathbb{B}^m \hookrightarrow \left(\overline{\mathbb{B}^m},g_0\right)$ is totally geodesic. Since the position vector field X is conformal on $\left(\overline{\mathbb{B}^m},\langle\,,\,\rangle\right)$, it is conformal on $\left(\overline{\mathbb{B}^m},g_0\right)$ and, since $g_0|_{\partial\overline{\mathbb{B}^m}} = \langle\,,\,\rangle_{\partial\overline{\mathbb{B}^m}}$, $X|_{\partial\overline{\mathbb{B}^m}}$ is the outward unit normal with respect to g_0.

Since g is conformally related to $\langle\,,\,\rangle$, it is conformally related to g_0. Let $g = \varphi^{-2}g_0$, $\varphi > 0$, $\varphi \in C^\infty\left(\overline{\mathbb{B}^m}\right)$. First we show that g is Einstein. Indeed, we are in the assumptions of Lemma 2.20 so that, with the same notation, we have the validity of (2.75). On the other hand, on $\partial\mathbb{B}^m$,

$$X = \varphi^{-1}\nu$$

where ν is the outward unit normal with respect to g. From Lemma 2.18,

$$\int_{\mathbb{B}^m} \varphi^{-1} T(\nu,\nu) = 0,$$

and constancy of h and (2.75) give

$$\int_M \varphi^{-1} |T|^2 \equiv 0,$$

that is, $T \equiv 0$ on $\left(\overline{\mathbb{B}^m}, g\right)$ or, in other words, g is an Einstein metric. If $m = 2$ we are done; for $m \geq 3$, to complete the proof we consider the decomposition of the curvature tensor R^i_{jkl} given in (2.27) that we rewrite as

$$R^i_{jkl} = W^i_{jkl} + \frac{1}{m-2}(T_{ik}\delta_{jl} - T_{il}\delta_{jk} + T_{jl}\delta_{ik} - T_{jk}\delta_{il})$$
$$+ \frac{S}{m(m-1)}(\delta_{ik}\delta_{jl} - \delta_{il}\delta_{jk})$$

with respect to a local o.n. coframe for g. But g is conformally related to the Euclidean metric and therefore it is conformally flat. Furthermore, g is Einstein and hence $T \equiv 0$ Thus the above formula becomes

$$R^i_{jkl} = \frac{S}{m(m-1)}(\delta_{ik}\delta_{jl} - \delta_{il}\delta_{jk})$$

and $\left(\overline{\mathbb{B}^m}, g\right)$ has constant sectional curvature. $\qquad\square$

Chapter 3

General nonexistence results

The aim of this chapter is to prove a number of very general nonexistence results for Yamabe-type inequalities of the form

$$\Delta u + a(x)u - b(x)u^\sigma \geq 0$$

on complete, noncompact, Riemannian manifolds. Loosely speaking, triviality of the solutions is obtained under assumptions focusing on the nonnegativity of the spectral radius for the related Schrödinger operator and L^p-properties of the solutions themselves. The spectral condition is exploited and generalized *via* the existence on M of a positive solution φ of a differential inequality of the form

$$\Delta\varphi + Ha(x)\varphi \leq -K\frac{|\nabla\varphi|^2}{\varphi}$$

with H, K parameters satisfying $H > 0$, $K > -1$. The function φ is used in two different ways: in Theorem 3.2 one uses it to obtain an integral inequality involving u and its gradient from which one concludes that u is constant, and therefore necessarily identically zero; in a second group of results, we combine φ with the supposed solution u to give rise to a diffusion-type differential inequality for which we prove a Liouville theorem.

Note that the limiting case $K = -1$ amounts to providing a solution on M of the Poisson equation

$$\Delta\psi + a(x) = 0$$

and it will be considered in Section 3.2 below. In the last section we give a refined version of Theorem 3.2 by using a conformal transformation of the metric in case $\sigma \geq \frac{m+2}{m-2}$ (see Proposition 3.10 and Theorem 3.11). As expected, the above geometrical limitation on σ plays an essential role with respect to the conformal transformation.

3.1   Some spectral considerations

First we recall some definitions and notation. From now on $(M, \langle \, , \, \rangle)$ will always denote a complete, noncompact, connected Riemannian manifold. Let $a(x) \in C^0(M)$ and for $\Omega \subset M$, a bounded domain, define

$$\lambda_1^{L_H}(\Omega) = \inf_{\varphi \in C_0^\infty(\Omega), \varphi \not\equiv 0} \frac{\int_\Omega |\nabla \varphi|^2 - Ha(x)\varphi^2}{\int_\Omega \varphi^2} \tag{3.1}$$

so that L_H denotes the operator

$$L_H = \Delta + Ha(x), \quad H \in (0, +\infty). \tag{3.2}$$

If Ω has sufficiently regular boundary, $\lambda_1^{L_H}(\Omega)$ is achieved (see, e.g., [Dav95], [GT01]) by the nonzero solutions of the Dirichlet problem

$$\begin{cases} \Delta u + Ha(x)u + \lambda_1^{L_H}(\Omega)u = 0 & \text{on } \Omega, \\ u \equiv 0 & \text{on } \partial\Omega. \end{cases} \tag{3.3}$$

We then define the first eigenvalue of L_H on M as

$$\lambda_1^{L_H}(M) = \inf_\Omega \lambda_1^{L_H}(\Omega), \tag{3.4}$$

where Ω runs over all bounded domains in M. Observe that, due to the monotonicity of $\lambda_1^{L_H}$ with respect to Ω, that is

$$\Omega_1 \subseteq \Omega_2 \quad \text{implies} \quad \lambda_1^{L_H}(\Omega_1) \geq \lambda_1^{L_H}(\Omega_2), \tag{3.5}$$

we have

$$\lambda_1^{L_H}(M) = \lim_{R \to +\infty} \lambda_1^{L_H}(B_R). \tag{3.6}$$

Note that in case $\overset{\circ}{\Omega_2 \backslash \Omega_1} \neq \emptyset$, that is the interior of $\Omega_2 \backslash \Omega_1$ is nonempty, then the inequality in (3.5) becomes strict. Of course (3.5) is a trivial consequence of the variational characterization (3.1), while to prove the last statement we can proceed in an elementary way by using Picone's identity, (3.7) below, as we are now going to show (see for instance [Swa75]). Let $u, v \in C^2(\Omega)$, $\Omega \subset M$ a bounded domain, and suppose $v \neq 0$ on Ω. A direct computation yields

$$0 \leq v^2 \left| \nabla \left(\frac{u}{v} \right) \right|^2 = \left| \nabla u - \frac{u}{v} \nabla v \right|^2 = |\nabla u|^2 - \left\langle \nabla \left(\frac{u^2}{v} \right), \nabla v \right\rangle. \tag{3.7}$$

In particular, $\left| \nabla u - \frac{u}{v} \nabla v \right| \equiv 0$ on Ω if and only if $u = Cv$ for some $C \in \mathbb{R}$.

Let now u and v be eigenfunctions on Ω_1 and Ω_2 respectively ($\Omega_1 \subseteq \Omega_2$), with eigenvalues $\lambda_1^{L_H}(\Omega_1)$ and $\lambda_1^{L_H}(\Omega_2)$. Thus

$$\begin{cases} \Delta u + Ha(x)u + \lambda_1^{L_H}(\Omega_1)u = 0 & \text{on } \Omega_1, \\ u \equiv 0 & \text{on } \partial\Omega_1, \end{cases} \tag{3.8}$$

$$\begin{cases} \Delta v + Ha(x)v + \lambda_1^{LH}(\Omega_2)v = 0 & \text{on } \Omega_2, \\ v \equiv 0 & \text{on } \partial\Omega_2. \end{cases} \qquad (3.9)$$

Note that we can suppose $v > 0$ on Ω_2. Observing that, by the divergence theorem, $\forall f, g \in C^2(\Omega)$,

$$\int_{\partial\Omega} f\, \langle \nabla g, \nu \rangle = \int_\Omega fL_H g + \int_\Omega \langle \nabla f, \nabla g \rangle - \int_\Omega fHa(x)g,$$

with ν the outward unit normal to $\partial\Omega$, integrating (3.7) on Ω_1 and using (3.8) and (3.9) gives

$$0 \le \int_{\Omega_1} |\nabla u|^2 - \int_{\Omega_1} \left\langle \nabla\left(\frac{u^2}{v}\right), \nabla v \right\rangle = \int_{\Omega_1} |\nabla u|^2 + \int_{\Omega_1} \frac{u^2}{v} L_H v - \int_{\Omega_1} Ha(x)u^2$$

$$= \int_{\Omega_1} |\nabla u|^2 - Ha(x)u^2 - \int_{\Omega_1} \lambda_1^{LH}(\Omega_2)u^2 = \left[\lambda_1^{LH}(\Omega_1) - \lambda_1^{LH}(\Omega_2) \right] \int_{\Omega_1} u^2.$$

We now reason by contradiction by assuming $\overset{\circ}{\Omega_2}\backslash\Omega_1 \ne \emptyset$ and $\lambda_1^{LH}(\Omega_1) = \lambda_1^{LH}(\Omega_2)$. From the above inequalities it follows that, on the connected component of Ω_1, $u = Cv$ for some $C \in \mathbb{R}$ (depending on the component). Choose one of the component, say $\widetilde{\Omega}_1$, with $\partial\widetilde{\Omega}_1 \cap \overset{\circ}{\Omega_2} \ne \emptyset$ (this is possible since $\overset{\circ}{\Omega_2}\backslash\Omega_1 \ne \emptyset$). Since $u \equiv 0$ on $\partial\Omega_1$, we have that $v \equiv 0$ on $\partial\widetilde{\Omega}_1 \cap \Omega_2$, contradiction.

For $H = 1$ we simply write L instead of L_1.

It often happens, in geometrical problems, that the coefficient $a(x)$ of L is related to the Ricci curvature of M: precisely it is a lower bound for the radial Ricci curvature $\mathrm{Ric}\,(\nabla r, \nabla r)$ once we fix an origin $o \in M$, that is,

$$\mathrm{Ric}(\nabla r, \nabla r) \ge -(m-1)A(r) \qquad (3.10)$$

and $a(x) \ge A(r(x))$ on M. We observe that the condition $a(x) \le 0$, or $\mathrm{Ric}(\nabla r, \nabla r) \ge 0$ if equation (3.10) holds, implies (but it is *not* implied by)

$$\lambda_1^{LH}(M) \ge 0, \qquad (3.11)$$

and therefore a request of the type of this latter may be interpreted as a weakening of the assumption of nonnegative Ricci curvature. It is also worth observing that (3.11) may be satisfied even if the entire sectional curvature is strictly negative. For instance, on $\mathbb{H}^m_{-H^2}$ (the hyperbolic space of constant sectional curvature $-H^2 < 0$) let $a(x) = A(r(x))$ with $A(t) \in C^\infty([0, +\infty))$, positive, nonincreasing and such that, for some $\varepsilon > 0$,

$$A(t) \begin{cases} = \frac{(m-1)^2}{4} H^2 \coth(Ht) & \text{on } [\varepsilon, +\infty), \\ \le \frac{(m-1)^2}{4} H^2 \coth(Ht) & \text{on } [0, \varepsilon). \end{cases} \qquad (3.12)$$

Then, the problem

$$\begin{cases} \left(\sinh^{m-1}(Ht)\alpha'\right)' + A(t)\sinh^{m-1}(Ht)\alpha = 0, \ \ H > 0, \ \text{on } [0,+\infty), \\ \alpha'(0), \ \alpha(0) > 0 \end{cases}$$

has a solution $\alpha \in C^2([0,+\infty))$ satisfying (see Proposition 7.16)

$$\alpha(t) \geq C\, t\, e^{-\frac{m-1}{2}Ht} \ \text{on } [0,+\infty)$$

for some appropriate constant $C > 0$. It follows that

$$u(x) = \alpha(r(x))$$

is a positive, C^2-solution on M of

$$\Delta u + a(x)u = 0 \tag{3.13}$$

and, by a result of Fisher-Colbrie and Schoen (see [FCS80], [PRS05c]),

$$\lambda_1^L(M) \geq 0. \tag{3.14}$$

We also note that, as shown in the aforementioned papers, the reverse implication is true, that is, the validity of (3.14) implies the existence of a *positive* solution of equation (3.13). Thus (3.11) is a *genuine relaxation of the assumption* $\mathrm{Ric} \geq 0$ when the coefficient of the linear term in L provides a lower bound for Ric.

We now relate (3.11) with the assumption on the existence of a positive solution φ of

$$\Delta\varphi + Ha(x)\varphi \leq -K\frac{|\nabla\varphi|^2}{\varphi}, \quad K > -1, H > 0. \tag{3.15}$$

As already remarked, (3.11) yields the validity of (3.15) for some $\varphi \in C^\infty(M)$, $\varphi \neq 0$, with $K = 0$ and even with the equality sign. Thus if (3.11) holds, then (3.15) holds with $K \leq 0$. On the other hand, assume that φ is a positive C^2-solution of (3.15). Let $\psi \in C_0^\infty(M)$ and consider the vector field $\psi^2\nabla\log\varphi$. We compute its divergence and we use (3.15), Schwarz's and Young's inequalities to obtain

$$\mathrm{div}(\psi^2\nabla\log\varphi) = 2\frac{\psi}{\varphi}\langle\nabla\psi,\nabla\varphi\rangle + \frac{\psi^2}{\varphi}\left(\Delta\varphi - \frac{|\nabla\varphi|^2}{\varphi}\right)$$

$$\leq \frac{\psi^2}{\varphi}\left(-Ha(x)\varphi - (K+1)\frac{|\nabla\varphi|^2}{\varphi}\right) + \frac{\psi^2}{\varphi^2}|\nabla\varphi|^2 + |\nabla\psi|^2.$$

Thus, integrating,

$$\int_M |\nabla\psi|^2 - Ha(x)\psi^2 \geq K\int_M \psi^2\frac{|\nabla\varphi|^2}{\varphi^2},$$

and from the variational characterization of the bottom of the spectrum we conclude that

$$\text{if } K \geq 0, \quad \text{then } \lambda_1^{L_H}(M) \geq 0.$$

The following simple proposition (see [BRS98]) motivates the nonexistence results of the next section.

Proposition 3.1. *Let $(M, \langle , \rangle)$ be a complete manifold. $a(x), b(x) \in C^0(M)$ and suppose that*

$$\text{i) } b(x) \geq 0; \qquad \text{ii) } b(x) \not\equiv 0. \tag{3.16}$$

Assume that for some $\sigma \in \mathbb{R}$ there exists a positive C^2-solution $u \in L^2(M)$ of the differential inequality

$$\Delta u + a(x)u - b(x)u^\sigma \geq 0. \tag{3.17}$$

Then $\lambda_1^L(M) < 0$, with $L = \Delta + a(x)$.

Proof. We reason by contradiction and assume that $\lambda_1^L(M) \geq 0$. We fix a geodesic ball B_R and we consider a smooth cut-off function φ such that $0 \leq \varphi \leq 1$ and

$$\varphi \equiv 1 \text{ on } B_R, \quad \text{supp } \varphi \subset B_{R+1}, \quad |\nabla\varphi| \leq 2, \tag{3.18}$$

where $\text{supp } \varphi$ denotes the support of the function φ, i. e. the closure of the set where $\varphi \neq 0$. Let u be a positive solution of (3.17) and note that, according to the variational characterization of $\lambda_1^L(B_{R+1})$, we have

$$\lambda_1^L(B_{R+1}) \leq \frac{\int_{B_{R+1}} |\nabla(u\varphi)|^2 - a(x)u^2\varphi^2}{\int_{B_{R+1}} u^2\varphi^2}. \tag{3.19}$$

Furthermore $\lambda_1^L(M) \geq 0$ and the monotonicity property of eigenvalues yields $\lambda_1^L(B_{R+1}) \geq 0$. Next we consider the vector field $W = u\varphi^2\nabla u$. Using the differential inequality (3.17) and computing we obtain

$$\text{div } W \geq -a(x)u^2\varphi^2 + b(x)u^{\sigma+1}\varphi^2 + |\nabla(u\varphi)|^2 - u^2|\nabla\varphi|^2.$$

Hence, by (3.19) and the divergence theorem,

$$0 \geq \lambda_1^L(B_{R+1})\int_{B_{R+1}} u^2\varphi^2 - \int_{B_{R+1}} u^2|\nabla\varphi|^2 + \int_{B_{R+1}} b(x)u^{\sigma+1}\varphi^2.$$

Rearranging and using (3.16) i) and (3.18) we obtain

$$\lambda_1^L(B_{R+1})\int_{B_{R+1}} u^2 + \int_{B_{R+1}} b(x)u^{\sigma+1} \leq 4\int_{B_{R+1}\setminus B_R} u^2.$$

Now let $R \to +\infty$ and use $u \in L^2(M)$ to finally have

$$\lambda_1^L(M)\int_M u^2 + \int_M b(x)u^{\sigma+1} \leq 0.$$

This contradicts $\lambda_1^L(M) \geq 0$ and (3.16). $\qquad\qquad\square$

3.1.1 The main nonexistence result

The next theorem, which is the main nonexistence result of the chapter, has been proved in [BRS98].

Theorem 3.2. *Let* $\sigma > 1$ *and* $a(x), b(x) \in C^0(M)$ *satisfy*

$$b(x) > 0 \ \ on \ M; \quad a(x) \leq Cb(x) \ \ for \ r(x) \gg 1 \tag{3.20}$$

and some constant $C > 0$. *Given* $\alpha > 0$ *and* $K > -1$, *suppose that, for some*

$$H \geq \frac{(1+\alpha)^2}{4\alpha(K+1)} \tag{3.21}$$

there exists $\varphi > 0$ *satisfying*

$$\Delta\varphi + Ha(x)\varphi \leq -K\frac{|\nabla\varphi|^2}{\varphi} \quad on \ M. \tag{3.22}$$

Then, the differential inequality

$$\Delta u + a(x)u - b(x)u^\sigma \geq 0 \tag{3.23}$$

has no bounded, nonnegative, nonidentically null C^2-*solutions* u *satisfying*

$$\begin{cases} \text{i)} \quad a(x)u^{1+\alpha} \quad \in L^1(M); \\ \text{ii)} \left\{\int_{\partial B_r} u^{1+\alpha}\right\}^{-1} \notin L^1(+\infty). \end{cases} \tag{3.24}$$

Remark. Note that if $a(x)$ is bounded, then (3.24) i), ii) are satisfied whenever $u^{1+\alpha} \in L^1(M)$.

Remark. We recall that a sufficient condition for the validity of (3.24) ii) is

$$\frac{r}{\int_{B_r} u^{1+\alpha}} \notin L^1(+\infty) \tag{3.25}$$

(see [RS01], Proposition 1.3).

Proof. We reason by contradiction and we assume that $u \in C^2(M)$ is a bounded, nonnegative, non-identically null solution of (3.23) satisfying (3.24). Because of (3.20) we can suppose that

$$a(x) \leq \widetilde{C}b(x) \quad on \ M$$

for some appropriate constant $\widetilde{C} > 0$. We define

$$A = \max\left\{\sup_M u, \frac{1}{2}\widetilde{C}^{\frac{1}{\sigma-1}}, 1\right\}$$

and we set

$$v(x) = \frac{1}{2A} u(x).$$

Then, $\sup_M v \le \frac{1}{2}$ and v satisfies

$$\Delta v + a(x)v \ge \bar{b}(x)v^\sigma, \tag{3.26}$$

with

$$\bar{b}(x) = (2A)^{\sigma-1} b(x) \ge \tilde{C}b(x) \ge a(x) \quad \text{on } M.$$

Thus, it suffices to show that, if $a(x) \le \bar{b}(x)$ on M, then (3.26) has no nonnegative, nontrivial solutions u satisfying (3.24) and the further requirement

$$\text{iii)} \quad \sup_M u < 1.$$

We proceed by contradiction and we divide the argument in four steps.

Step 1. We claim that there exists a sequence $\{r_k\} \uparrow +\infty$ such that

$$\lim_{k \to +\infty} \int_{\partial B_{r_k}} |u^\alpha \langle \nabla u, \nabla r \rangle| = 0. \tag{3.27}$$

Indeed, let $\varepsilon > 0$. Define $W_\varepsilon = (u+\varepsilon)^\alpha \nabla u$: compute its divergence using (3.26) and apply the divergence theorem on the geodesic ball B_r to obtain

$$\int_{\partial B_r} (u+\varepsilon)^\alpha \langle \nabla u, \nabla r \rangle - \alpha \int_{B_r} (u+\varepsilon)^{\alpha-1} |\nabla u|^2 \tag{3.28}$$

$$\ge \int_{B_r} (\bar{b}(x)u^\sigma - a(x)u)(u+\varepsilon)^\alpha \ge \int_{B_r} a(x)(u^\sigma - u)(u+\varepsilon)^\alpha$$

where, in the last inequality, we have been using $\bar{b}(x) \ge a(x)$. Letting $\varepsilon \downarrow 0^+$, the last term on the right-hand side of (3.28) converges to

$$\int_{B_r} a(x)(u^\sigma - u)u^\alpha$$

and the first integral on the left-hand side tends to

$$\int_{\partial B_r} u^\alpha \langle \nabla u, \nabla r \rangle.$$

Thus, by B. Levi's monotone convergence theorem,

$$\int_{B_r} (u+\varepsilon)^{\alpha-1} |\nabla u|^2 \longrightarrow \int_{B_r} u^{\alpha-1} |\nabla u|^2 < +\infty$$

as $\varepsilon \downarrow 0$. In particular $u^{\alpha-1} |\nabla u|^2 \in L^1_{loc}(M)$. Moreover,

$$\int_{\partial B_r} u^\alpha \langle \nabla u, \nabla r \rangle - \alpha \int_{B_r} u^{\alpha-1} |\nabla u|^2 \ge \int_{B_r} a(x)(u^\sigma - u)u^\alpha.$$

Now, since $u < 1$, $\sigma > 1$, (3.24) i) implies that $a(x)(u^{\sigma+\alpha} - u^{1+\alpha}) \in L^1(M)$ and, letting $r \to +\infty$, we obtain

$$\liminf_{r \to +\infty} \left\{ \int_{\partial B_r} u^\alpha \langle \nabla u, \nabla r \rangle - \alpha \int_{B_r} u^{\alpha-1} |\nabla u|^2 \right\} = B > -\infty.$$

We define

$$G(r) = \int_{B_r} u^{\alpha-1} |\nabla u|^2.$$

Since $u^{\alpha-1} |\nabla u|^2 \in L^1_{loc}(M)$, it follows from the co-area formula (1.87) that $G(r)$ is absolutely continuous and that

$$G'(r) = \int_{\partial B_r} u^{\alpha-1} |\nabla u|^2$$

is defined almost everywhere and is locally L^1. Assume by contradiction that $G(r) \to +\infty$ as $r \to +\infty$; then, for large enough $r > R$,

$$\int_{\partial B_r} u^\alpha \langle \nabla u, \nabla r \rangle \geq \frac{\alpha}{2} \int_{B_r} u^{\alpha-1} |\nabla u|^2,$$

and therefore, by the Hölder inequality,

$$\left\{ \frac{\alpha}{2} G(r) \right\}^2 \leq \left\{ \int_{\partial B_r} u^\alpha \langle \nabla u, \nabla r \rangle \right\}^2 \leq G'(r) \int_{\partial B_r} u^{\alpha+1}.$$

We can also suppose to have chosen R so large that $G(r) > 0$ for $r \geq R$. From the above we obtain

$$\frac{G'(r)}{G^2(r)} \geq \frac{\alpha^2}{4} \left\{ \int_{\partial B_r} u^{\alpha+1} \right\}^{-1}$$

and integrating over $[R, r)$, we get

$$-\frac{1}{G(r)} + \frac{1}{G(R)} \geq \frac{\alpha^2}{4} \int_R^r \left\{ \int_{\partial B_t} u^{\alpha+1} \right\}^{-1} dt.$$

Whence, letting $r \to +\infty$, we contradict assumption (3.24) ii). Hence $u^{\alpha-1} |\nabla u|^2 \in L^1(M)$ and therefore

$$\int_{\partial B_r} u^{\alpha-1} |\nabla u|^2 \in L^1((0, +\infty)).$$

It follows that there exists a sequence $\{r_k\} \uparrow +\infty$ such that

$$\lim_{k \to +\infty} \left(\int_{\partial B_{r_k}} u^{\alpha+1} \right) \left(\int_{\partial B_{r_k}} u^{\alpha-1} |\nabla u|^2 \right) = 0.$$

By the Hölder inequality,

$$\int_{\partial B_{r_k}} |u^\alpha \langle \nabla u, \nabla r \rangle| \le \left\{ \int_{\partial B_{r_k}} u^{\alpha+1} \right\}^{\frac{1}{2}} \left\{ \int_{\partial B_{r_k}} u^{\alpha-1} |\nabla u|^2 \right\}^{\frac{1}{2}} \to 0 \text{ as } k \to +\infty,$$

which implies the claim (3.27).

Step 2. Let $\varphi > 0$ be a solution of (3.22), i.e.,

$$\Delta\varphi + Ha(x)\varphi \le -K\frac{|\nabla\varphi|^2}{\varphi}.$$

We claim that

$$u^{\alpha+1}\varphi^{-2}|\nabla\varphi|^2 \in L^1(M).$$

Indeed, let $V = u^{\alpha+1}\varphi^{-1}\nabla\varphi$; since $K > -1$ we can choose $\theta > 0$ such that $K + 1 - \frac{1}{2\theta} > 0$. Using (3.22), a computation yields

$$\operatorname{div} V = (\alpha + 1)u^\alpha\varphi^{-1}\langle \nabla\varphi, \nabla u \rangle + u^{\alpha+1}\varphi^{-1}\Delta\varphi - u^{\alpha+1}\varphi^{-2}|\nabla\varphi|^2 \qquad (3.29)$$

$$\le 2\theta(\alpha+1)^2 u^{\alpha-1}|\nabla u|^2 + \frac{1}{2\theta}u^{\alpha+1}\varphi^{-2}|\nabla\varphi|^2 - Ha(x)u^{\alpha+1}$$

$$- (K+1)u^{\alpha+1}\varphi^{-2}|\nabla\varphi|^2$$

$$= 2\theta(\alpha+1)^2 u^{\alpha-1}|\nabla u|^2 - Ha(x)u^{\alpha+1} + \left(\frac{1}{2\theta} - K - 1 \right)u^{\alpha+1}\varphi^{-2}|\nabla\varphi|^2,$$

and applying the divergence theorem

$$\int_{\partial B_r} u^{\alpha+1}\varphi^{-1}\langle \nabla\varphi, \nabla r \rangle + \left(K + 1 - \frac{1}{2\theta} \right)\int_{B_r} u^{\alpha+1}\varphi^{-2}|\nabla\varphi|^2$$

$$\le -H\int_{B_r} a(x)u^{\alpha+1} + 2\theta(\alpha+1)^2\int_{B_r} u^{\alpha-1}|\nabla u|^2 \le B \in \mathbb{R}$$

where, in the last inequality, we have used the fact that $a(x)u^{\alpha+1}, u^{\alpha-1} \in L^1(M)$.

Setting for the ease of notation

$$F(r) = \left(K + 1 - \frac{1}{2\theta} \right)\int_{B_r} u^{\alpha+1}\varphi^{-2}|\nabla\varphi|^2,$$

the above inequality can be written in the form

$$\int_{\partial B_r} u^{\alpha+1}\varphi^{-1}\langle \nabla\varphi, \nabla r \rangle \le B - F(r).$$

Now assume by contradiction that $F(r) \to +\infty$ as $r \to +\infty$. Thus, for each $r \ge R$ sufficiently large,

$$\int_{\partial B_r} u^{\alpha+1}\varphi^{-1}\langle \nabla\varphi, \nabla r \rangle \le -\frac{1}{2}F(r).$$

On the other hand, by the Schwarz inequality,

$$F(r) \leq -2 \int_{\partial B_r} u^{\alpha+1} \varphi^{-1} \langle \nabla \varphi, \nabla r \rangle$$

$$\leq \left\{ \int_{\partial B_r} u^{\alpha+1} \right\}^{1/2} \left\{ 4 \int_{\partial B_r} u^{\alpha+1} \varphi^{-2} |\nabla \varphi|^2 \right\}^{1/2},$$

that is,

$$F(r) \leq \left\{ \int_{\partial B_r} u^{\alpha+1} \right\}^{1/2} \{8F'(r)\}^{1/2}.$$

Thus, squaring,

$$\frac{F'(r)}{F(r)^2} \geq \frac{1}{8} \left\{ \int_{\partial B_r} u^{\alpha+1} \right\}^{-1}, \quad r \geq R.$$

Integrating this latter over $[R, r]$ gives

$$F(R)^{-1} - F(r)^{-1} \geq \frac{1}{8} \int_R^r \left\{ \int_{\partial B_t} u^{\alpha+1} \right\}^{-1} dt,$$

thus, letting $r \to +\infty$, we contradict (3.24) ii).

Step 3. We note that

$$\lim_{r \to +\infty} \int_{\partial B_r} u^{\alpha+1} \varphi^{-1} \langle \nabla \varphi, \nabla r \rangle = 0.$$

Indeed, because of Step 2, there exists a sequence $\{r_k\} \uparrow +\infty$ such that

$$\lim_{k \to +\infty} \left\{ \int_{\partial B_{r_k}} u^{\alpha+1} \right\} \left\{ \int_{\partial B_{r_k}} u^{\alpha+1} \varphi^{-2} |\nabla \varphi|^2 \right\} = 0$$

and therefore,

$$\left\{ \int_{\partial B_{r_k}} u^{\alpha+1} \varphi^{-1} \langle \nabla \varphi, \nabla r \rangle \right\}^2$$

$$\leq \left\{ \int_{\partial B_{r_k}} u^{\alpha+1} \right\} \left\{ \int_{\partial B_{r_k}} u^{\alpha+1} \varphi^{-2} |\nabla \varphi|^2 \right\} \to 0 \quad \text{as } k \to +\infty.$$

It remains to show that the desired limit exists. According to (3.22), (3.29) and the divergence theorem, we have

$$\int_{\partial B_r} u^{\alpha+1} \varphi^{-1} \langle \nabla \varphi, \nabla r \rangle = (\alpha + 1) \int_{B_r} u^\alpha \varphi^{-1} \langle \nabla \varphi, \nabla u \rangle \qquad (3.30)$$

$$- H \int_{B_r} u^{\alpha+1} a(x) - \int_{B_r} u^{\alpha+1} \varphi^{-2} |\nabla \varphi|^2.$$

Thus, we are reduced to showing that each of the summands on the right-hand side of (3.30) has a finite limit, as $r \to +\infty$. This easily follows from assumption (3.24) i), and from the following fact from Step 1:

$$\left| u^\alpha \varphi^{-1} \langle \nabla\varphi, \nabla u \rangle \right| \le 2u^{\alpha-1}|\nabla u|^2 + \frac{1}{2}u^{\alpha+1}\varphi^{-2}|\nabla\varphi|^2 \in L^1(M).$$

Step 4. Fix $\varepsilon > 0$ so small that $\sup_M u + \varepsilon < 1$, and define the vector field

$$Z_\varepsilon = H^{-1}\frac{u^{\alpha+1}}{\varphi}\nabla\varphi - \frac{(u+\varepsilon)^\alpha}{1-(u+\varepsilon)^{\sigma-1}}\nabla u.$$

Then, using (3.22), (3.26) and $\bar{b}(x) \ge a(x)$, we have

$$\text{div } Z_\varepsilon = H^{-1}(\alpha+1)u^\alpha\varphi^{-1}\langle\nabla\varphi,\nabla u\rangle + H^{-1}u^{\alpha+1}\varphi^{-1}\Delta\varphi$$

$$- H^{-1}u^{\alpha+1}\varphi^{-2}|\nabla\varphi|^2 - \alpha\frac{(u+\varepsilon)^{\alpha-1}}{1-(u+\varepsilon)^{\sigma-1}}|\nabla u|^2$$

$$- (\sigma-1)\frac{(u+\varepsilon)^{\alpha+\sigma-2}}{[1-(u+\varepsilon)^{\sigma-1}]^2}|\nabla u|^2 - \frac{(u+\varepsilon)^\alpha}{1-(u+\varepsilon)^{\sigma-1}}\Delta u$$

$$\le H^{-1}(\alpha+1)u^\alpha\varphi^{-1}\langle\nabla\varphi,\nabla u\rangle - H^{-1}(K+1)u^{\alpha+1}\varphi^{-2}|\nabla\varphi|^2$$

$$- a(x)u^{\alpha+1} - \alpha\frac{(u+\varepsilon)^{\alpha-1}}{1-(u+\varepsilon)^{\sigma-1}}|\nabla u|^2 - (\sigma-1)\frac{(u+\varepsilon)^{\alpha+\sigma-2}}{[1-(u+\varepsilon)^{\sigma-1}]^2}|\nabla u|^2$$

$$- \frac{(u+\varepsilon)^\alpha}{1-(u+\varepsilon)^{\sigma-1}}\{\bar{b}(x)u^\sigma - a(x)u\} \le$$

$$\le H^{-1}(\alpha+1)\varphi^{-1}u^\alpha\langle\nabla\varphi,\nabla u\rangle - H^{-1}(K+1)u^{\alpha+1}\varphi^{-2}|\nabla\varphi|^2$$

$$- \alpha\frac{(u+\varepsilon^{\alpha-1}}{1-(u+\varepsilon)^{\sigma-1}}|\nabla u|^2 - (\sigma-1)\frac{(u+\varepsilon)^{\alpha+\sigma-2}}{[1-(u+\varepsilon)^{\sigma-1}]^2}|\nabla u|^2$$

$$+ \frac{(u+\varepsilon)^\alpha}{1-(u+\varepsilon)^{\sigma-1}}a(x)(1-u^{\sigma-1})u - a(x)u^{\alpha+1}.$$

Integrating over B_r and applying the divergence theorem gives

$$H^{-1}\int_{\partial B_r} u^{\alpha+1}\varphi^{-1}\langle\nabla\varphi,\nabla r\rangle - \int_{\partial B_r}\frac{(u+\varepsilon)^\alpha}{1-(u+\varepsilon)^{\sigma-1}}\langle\nabla u,\nabla r\rangle \le I + II + III,$$

where we have set

$$I = \int_{B_r} H^{-1}\{(\alpha+1)\varphi^{-1}u^\alpha\langle\nabla\varphi,\nabla u\rangle - (K+1)u^{\alpha+1}\varphi^{-2}|\nabla\varphi|^2\};$$

$$II = \int_{B_r} -\alpha\frac{(u+\varepsilon)^{\alpha-1}}{1-(u+\varepsilon)^{\sigma-1}}|\nabla u|^2 - (\sigma-1)\frac{(u+\varepsilon)^{\alpha+\sigma-2}}{[1-(u+\varepsilon)^{\sigma-1}]^2}|\nabla u|^2;$$

$$III = \int_{B_r} a(x)\left\{u^{\alpha+1} - \frac{(u+\varepsilon)^\alpha}{1-(u+\varepsilon)^{\sigma-1}}(1-u^{\sigma-1})u\right\}.$$

Letting $\varepsilon \downarrow 0^+$ and applying the dominated convergence theorem, the left-hand side of the above converges to

$$H^{-1} \int_{\partial B_r} u^{\alpha+1}\varphi^{-1} \langle \nabla\varphi, \nabla r \rangle - \int_{\partial B_r} \frac{u^\alpha}{1 - u^{\sigma-1}} \langle \nabla u, \nabla r \rangle$$

while

$$II \to - \int_{B_r} \alpha \frac{u^{\alpha-1}}{1 - u^{\sigma-1}} |\nabla u|^2 + (\sigma - 1) \frac{u^{\alpha+\sigma-2}}{[1 - u^{\sigma-1}]^2} |\nabla u|^2$$

and finally

$$III \to 0.$$

Then, using

$$H^{-1} \frac{(\alpha+1)^2}{4(K+1)} \leq \alpha,$$

we obtain

$$H^{-1} \int_{\partial B_r} u^{\alpha+1}\varphi^{-1} \langle \nabla\varphi, \nabla r \rangle - \int_{\partial B_r} \frac{u^\alpha}{1 - u^{\sigma-1}} \langle \nabla u, \nabla r \rangle$$

$$\leq -H^{-1} \int_{B_r} \left\{ (K+1)\varphi^{-2}u^{\alpha+1}|\nabla\varphi|^2 + \frac{(\alpha+1)^2}{4(K+1)} u^{\alpha-1}|\nabla u|^2 \right.$$
$$\left. - (\alpha+1) \langle \nabla\varphi, \nabla u \rangle \varphi^{-1}u^\alpha \right\}$$

$$- \int_{B_r} \left\{ \alpha \frac{u^{\alpha-1}}{1 - u^{\sigma-1}} + (\sigma-1) \frac{u^{\alpha+\sigma-2}}{[1 - u^{\sigma-1}]^2} - H^{-1} \frac{(\alpha+1)^2}{4(K+1)} u^{\alpha-1} \right\} |\nabla u|^2$$

$$\leq -H^{-1} \int_{B_r} u^{\alpha-1} \left| \frac{\alpha+1}{2\sqrt{K+1}} \nabla u - \frac{u}{\varphi} \sqrt{K+1}\nabla\varphi \right|^2$$

$$- \alpha \int_{B_r} u^{(\alpha-1)+(\sigma-1)} \frac{1 + \frac{\sigma-1}{\alpha} - u^{\sigma-1}}{(1 - u^{\sigma-1})^2} |\nabla u|^2.$$

Now, letting $r \to +\infty$ along the sequence $\{r_k\}$ constructed in Step 1 and using Step 3 yields

$$0 \leq H^{-1} \int_{B_r} u^{\alpha-1} \left| \frac{\alpha+1}{2\sqrt{K+1}} \nabla u - \frac{u}{\varphi} \sqrt{K+1}\nabla\varphi \right|^2$$

$$- \alpha \int_{B_r} u^{\alpha+\sigma-2} \frac{1 + \frac{\sigma-1}{\alpha} - u^{\sigma-1}}{(1 - u^{\sigma-1})^2} |\nabla u|^2 \leq 0.$$

Let now $A \neq \emptyset$ be a connected component of the set

$$\{x \in M : u(x) > 0\}.$$

The above forces $\nabla u = 0$ on A and thus $u = c_1 > 0$ on A. In turn this implies $\nabla\varphi = 0$ on A and $\varphi = c_2 > 0$ on A. But, on all of M, hence on A,

$$\Delta\varphi + Ha(x)\varphi \le 0,$$

then

$$Ha(x)c_2 \le 0.$$

Because $a(x)c_1 - b(x)c_1^\sigma \ge 0$ implies $a(x)c_1 \ge 0$, it follows that $a(x) = 0$ on A and therefore, since

$$0 = \Delta u \ge -a(x)u + \bar{b}(x)u^\sigma = \bar{b}(x)c_1^\sigma \quad \text{on } A$$

we have $\bar{b}(x) = 0$ on A contradicting (3.20). □

Before commenting on Theorem 3.2 we prove the next result (see [PRS10]) in the same spirit.

Theorem 3.3. *Let $a(x), b(x) \in C^0(M)$ and assume that $b(x) \ge 0$. Let $H > 0$, $K > -1$, $A \in \mathbb{R}$ be constants satisfying*

$$\max\{0, A\} \le H(K+1) - 1, \tag{3.31}$$

and suppose that there exists a positive C^2-solution φ of the differential inequality

$$\Delta\varphi + Ha(x)\varphi \le -K\frac{|\nabla\varphi|^2}{\varphi} \quad \text{on } M. \tag{3.32}$$

Then the differential inequality

$$u\Delta u + a(x)u^2 - b(x)u^{\sigma+1} \ge -A|\nabla u|^2, \quad \sigma \ge 1, \tag{3.33}$$

has no nonnegative C^2-solutions u on M satisfying

$$\operatorname{supp} u \cap \{x \in M : b(x) > 0\} \ne \emptyset \tag{3.34}$$

and

$$\left\{\int_{\partial B_r} \varphi^{\frac{\beta+1}{H}(2-p)} u^{p(\beta+1)}\right\}^{-1} \notin L^1(+\infty) \tag{3.35}$$

for some $p > 1$ and $\max\{0, A\} \le \beta \le H(K+1) - 1$.

Remark. Note that if $p = 2$, then the nonintegrability assumption (3.35) involves u alone and reduces to

$$\left\{\int_{\partial B_r} u^{2(\beta+1)}\right\}^{-1} \notin L^1(+\infty). \tag{3.36}$$

Proof. Let $u \geq 0$ be a solution of (3.33) on M satisfying (3.34) and (3.35). Fix $\varepsilon > 0$, $\alpha \in \mathbb{R}$ and set

$$v = \varphi^{-\alpha}(u^2 + \varepsilon)^{\frac{\beta+1}{2}}.$$

A straightforward computation that uses (3.32) and (3.33) yields

$$v \operatorname{div}(\varphi^{2\alpha} \nabla v) \geq \alpha(K - \alpha + 1)(u^2 + \varepsilon)^{\beta+1}\frac{|\nabla \varphi|^2}{\varphi^2} + (\beta + 1)(u^2 + \varepsilon)^{\beta} b(x) u^{\sigma+1}$$

$$+ a(x)(u^2 + \varepsilon)^{\beta+1}\left[\alpha H - (\beta + 1)\frac{u^2}{u^2 + \varepsilon}\right]$$

$$+ (\beta + 1)(u^2 + \varepsilon)^{\beta}\left[1 - A + (\beta - 1)\frac{u^2}{u^2 + \varepsilon}\right]|\nabla u|^2.$$

We choose $\alpha = H^{-1}(\beta + 1)$, so that our assumptions on β, H and K give $0 < \alpha \leq K + 1$. Therefore $\alpha(K - \alpha + 1) \geq 0$, and using $b(x) \geq 0$ and $\beta + 1 \geq 0$ we deduce that

$$v \operatorname{div}(\varphi^{2\alpha} \nabla v) \geq \varepsilon(\beta + 1) a(x)(u^2 + \varepsilon)^{\beta} \tag{3.37}$$

$$+ (\beta + 1)(u^2 + \varepsilon)^{\beta}\left[1 - A + (\beta - 1)\frac{u^2}{u^2 + \varepsilon}\right]|\nabla u|^2.$$

Let $r(t) \in C^1(\mathbb{R})$ and $s(t) \in C^0(\mathbb{R})$ satisfy the conditions

$$i)\, r(v) \geq 0, \quad ii)\, r(v) + vr'(v) \geq s(v) > 0 \text{ on } [0, +\infty) \tag{3.38}$$

and consider the vector field

$$Z = vr(v)\varphi^{2\alpha}.$$

For fixed t and $\delta > 0$ let also ψ_δ be the Lipschitz function defined by

$$\psi_\delta(x) = \begin{cases} 1, & \text{if } r(x) \leq t, \\ \frac{t + \delta - r(x)}{\delta}, & \text{if } t < r(x) < t + \delta, \\ 0, & \text{if } r(x) \geq t + \delta. \end{cases}$$

Using (3.37), (3.38) and the definition of ψ_δ we compute

$$\operatorname{div}(\psi_\delta) = \psi_\delta \operatorname{div} Z + \langle \nabla \psi_\delta, Z \rangle$$

$$\geq (\beta + 1)r(v)(u^2 + \varepsilon)^{\beta}\left\{\varepsilon a(x) + \left[1 - A + (\beta - 1)\frac{u^2}{u^2 + \varepsilon}\right]|\nabla u|^2\right\}\chi_{B_t}$$

$$+ s(v)\varphi^{2\alpha}|\nabla v|^2\chi_{B_t} + \frac{1}{\delta}\langle \nabla r, Z \rangle\,\chi_{\bar{B}_{t+\delta}\setminus B_t}$$

whence, integrating and using the divergence theorem and Cauchy-Schwarz inequality we obtain

$$\int_{B_t} (\beta+1)r(v)(u^2+\varepsilon)^\beta \left\{ \varepsilon r(x) + \left[1 - A + (\beta-1)\frac{u^2}{u^2+\varepsilon} \right] |\nabla u|^2 \right\}$$
$$+ \int_{B_t} s(v)\varphi^{2\alpha}|\nabla v|^2 \leq \frac{1}{\delta}\int_{B_{t+\delta}\setminus B_t} |Z|.$$

By Hölder's inequality the integral on the right-hand side is bounded above by

$$\left\{ \frac{1}{\delta}\int_{B_{t+\delta}\setminus B_t} \varphi^{2\alpha}\frac{r(v)^2}{s(v)}v^2 \right\}^{\frac{1}{2}} \left\{ \frac{1}{\delta}\int_{B_{t+\delta}\setminus B_t} \varphi^{2\alpha}s(v)|\nabla v|^2 \right\}^{\frac{1}{2}}.$$

Inserting this into the above inequality, letting $\delta \downarrow 0^+$ and using the co-area formula (1.87) we deduce that

$$\int_{B_t} (\beta+1)r(v)(u^2+\varepsilon)^\beta \left\{ \varepsilon\, a(x) + \left[1 - A + (\beta-1)\frac{u^2}{u^2+\varepsilon} \right] |\nabla u|^2 \right\} \qquad (3.39)$$
$$+ s(v)\varphi^{2\alpha}|\nabla v|^2 \leq \left\{ \varphi^{2\alpha}\frac{r(v)^2}{s(v)}v^2 \right\}^{\frac{1}{2}} \left\{ \varphi^{2\alpha}s(v)|\nabla v|^2 \right\}^{\frac{1}{2}}.$$

As $\varepsilon \downarrow 0$, $v = v_\varepsilon \to v_0 = \varphi^{-\alpha}u^{\beta+1}$, therefore, using the dominated convergence theorem in (3.39) we get

$$\int_{B_t} s(v_0)\varphi^{2\alpha}|\nabla v_0|^2 + (\beta+1)(\beta-A)\int_{B_t} r(v_0)u^{2\beta}|\nabla u|^2 \qquad (3.40)$$

$$\leq \left\{ \varphi^{2\alpha}\frac{r(v_0)^2}{s(v_0)}v_0^2 \right\}^{\frac{1}{2}} \left\{ \varphi^{2\alpha}s(v_0)|\nabla v_0|^2 \right\}^{\frac{1}{2}}. \qquad (3.41)$$

Define

$$h(t) = \int_{B_t} \varphi^{2\alpha}s(v_0)|\nabla v_0|^2;$$

by the co-area formula (1.87) h is Lipschitz and

$$h'(t) = \int_{\partial B_t} \varphi^{2\alpha}s(v_0)|\nabla v_0|^2,$$

and noting that the coefficient of the second integral on the left-hand side of (3.40) is nonnegative by the conditions imposed on β, we obtain

$$h(t) \leq \left\{ \int_{\partial B_t} \varphi^{2\alpha}\frac{r(v_0)^2}{s(v_0)}v_0^2 \right\}^{\frac{1}{2}} [h'(t)]^{\frac{1}{2}}. \qquad (3.42)$$

Our aim is to show that under assumption (3.35) v_0 is constant. We reason by contradiction, then there exists R_0 such that $h(t) > 0 \, \forall t \geq R_0$ and therefore the right-hand side of (3.42) is positive for $t \geq R_0$. Dividing through by $h(t)$, squaring and integrating the resulting differential inequality between R and r with $R_0 \leq R \leq r$ yields

$$h(R)^{-1} \geq h(R)^{-1} - h(r)^{-1} \geq \int_R^r \left\{ \varphi^{2\alpha} \frac{r(v_0)^2}{s(v_0)} v_0^2 \right\}^{-1} dt. \qquad (3.43)$$

We choose a sequence f of functions

$$r_n(t) = \left(t^2 + \frac{1}{n} \right)^{\frac{p-2}{2}}, \quad s_n(t) = \min \{ p-1, 1 \} \, r_n(t) \ \ \forall n \in \mathbb{N}, p > 1.$$

Since condition (3.38) holds for all n, so does (3.43), hence, letting $n \to +\infty$ and using the Lebesgue and monotone convergence theorems we deduce that there exists $c > 0$ depending only on p such that

$$\left\{ \int_{B_R} v_0^{p-2} \varphi^{2\alpha} |\nabla v_0|^2 \right\}^{-1} \geq C \int_R^r \left\{ \int_{\partial B_t} \varphi^{2\alpha} v_0^p \right\}^{-1} dt. \qquad (3.44)$$

Now, recalling that $\alpha = (\beta + 1)/H$ and the definition of v_0, we have

$$\varphi^{2\alpha} v_0^p = \varphi^{(2-p)(\beta+1)/H} u^{2(\beta+1)}$$

and the required contradiction is obtained by letting $r \to +\infty$ and using assumption (3.35). Thus v_0 is constant, and we deduce that there exists a constant $C \geq 0$ such that

$$u^H = C\varphi.$$

Since u is not identically zero by (3.34), $C > 0$ and u is strictly positive on M. We insert the expression of φ in terms of u in (3.32), divide by $CH u^{H-2}$ and subtract the result from (3.33) to obtain

$$[A - H(K+1)]|\nabla u|^2 \geq b(x)u^{\sigma+1}.$$

Since the coefficient of $|\nabla u|^2$ is nonpositive, by (3.31) we conclude that

$$b(x)u^{\sigma+1} \leq 0$$

which contradicts (3.34).   □

Remark. Observe that the above proof actually shows that if $A < H(K+1) - 1$, then $\nabla u = 0$, so that u, and therefore φ, are necessarily constant. It follows from (3.32) and (3.33) that $0 \geq a(x)u \geq b(x)u^{\sigma+1} \geq 0$, so that, if $a(x)$ does not vanish identically, then $u \equiv 0$ without any strict positivity assumption on $b(x)$. On the other hand, if $A = H(K+1) - 1$, then the conclusion depends on the fact that $b(x)$ is positive somewhere.

Remark. We also note that if u is assumed to be strictly positive, then the conclusion of Theorem 3.3 holds, with a much easier proof, if we assume that

$$\max\{-1, A\} \leq H(K+1) - 1$$

and that

$$\beta > -1, \ \beta \geq A, \ \beta \leq H(K+1) - 1.$$

We observe that in both Theorems 3.2 and 3.3 we require some integrability type condition for the positive (nonnegative) solution u of (3.23)

$$\Delta u + a(x)u \geq b(x)u^\sigma, \tag{3.23}$$

respectively (3.24) ii), that is

$$\left\{ \int_{\partial B_r} u^{\alpha+1} \right\}^{-1} \notin L^1(+\infty) \quad \text{for some } \alpha > 0$$

and (3.36), that is

$$\left\{ \int_{\partial B_r} u^{2(1+\beta)} \right\}^{-1} \notin L^1(+\infty) \quad \text{for some } \max\{0, A\} \leq \beta \leq H(K+1) - 1.$$

Sometimes this type of information is already contained in the differential inequality. This is the content of the next

Proposition 3.4. *Let* $(M, \langle\,,\,\rangle)$ *be a complete Riemannian manifold and let* $a(x)$, $b(x) \in C^0(M)$ *with* $b(x) > 0$ *on* M. *Assume that* $u \geq 0$ *is a* C^2-*solution of the differential inequality*

$$u\Delta u + a(x)u^2 - b(x)u^{\sigma+1} \geq -A|\nabla u|^2 \tag{3.45}$$

for $A \leq 1$ *and* $\sigma > 1$. *Then, for every* $p \geq 1$, $p > A + 2$ *there exist constants* $C_1, C_2 > 0$ *which depend only on* p, σ *and* $R_0 > 0$ *sufficiently large, such that,* $\forall R \geq R_0$,

$$\int_{B_R} b(x)u^{p+\sigma-2} \leq C_1 R^{-2\frac{p+\sigma-2}{\sigma-1}} \int_{B_{2R}} b(x)^{-\frac{p-1}{\sigma-1}} + C_2 \int_{B_{2R}} \left(\frac{a_+(x)}{b(x)} \right)^{\frac{p-1}{\sigma-1}} a_+(x), \tag{3.46}$$

where a_+ *denotes the positive part of* a.

Proof. First we observe that we may suppose that $u \not\equiv 0$, for otherwise there is nothing to prove. Thus, there exists $R_0 > 0$ such that $u \not\equiv 0$ on $B_R, \forall R \geq R_0$. Next, $\forall R \geq R_0$ let $\psi = \psi_R : M \to [0, 1]$ be a smooth cut-off function such that

$$\psi \equiv 1 \text{ on } B_R, \quad \psi \equiv 0 \text{ on } M \backslash B_{2R}, \quad |\nabla \psi| \leq \frac{C}{R} \psi^{\frac{p-1}{p+\sigma-2}} \text{ on } B_{2R} \tag{3.47}$$

for some $C > 0$ which depends only on σ and p. Note that this is possible since the exponent

$$\frac{p-1}{p+\sigma-2} < 1.$$

Having fixed $\varepsilon > 0$, we let W be the vector field defined by

$$W = \psi^2(u+\varepsilon)^{p-3}u\nabla u;$$

a computation that uses (3.45) yields

$$\operatorname{div} W \geq \psi^2(u+\varepsilon)^{p-3}\left\{-a(x)u^2 + b(x)u^{\sigma+1} + \left(1 - A - (p-3)\frac{u}{u+\varepsilon}\right)|\nabla u|^2\right\}$$

$$+ 2\psi(u+\varepsilon)^{p-3}u\,\langle\nabla u, \nabla\psi\rangle.$$

We estimate the last term on the right-hand side using Cauchy-Schwarz inequality and Young inequality $2ab \leq \lambda a^2 + \lambda^{-1}b^2$ with $\lambda = p - 2 - A > 0$ to obtain

$$\operatorname{div} W \geq \psi^2(u+\varepsilon)^{p-3}\left\{-a(x)u^2 + b(x)u^{\sigma+1}\right\} - \frac{1}{p-2-A}u(u+\varepsilon)^{p-2}|\nabla\psi|^2.$$

We integrate the above inequality, apply the divergence theorem, rearrange, let $\varepsilon \to 0^+$ and use the dominated convergence theorem in this order, to deduce that

$$\int_{B_{2R}} b(x)\psi^2 u^{p+\sigma-2} \leq \frac{1}{p-2-A}\int_{B_{2R}} u^{p-1}|\nabla\psi|^2 + \int_{B_{2R}} \psi^2 a_+(x)u^{p-1}. \quad (3.48)$$

If $p = 1$, the conclusion follows immediately using (3.47). If $p > 1$, we denote by I and II the two integrals on the right-hand side of (3.48), and use Hölder inequality with conjugate exponents

$$\frac{p+\sigma-2}{p-1} > 1 \quad \text{and} \quad \frac{p+\sigma-2}{\sigma-1}$$

and the assumption that $b(x) > 0$ to estimate

$$I \leq \left\{\int_{B_{2R}} b(x)\psi^2 u^{p+\sigma-2}\right\}^{\frac{p-1}{p+\sigma-2}}\left\{\int_{B_{2R}} \psi^{-2\frac{p-1}{\sigma-1}}b(x)^{-\frac{p-1}{\sigma-1}}|\nabla\psi|^{2\frac{p+\sigma-2}{\sigma-1}}\right\}^{\frac{\sigma-1}{p+\sigma-2}}$$

and

$$II \leq \left\{\int_{B_{2R}} b(x)\psi^2 u^{p+\sigma-2}\right\}^{\frac{p-1}{p+\sigma-2}}\left\{\int_{B_{2R}} \psi^2 a_+(x)^{\frac{p+\sigma-2}{\sigma-1}}b(x)^{-\frac{p-1}{\sigma-1}}\right\}^{\frac{\sigma-1}{p+\sigma-2}}.$$

Inserting into (3.48), noting that the integral on the left-hand side is strictly positive by the choice of R, and simplifying, we obtain

$$\int_{B_{2R}} b(x)\psi^2 u^{p+\sigma-2} \leq \left\{\frac{1}{p-2-A}\left[\int_{B_{2R}} \psi^{-2\frac{p-1}{\sigma-1}}b(x)^{-\frac{p-1}{\sigma-1}}|\nabla\psi|^{2\frac{p+\sigma-2}{\sigma-1}}\right]^{\frac{\sigma-1}{p+\sigma-2}}\right.$$

$$\left. + \left[\int_{B_{2R}} \psi^2 a_+(x)^{\frac{p+\sigma-2}{\sigma-1}}b(x)^{-\frac{p-1}{\sigma-1}}\right]^{\frac{\sigma-1}{p+\sigma-2}}\right\}.$$

The required conclusion follows again using (3.47) and the elementary inequality $(a+b)^\tau \leq 2^\tau (a^\tau + b^\tau)$ valid for $a, b, \tau \geq 0$. $\qquad\square$

For the sake of illustration let us consider the case

$$b(x) > 0, \quad b(x) \geq \frac{C}{r(x)^\mu} \quad \text{for } r(x) \gg 1 \tag{3.49}$$

and some constants $C > 0, \mu \geq 0$. Then, from (3.46) we have

$$\int_{B_R} u^{p+\sigma-2} \leq C_1 R^{\mu - 2\frac{p+\sigma-2}{\sigma-1} + \frac{p-1}{\sigma-1}\mu} \operatorname{vol}(B_{2R}) + C_2 \int_{B_{2R}} \frac{a_+(x)^{\frac{p+\sigma-2}{\sigma-1}}}{b(x)} a_+(x). \tag{3.50}$$

In particular, if $\frac{a_+(x)}{b(x)}$ is bounded above and $a_+(x) \in L^1(M)$,

$$\int_{B_{2R}} u^{p+\sigma-2} \leq C_2 + C_1 R^{\frac{\mu-2}{\sigma-1}(p+\sigma-2)} \operatorname{vol}(B_{2R}). \tag{3.51}$$

We observe now that (3.24) or (3.36) are implied by

$$\frac{r}{\int_{B_r} u^\xi} \notin L^1(+\infty), \quad \xi \geq 0$$

and from (3.51)

$$\left\{ \int_{\partial B_r} u^{p+\sigma-2} \right\}^{-1} \notin L^1(+\infty)$$

in case

$$C_1 r^{\frac{\mu-2}{\sigma-1}(p+\sigma-2)} \operatorname{vol}(B_{2R}) \leq C r^2 \log r \quad \text{for } r \gg 1,$$

in other words if

$$\operatorname{vol}(B_{2r}) \leq C r^{2 - \frac{\mu-2}{\sigma-1}(p+\sigma-2)} \log r \quad \text{for } r \gg 1.$$

This is a slow growth for $\operatorname{vol}(B_r)$.

This somehow forces us to find a better way to guarantee integrability and this will be achieved with upper *a priori* estimates which however are obtained under lower bound type restrictions on the Ricci tensor. We shall return to this in Chapter 4.

Using Proposition 3.4 we obtain the following

Theorem 3.5. *Let* $(M, \langle\,,\,\rangle)$ *be a complete Riemannian manifold, and let* $a(x)$, $b(x) \in C^0(M)$ *where* $b(x) > 0$ *on* M *and*

$$b(x) \geq \frac{C}{r(x)^\mu} \tag{3.52}$$

for $r(x) \gg 1$ and for some constants $C > 0$ and $0 \le \mu \le 2$. Assume that

(i) $\sup_{M} \dfrac{a_+(x)}{b(x)} < +\infty$ *and* (ii) $\displaystyle\int_{B_r} a_+(x) = O(r^{2-\mu} \log r)$ *as* $r \to +\infty$,   (3.53)

and that, for some $H \ge 1$, the operator $L_H = \Delta + Ha(x)$ satisfies

$$\lambda_1^{L_H}(M) \ge 0. \tag{3.54}$$

Finally, let A and σ be such that $A \le 1$, $A < H - 1$, $1 < \sigma \le 2H + 1$ and $\sigma < 2H - A$ and assume that

$$\mathrm{vol}\, B_r = O\!\left(r^{2+(2-\mu)\frac{2H}{\sigma-1}} \log r\right) \quad \text{as } r \to +\infty. \tag{3.55}$$

Then the only nonnegative C^2 solution u of the differential inequality

$$u\Delta u + a(x)u^2 - b(x)u^{\sigma+1} \ge -A|\nabla u|^2 \tag{3.56}$$

is $u \equiv 0$.

Proof. If we set $p = 2H + 2 - \sigma$, the conditions imposed on the parameters imply that p satisfies the assumptions listed in the statement of Proposition 3.4. The lemma and condition (3.53) (i) show that there exist constants $C_i > 0$ such that

$$\int_{B_r} b(x)u^{2H} \le C_1 r^{-\frac{4H}{\sigma-1}} \int_{B_{2r}} b(x)^{1-\frac{2H}{\sigma-1}} + C_2 \int_{B_{2r}} a_+(x) \tag{3.57}$$

for $r > 0$ sufficiently large. We use condition (3.52) to estimate from below the integral on the left-hand side. On the other hand, since $\sigma < 2H + 1$, we may again use condition (3.52) to estimate from above the first integral on the right-hand side, and (3.53) (ii) to estimate from above the second integral, and deduce that, for r sufficiently large,

$$\int_{B_r} u^{2H} \le C\!\left(r^{(\mu-2)\frac{2H}{\sigma-1}} \mathrm{vol}\, B_{2r} + r^2 \log r\right),$$

whence, using the volume growth condition (3.57) we conclude that

$$\int_{B_r} u^{2H} \le C r^2 \log r \quad \text{for } r \gg 1.$$

This immediately implies that, for r large,

$$\frac{r}{\int_{B_r} u^{2H}} \ge C\frac{1}{r \log r} \notin L^1(+\infty),$$

which in turn yields (see, e.g., [RS01] Proposition 1.3)

$$\frac{1}{\int_{\partial B_r} u^{2H}} \notin L^1(+\infty).$$

We may therefore apply Theorem 3.3 with $K = 0$ and $\beta = H - 1$ to deduce that $\mathrm{supp}\, u = \emptyset$, that is, $u \equiv 0$. $\square$

Remark. The argument used in the proof shows that the condition that $\frac{a_+}{b}$ is bounded above may be removed provided we replace (3.57) with

$$\int_{B_r} a_+^{\frac{2H}{\sigma-1}} = O\left(r^{2-\mu\frac{2H}{\sigma-1}} \log r\right) \quad \text{as } r \to +\infty. \tag{3.58}$$

Note that since the integral on the left-hand side is a nondecreasing function of r this also imposes the further restriction $\mu \leq (\sigma - 1)/H$, with corresponding restrictions being imposed on the range of the other parameters.

3.2 The endpoint case $K = -1$ and the Poisson equation

Theorem 3.3 does not cover the "endpoint" case where $K = -1$ in (3.22), which we are going to consider presently. We therefore assume that there exists a positive solution φ of

$$\Delta\varphi + Ha(x)\varphi \leq \frac{|\nabla\varphi|^2}{\varphi}, \quad H > 0. \tag{3.59}$$

If u is a C^2 solution of (3.33) with $\sigma \geq 0$, we define $v = \varphi^{-\gamma}u$, $\gamma \geq 0$. A computation that uses (3.59), (3.33) and Young inequality yields

$$
\begin{aligned}
v\Delta v &= (\gamma H - 1)a(x)\varphi^{-2\gamma}u^2 + b(x)\varphi^{-2\gamma}u^{\sigma+1} \\
&\quad + \gamma^2\varphi^{-2\gamma-2}u^2|\nabla\varphi|^2 - A\varphi^{-2\gamma}|\nabla u|^2 - 2\gamma\varphi^{-2\gamma-1}u\langle\nabla u, \nabla\varphi\rangle \\
&\geq (\gamma H - 1)a(x)\varphi^{-2\gamma}u^2 + b(x)\varphi^{-2\gamma}u^{\sigma+1} \\
&\quad + \gamma^2(1 - \frac{1}{\varepsilon})\varphi^{-2\gamma-2}u^2|\nabla\varphi|^2 - (A+\varepsilon)\varphi^{-2\gamma}|\nabla u|^2.
\end{aligned}
$$

Choosing $\gamma = 1/H$ and $\varepsilon = -A$, the right-hand side reduces to

$$b(x)\varphi^{-2/H}u^{\sigma+1} + \frac{1}{H^2}(1 + \frac{1}{A})\varphi^{-2/H-2}u^2|\nabla\varphi|^2,$$

and we easily deduce that, if $A \leq -1$, then the function v satisfies

$$\Delta v \geq b(x)\varphi^{(\sigma-1)/H}v^\sigma. \tag{3.60}$$

Using (3.60) we obtain the following version of Theorem 3.3.

Theorem 3.6. *Let $a(x)$, $b(x) \in C^0(M)$, with $b(x) \geq 0$, and assume that φ is a positive C^2 solution of (3.59) satisfying*

$$\varphi(x) \geq Cr(x)^{1/\delta} \tag{3.61}$$

for $r(x) \gg 1$, and some constants $C > 0$ and $\delta > 0$. Then the differential inequality

$$u\Delta u + a(x)u^2 - b(x)u^{\sigma+1} \geq -A|\nabla u|^2,$$

with $\sigma \geq 0$ and $A \leq -1$, has no nonnegative C^2 solution satisfying

$$\operatorname{supp} u \cap \{x : M : b(x) > 0\} \neq \emptyset \tag{3.62}$$

and

$$\frac{r^{\delta p}}{\int_{\partial B_r} u^p} \notin L^1(+\infty) \tag{3.63}$$

for some $p > 1$.

Proof. According to (3.60) above, the function $v = \varphi^{-1/H} u$ is sub-harmonic. Further, (3.61) and (3.63) imply that

$$\left(\int_{\partial B_r} v^p \right)^{-1} \notin L^1(++\infty).$$

An application of Theorem B in [RS01] shows that v is constant. The conclusion now follows as in the proof of Theorem 3.3. □

Note that, even in the case of Theorem 3.6, if $A < -1$, then the conclusion can be strengthened to assert that every nonnegative solution of (3.6) vanishes identically, unless $a(x) = b(x) \equiv 0$.

In applying Theorem 3.6 it is of course crucial to be able to find positive solutions of (3.59) satisfying the asymptotic lower bound (3.61). By contrast, in order to apply Theorem 3.3 one needs a positive solution of (3.22), whose existence, in typical applications like the one exemplified by Theorem 3.5 above, is guaranteed by means of assumptions on the spectrum of a suitable operator.

Observe now that if v is a solution of the Poisson equation

$$\Delta v = a(x),$$

then the function $\varphi = e^{-v}$ is a positive solution of

$$\Delta \varphi + a(x)\varphi = \frac{|\nabla \varphi|^2}{\varphi},$$

and furthermore, an upper bound for v yields a lower bound for φ.

The Poisson equation on complete Riemannian manifolds has been extensively studied using heat kernel techniques to obtain bounds on the Green kernel. To illustrate an application of Theorem 3.6, we consider the elementary case where the positive part of $a(x)$ is integrable. We first recall that a manifold is said to be *parabolic* if every positive bounded-above subharmonic function is constant, and *nonparabolic* otherwise (see [Gri99], [PRS05b]); we also recall that the definition of non-parabolicity is equivalent to the existence of a positive Green kernel (see again [Gri99]). Then we have the following lemma (see, e.g., the proof of Theorem 3.2 in [NST01]).

Lemma 3.7. *Let $(M, \langle \, , \, \rangle))$ be a complete, nonparabolic manifold, and let $\rho \in C^{0,\alpha}(M) \cap L^1(M)$, $0 \leq \alpha < 1$, be a nonnegative function. Then, there exists a solution $v \in C^2$ of the Poisson equation*

$$\Delta v = \rho$$

satisfying $v \leq 0$.

Proof. Let $G(x, y)$ be the Green kernel, i.e., the minimal positive fundamental solution of the Laplacian, which exists by the assumption that M is nonparabolic. The Green kernel is symmetric and, if $\psi \in C_c^\infty(M)$, then the function $u(x) = -\int_M G(x, y)\psi(y)$ is smooth and satisfies $\Delta u = \psi$.

We claim that if $\rho \in C^0(M) \cap L^1(M)$, then the function $v = -\int_M G(x, y)\rho(y)$ is well defined and locally bounded. Assuming the claim, for every $\psi \in C_0^\infty(M)$ we have

$$\int_M v \Delta \psi = -\int_M \psi(x) \int_M \Delta G(x, y)\rho(y)$$

$$= -\int_M \rho(y) \int_M G(x, y)\Delta \psi(x) = \int_M \rho(y)\psi(y),$$

so that v satisfies the Poisson equation in distributional sense, and therefore, by standard elliptic regularity (see [Aub98]), Theorem 3.55), it is a classical solution. Clearly, v is nonpositive.

To prove the claim, fix $R > 0$ and for every $x \in B_R$ we write

$$\int_M G(x, y)\rho(y) = \int_{B_{2R}} G(x, y)\rho(y) + \int_{M \setminus B_{2R}} G(x, y)\rho(y). \qquad (3.64)$$

Since $G(x, y)$ is locally integrable uniformly for $x \in B_R$, the first integral on the right-hand side is bounded above by a constant independent of $x \in B_R$.

On the other hand, by the local Harnack inequality there exists a constant C independent of $x \in B_R$ and such that

$$G(x, y) \leq CG(o, y) \quad \text{for every } y \in M \setminus B_{2R}.$$

Moreover,

$$\sup_{M \setminus B_{2R}} G(o, y) < +\infty.$$

Indeed, let Ω_n be an exhaustion of M by open sets containing o and with smooth boundary and let G_n by the Green kernel of Ω_n and recall that, by the standard construction of the Green kernel $G(x, y)$, $G_n(x, y) \to G(x, y)$ locally uniformly in $M \setminus \{x\}$. Let $C > \sup_{\partial B_{2R}} G(o, y)$, then, for every sufficiently large n we have $C > G(o, y) \geq G_n(o, y)$ for $y \in \partial B_{2R}$ and clearly $C > G_n(o, y) = 0$ for $y \in \partial \Omega_n$. Thus, by the comparison principle, $C > G_n(o, y)$ in $\Omega \setminus B_{2R}$, whence, letting

$n \to +\infty$, $G(o, y) \leq C$ for $y \in M \setminus B_{2R}$. It follows that there exists a constant C' independent of $x \in B_R$ and $y \in M \setminus B_{2R}$ such that

$$G(x, y) \leq C'.$$

Since ρ is integrable, this implies that the second integral on the right-hand side of (3.64) is also bounded independently of $x \in B_R$, as required to complete the proof of the claim. $\qquad\square$

Corollary 3.8. *Let* $(M, \langle\,,\,\rangle)$ *be a complete, nonparabolic manifold, let the functions* $a(x) \in C^{0,\alpha}(M)$, *and* $b(x) \in C^0(M)$ *satisfy* $b(x) > 0$ *and*

$$a_+(x) \in L^1(M), \tag{3.65}$$

and suppose that for some constants $\sigma > 1$, $A \in (-\infty, -1]$ *and* μ, p, q *satisfying*

$$q > \max\{1, 3 - \sigma\}, \quad 0 \leq \mu \leq 2\frac{\sigma - 1}{\sigma + q - 2}, \quad p > \frac{q + \sigma - 2}{\sigma - 1},$$

we have

$$\int_{B_r} a_+(x)^p = O\left(r^{[2(\sigma-1)-\mu(q+\sigma-2)]\frac{p-1}{q-1}}\right) \quad as \ \ r \to +\infty \tag{3.66}$$

$$\mathrm{vol}\, B_r = O\left(r^{2+(2-\mu)\frac{\sigma+q-2}{\sigma-1}}\right) \quad as \ \ r \to +\infty \tag{3.67}$$

$$b(x) \geq \frac{C}{r(x)^\mu} \quad for \ \ r(x) \gg 1. \tag{3.68}$$

Then there are no nonnegative, nonidentically zero $C^2(M)$ *solutions of the differential inequality*

$$u\Delta u + a(x)u^2 \geq b(x)u^{\sigma+1} - A|\nabla u|^2 \quad on \ M. \tag{3.69}$$

Proof. Since a_+ is integrable, by Lemma 3.7 and the preceding discussion there exists a solution $\varphi \geq 1$ of

$$\Delta\varphi + a_+\varphi = \frac{|\nabla\varphi|^2}{\varphi},$$

and φ is a solution of the differential inequality (3.59).

Now, let u be a nonnegative solution of (3.69). Noting that $q > 1$, and $A \leq -1$ imply $q > A + 2$, applying Proposition 3.4 and using the lower bound (3.68) for $b(x)$ we obtain

$$\int_{B_r} u^{q+\sigma-2} \leq C_1 r^{(\mu-2)\frac{q+\sigma-2}{\sigma-1}} \mathrm{vol}\, B_{2r} + C_2 r^{\mu\frac{q+\sigma-2}{\sigma-1}} \int_{B_{2r}} a_+(x)^{\frac{\sigma+q-2}{\sigma-1}}. \tag{3.70}$$

We claim that (3.66) implies

$$\int_{B_{2r}} a_+(x)^{\frac{\sigma+q-1}{\sigma-1}} = O\left(r^{2-\mu\frac{\sigma+q-2}{\sigma-1}}\right) \quad as \ \ r \to +\infty, \tag{3.71}$$

which, together with the volume growth assumption (3.67) yields

$$\int_{B_r} u^{q+\sigma-2} = O(r^2) \quad \text{as} \quad r \to +\infty.$$

As in the proof of Theorem 3.5, it follows (see, e.g., [RS01] Proposition 1.3) that u satisfies condition (3.63) with $\delta = 0$ and exponent $q + \sigma - 2$ which is greater than 1 by the conditions on q. Thus, Theorem 3.6 (with $a(x)$ replaced by $a_+(x)$) applies and u vanishes identically.

To conclude it remains to prove the claim. To this end, we set $p' = (q + \sigma - 2)/(\sigma - 1)$, and apply Hölder inequality with conjugate exponents $(p-1)/(p-p')$ and $(p-1)/(p'-1)$ to estimate

$$\int_{B_r} a_+(x)^{p'} = \int_{B_r} a_+(x)^{\frac{p-p'}{p-1}+p\frac{p'-1}{p-1}}$$

$$\leq \left(\int_{B_{2r}} a_+(x) \right)^{\frac{p-p'}{p-1}} \left(\int_{B_{2r}} a_+(x)^p \right)^{\frac{p'-1}{p-1}}$$

$$= O\left(r^{[2(\sigma-1)-\mu(q+\sigma-2)]\frac{p'-1}{q-1}} \right) = O\left(r^{2-\mu\frac{q+\sigma-2}{\sigma-1}} \right),$$

as required. □

Remark. Assume that $b(x)$ satisfies the condition stated in the corollary, with $\mu < \sigma - 1$, and that conditions (3.66) and (3.67) are replaced by

$$\int_{B_r} a_+(x)^{\frac{2}{\sigma-1}} = O\left(r^{2[1-\frac{\mu}{\sigma-1}]} \right) \quad \text{as} \quad r \to +\infty,$$

$$\text{vol} B_r = O\left(r^{2[1-\frac{\mu-2}{\sigma-1}]} \right) \quad \text{as} \quad r \to +\infty.$$

It follows from (3.70) above with $q + \sigma - 2 = 2$, that every nonnegative solution of

$$\Delta u + a(x)u - b(x)u^\sigma = 0 \tag{3.72}$$

satisfies

$$\int_{B_r} u^2 \leq Cr^2 \log r, \tag{3.73}$$

and the same estimate is clearly satisfied by the difference of two solutions. An application of Theorem 4.1 in [BRS98] shows that (3.72) has at most one positive solution. We remark in this respect that if we replace u^2 in (3.73) with u^p with $p > 2$, then the conclusion of Theorem 4.1 in [BRS98] fails, as the example described on pages 214-215 therein shows (see also Theorem 5.1 below).

Clearly, whenever a solution to the Poisson equation is available one can obtain a version of Corollary 3.8 above. According to a recent result by O. Munteanu and N. Sesum, [MS10], if $(M, \langle , \rangle)$ has Ricci curvature bounded from below by a negative constant, and its Laplace operator has positive bottom of the spectrum, and if ρ satisfies the decay condition

$$|\rho(x)| \leq \frac{C}{(1 + r(x))^\gamma}$$

for some $\gamma > 1$, then the Poisson equation

$$\Delta v = \rho(x)$$

has a solution satisfying

$$|v(x)| \leq A e^{Br(x)}$$

for some $A, B > 0$. Applying this result, and arguing as in the proof of Corollary 3.8 one obtains the following

Corollary 3.9. *Let $(M, \langle \, , \rangle)$ be a complete manifold with Ricci curvature bounded from below by a negative constant, and positive bottom of the spectrum. Let $a(x) \in C^{0,\alpha}(M)$, and $0 < b(x) \in C^0(M)$ and suppose that for some constants $\sigma > 1$, $A \in (-\infty, -1]$, $\mu > 0$, $p \geq 1$, $\gamma \in (1, 2]$ and $\lambda > 0$ satisfying*

$$p > A + 2, \quad 0 \leq \mu \leq 2\frac{\sigma - 1}{\sigma + p - 2}, \quad \gamma\frac{p + \sigma - 2}{\sigma - 1} \leq \lambda \leq 2 - (\mu - \gamma)\frac{p + \sigma - 2}{\sigma - 1},$$

we have

$$a_+(x) \leq \frac{C}{(1 + r(x))^\gamma}, \quad b(x) \geq \frac{C}{(1 + r(x))^\mu} \quad and \quad \mathrm{vol}\, B_r = O\left(r^\lambda\right).$$

Then there are no nonnegative, nonidentically zero $C^2(M)$ solutions of the differential inequality (3.69) on M.

3.3   A refined version of Theorem 3.2

We shall now relax the assumption $b(x) > 0$ on M in Theorem 3.2. This is related to geometry and first we need to derive the transformation law of the Laplace-Beltrami operator under a conformal change of the metric. We go back to Chapter 2 and we suppose to have a conformal change of the type

$$\widetilde{\langle \, , \rangle} = \varphi^2 \langle \, , \rangle, \quad \varphi > 0, \ \varphi \in C^\infty(M), \tag{3.74}$$

with the notation in Chapter 2,

$$\widetilde{\theta}^i = \varphi \theta^i \quad i = 1, \ldots, m = \dim M, \tag{3.75}$$

and, having set

$$d\varphi = \varphi_t \theta^t, \tag{3.76}$$

$$\widetilde{\theta}^i_j = \theta^i_j + \frac{\varphi_j}{\varphi}\theta^i - \frac{\varphi_i}{\varphi}\theta^j. \tag{3.77}$$

Let $u \in C^2(M)$. Then

$$du = \widetilde{u}_j\widetilde{\theta}^j = \widetilde{u}_j\varphi\theta^j = u_j\theta^j,$$

that is,

$$\widetilde{u}_j = u_j\varphi^{-1}. \tag{3.78}$$

Furthermore, using (3.75) and (3.77), (3.76)

$$\widetilde{u}_{jt}\widetilde{\theta}^t = d\widetilde{u}_j - \widetilde{u}_k\widetilde{\theta}^k_j = d(u_j\varphi^{-1}) - (u_k\varphi^{-1})\left(\theta^k_j + \frac{\varphi_j}{\varphi}\theta^k - \frac{\varphi_k}{\varphi}\theta^j\right)$$

$$= \varphi^{-1}du_j - u_j\varphi^{-2}\varphi_t\theta^t - \varphi^{-1}u_k\theta^k_j - \varphi^{-2}u_k\varphi_j\theta^k + \varphi^{-2}u_k\varphi_k\theta^j$$

$$= \varphi^{-1}u_{jt}\theta^t - \varphi^{-2}\left(u_j\varphi_t + u_t\varphi_j - \delta^j_t u_k\varphi_k\right)\theta^t$$

$$= \varphi^{-2}\left[u_{jt} - \varphi^{-1}\left(u_j\varphi_t + u_t\varphi_j - \delta^j_t u_k\varphi_k\right)\right]\widetilde{\theta}^t.$$

It follows that

$$\widetilde{u}_{jt} = \frac{1}{\varphi^2}u_{jt} - \frac{1}{\varphi^3}\left(u_j\varphi_t + u_t\varphi_j - \delta^j_t u_k\varphi_k\right). \tag{3.79}$$

Hence,

$$\widetilde{\mathrm{Hess}}(u) = \mathrm{Hess}(u) - \frac{1}{\varphi}(du \otimes d\varphi + d\varphi \otimes du) + \frac{1}{\varphi}\langle\nabla u, \nabla\varphi\rangle\langle\,,\,\rangle \tag{3.80}$$

and

$$\widetilde{\Delta}u = \frac{1}{\varphi^2}\Delta u + \frac{m-2}{\varphi^3}\langle\nabla u, \nabla\varphi\rangle. \tag{3.81}$$

Note the simple form that equation (3.81) takes if $m = 2$: in this case, harmonic functions do depend only on the conformal class of the metric.

Remark. Formula (3.81) shall also be used in Step 3 of the proof of Theorem 6.8.

Proposition 3.10. *Let* $(M, \langle\,,\,\rangle)$ *be a complete Riemannian manifold of dimension* $m \geq 3$ *and let* $b(x) \in C^0(M)$ *be such that*

$$b(x) > 0 \quad in \ M\backslash B_R \tag{3.82}$$

for some $R > 0$. *Let* v *be a positive solution of*

$$\Delta v + a(x)v - b(x)v^\sigma \geq 0 \tag{3.83}$$

with $a(x) \in C^0(M)$ *and* $\sigma \geq \frac{m+2}{m-2}$. *Then, there exists* $\varepsilon = \varepsilon(v) > 0$ *such that, if*

$$b(x) \geq -\varepsilon \quad on \ M, \tag{3.84}$$

then, there exist $\widetilde{b}(x) \in C^0(M)$ *and* $C > 0$ *satisfying*

$$\text{i) } \widetilde{b}(x) > 0 \ \text{ in } \ M, \quad \text{ii) } \widetilde{b}(x) = C^2 b(x) \ \text{ in } \ M \backslash B_{R+1} \tag{3.85}$$

and a positive solution u on M of

$$\Delta u + \frac{(m-2)(\sigma-1)}{4} a(x)u - \widetilde{b}(x)u^{\frac{m+2}{m-2}} \geq 0. \tag{3.86}$$

Furthermore, there exists $C_1 > 0$ such that

$$u(x) = C_1 v(x)^{\frac{(m-2)(\sigma-1)}{4}} \quad \text{for } r(x) \gg 1. \tag{3.87}$$

Proof. We set $\varphi = v^{\frac{\sigma-1}{2}}$ and we consider the conformally related metric $\widetilde{\langle\,,\,\rangle} = \varphi^2 \langle\,,\,\rangle$. Let ψ be a solution of the Dirichlet problem

$$\begin{cases} \widetilde{\Delta}\psi = \delta & \text{on } B_{R+2}, \\ \psi \equiv 0 & \text{on } \partial B_{R+2} \end{cases} \tag{3.88}$$

for some constant $\delta > 0$. We define

$$\beta(x) = \xi(x)\psi(x) + \beta_0 > 0 \quad \text{on } M \tag{3.89}$$

where $\beta_0 > \inf \psi$ is a constant that we will specify in the sequel and $\xi(x) \in C^\infty(M)$, $\xi : M \to [0,1]$ is a cut-off function such that

$$\xi(x) \equiv 1 \ \text{ on } B_R, \quad \xi(x) \equiv 0 \ \text{ on } M \backslash B_{R+1}. \tag{3.90}$$

Finally we set

$$u(x) = \varphi(x)^{\frac{m-2}{2}} \beta(x). \tag{3.91}$$

Using the transformation law (3.81) we now compute Δu. To simplify the writing we set

$$a = \frac{m-2}{2}, \quad b = \frac{\sigma-1}{2}, \tag{3.92}$$

and we observe that

$$\Delta\beta = \varphi^2 \widetilde{\Delta}\beta - (m-2)\varphi^{-1} \langle \nabla\beta, \nabla\varphi \rangle, \tag{3.93}$$

$$\Delta\varphi = bv^{b-1}\Delta v + b(b-1)v^{b-2}|\nabla v|^2. \tag{3.94}$$

We then have

$$\begin{aligned}
\Delta u &= \beta\Delta\varphi^a + \varphi^a\Delta\beta + 2a\varphi^{a-1} \langle \nabla\varphi, \nabla\beta \rangle \\
&= \beta a\varphi^{a-1}\Delta\varphi + \beta a(a-1)\varphi^{a-2}|\nabla\varphi|^2 + \varphi^{a+2}\widetilde{\Delta}\beta - (m-2)\varphi^{a-1}\langle\nabla\beta,\nabla\varphi\rangle \\
&\quad + 2a\varphi^{a-1}\langle\nabla\beta,\nabla\varphi\rangle \\
&= \beta ab\varphi^{a-1}v^{b-1}\Delta v + \beta ab\varphi^{a-1}(b-1)v^{b-2}|\nabla v| + \beta a(a-1)\varphi^{a-2}|\nabla\varphi|^2 \\
&\quad + (2a-m+2)\varphi^{a-1}\langle\nabla\beta,\nabla\varphi\rangle + \varphi^{a+2}\widetilde{\Delta}\beta \\
&\geq -\beta ab\varphi^{a-1}v^b a(x) + \beta ab\varphi^{a-1}v^{b-1+\sigma}b(x) + ab(b-1)\beta\varphi^{a-2}v^{2b-2}|\nabla v| \\
&\quad + (2a-m+2)\varphi^{a-1}\langle\nabla\beta,\nabla\varphi\rangle + \varphi^{a+2}\widetilde{\Delta}\beta.
\end{aligned}$$

Therefore,

$$\Delta u + aba(x)u \geq \varphi^{a+2}\widetilde{\Delta}\beta + ab\beta\varphi^{a-1}v^{b-1+\sigma}b(x) \tag{3.95}$$
$$+ ab[b-1+b(a-1)]\beta\varphi^{a-2}v^{2b-2}|\nabla v| \tag{3.96}$$
$$+ (2a-m+2)\varphi^{a-1}\langle\nabla\beta,\nabla\varphi\rangle$$
$$= \varphi^{a+2}\Big[\widetilde{\Delta}\beta + ab\beta v^{\sigma-1-2b}b(x)\Big] + ab(ab-1)\beta\varphi^{a-2}v^{2b-2}|\nabla v|^2$$
$$+ (2a-m+2)\varphi^{a-1}\langle\nabla\beta,\nabla\varphi\rangle.$$

Using (3.92) we get

$$\Delta u + \frac{(m-2)(\sigma-1)}{4}a(x)u \geq u^{\frac{m+2}{m-2}}\Big\{\beta^{-\frac{m+2}{m-2}}\Big[\widetilde{\Delta}\beta + \frac{(m-2)(\sigma-1)}{4}\beta b(x)\Big]\Big\}$$
$$+ \frac{(m-2)(\sigma-1)}{4}\Big[\frac{(m-2)(\sigma-1)}{4}-1\Big]\beta v^{\frac{(m-2)(\sigma-1)}{4}}|\nabla\log v|^2.$$

The choice $\sigma \geq \frac{m+2}{m-2}$ implies

$$\frac{(m-2)(\sigma-1)}{4} - 1 \geq 0.$$

Hence, having set

$$\widetilde{b}(x) = \beta^{-\frac{m+2}{m-2}}\Big[\widetilde{\Delta}\beta + \frac{(m-2)(\sigma-1)}{4}\beta b(x)\Big],$$

(3.95) implies the validity of (3.86). We also observe that, because of (3.89) and the choice of $\xi(x)$,

$$\widetilde{b}(x) = C^2 b(x) \quad \text{on } M\backslash B_{R+1},$$

for some constant $C > 0$. Furthermore, since

$$\inf_{\overline{B}_{R+2}\backslash B_R} b(x) > 0$$

we can choose β_0 sufficiently large so that

$$\widetilde{\Delta}\beta + \frac{(m-2)(\sigma-1)}{4}\beta b(x) > 0 \quad \text{on } \overline{B}_{R+2}\backslash B_R. \tag{3.97}$$

Finally, (3.89), (3.88), (3.84) imply that (3.97) holds on B_R provided

$$\frac{(m-2)(\sigma-1)}{4}(\|\psi\|_\infty + \beta_0)\varepsilon < \delta.$$

This proves that $\widetilde{b}(x) > 0$ on M. (3.87) follows from (3.89) and (3.91). $\qquad\square$

Remark. If in (3.83) we have the equality sign this does not hold in general in (3.86) unless $\sigma = \frac{m+2}{m-2}$. Note that we can also choose $\beta_0 \geq 1$ so that $u = \beta_0 v$ is large for $\sigma = \frac{m+2}{m-2}$. Hence $\sup_M v = v^* > 1$ implies $\sup_M u = u^* > 1$. This is used in Theorem 5.9.

Using Proposition 3.10 and Theorem 3.2 we obtain the following refined version.

Theorem 3.11. *Let* $a(x), b(x) \in C^0(M)$ *satisfy*

i) $b(x) \geq 0$ *on* M; ii) $b(x) > 0$ *for* $r(x) \gg 1$; iii) $a(x) \leq Cb(x)$ *for* $r(x) \gg 1$.

Let $\sigma \geq \frac{m+2}{m-2}$, $\alpha > 0$ *and assume the existence of* $\varphi > 0$ *satisfying*

$$L_{\sigma,\alpha}\varphi \leq -K\frac{|\nabla\varphi|^2}{\varphi}$$

on M *for* $K > -1$, *with* $L_{\sigma,\alpha} = \Delta + \frac{(m-2)(\sigma-1)(1+\alpha)^2}{16\alpha}a(x)$. *Then, the differential inequality*

$$\Delta u + a(x)u - b(x)u^\sigma \geq 0 \tag{3.98}$$

has no positive bounded solutions on M *satisfying*

$$a(x)u^{\frac{(m-2)(\sigma-1)(1+\alpha)}{4}} \in L^1(M) \tag{3.99}$$

and

$$\left\{\int_{\partial B_R} u^{\frac{(m-2)(\sigma-1)(1+\alpha)}{4}}\right\}^{-1} \notin L^1(+\infty). \tag{3.100}$$

Remark. Theorem 3.11 will be used in the proof of Theorem 4.8 below.

We would like to end the chapter with a further general nonexistence result which will be useful for the Yamabe problem once we will have obtained the *a priori* estimates from above in the next chapter; see Theorem 4.10. First we recall how the definition of $\lambda_1(\Omega)$ is extended in case Ω is not necessarily an open set. Thus, let S be an arbitrary bounded subset of M. We set

$$\lambda_1^L(S) = \sup \lambda_1^L(\Omega),$$

where the supremum is taken over all open bounded sets with smooth boundary Ω such that $S \subset \Omega$. Note that, by definition, if $S = \emptyset$, then $\lambda_1^L(S) = +\infty$. Finally, if S is an unbounded subset of M, we define

$$\lambda_1^L(S) = \inf \lambda_1^L(D \cap S),$$

where the infimum is taken over all bounded open sets D with smooth boundary. Note that the condition that $\lambda_1^L(S) > 0$ means that the set S is small in suitable spectral sense.

We are now ready to prove the following

Theorem 3.12. *Let $a(x), b(x) \in C^0(M)$ satisfy*

$$b(x) \geq 0 \quad on \ M$$

and

$$\lambda_1^L(\operatorname{supp} a_+) > 0 \tag{3.101}$$

with $L = \Delta + a(x)$. Then the differential equation

$$\Delta u + a(x)u - b(x)u^\sigma = 0, \quad \sigma > 1,$$

has no nontrivial nonnegative ground states.

Remark. Recall that a *ground state* for the above equation is an entire solution u satisfying

$$\lim_{x \to +\infty} u(x) = 0.$$

Proof. Let u be a nonnegative ground state. If u is not identically null it attains its absolute positive maximum at a point $x_o \in M$. We claim that $a(x_o) \geq 0$. Indeed, otherwise $a(x_o) < 0$ and $0 \geq \Delta u(x_o) = -a(x_o)u(x_o) + b(x_o)u(x_o)^\sigma \geq -a(x_o)u(x_o) > 0$, contradiction. Therefore, $x_o \in \operatorname{supp} a_+$. Using (3.101) we can find a sufficiently small bounded domain $\Omega \ni x_o$ such that

$$\lambda_1^L(\Omega) > 0.$$

Let φ be the positive corresponding eigenfunction, that is,

$$\begin{cases} \Delta\varphi + a(x)\varphi + \lambda_1^L(\Omega)\varphi = 0, & \varphi > 0 \text{ on } \Omega, \\ \varphi = 0 & \text{on } \partial\Omega. \end{cases}$$

Thus, in particular,

$$\Delta\varphi + a(x)\varphi = -\lambda_1^L(\Omega)\varphi \leq 0. \tag{3.102}$$

Next, we fix $0 < \gamma < u(x_o)$ sufficiently close to $u(x_o)$ such that the connected component containing x_o, Ω_o, of $\{x \in M : u(x) > \gamma\}$ is inside Ω. Note that u has an absolute maximum in Ω_o and

$$\Delta u + a(x)u = b(x)u^\sigma \geq 0 \quad \text{on } \Omega_o. \tag{3.103}$$

Since φ satisfies (3.102) and $\varphi > 0$ on Ω_o, by the generalized maximum principle (see [PW67], Section 5) $u = C\varphi$ for some positive constant C. From (3.102) and (3.103) we have

$$-\lambda_1^L(\Omega)u = b(x)u^\sigma \quad \text{on } \Omega_o.$$

Thus at x_o,

$$0 \leq b(x_o)u(x_o)^{\sigma-1} = -\lambda_1^L(\Omega) < 0,$$

contradiction. $\qquad\square$

Remark. If $a_+(x) \equiv 0$, $\operatorname{supp} a_+ = \emptyset$ and $\lambda_1^L(\emptyset) = +\infty$. Thus in case $a(x) \leq 0$ we have the validity of the conclusions of the theorem.

Chapter 4

A priori estimates

In this chapter we determine *a priori* estimates on the behavior at infinity of positive solutions of the equation

$$\Delta u + a(x)u - b(x)u^{\sigma} = 0, \quad \sigma > 1 \tag{4.1}$$

on M under assumptions on $a(x)$ and $b(x)$ related to the geometrical requirement

$$\text{Ric} \geq -(m-1)H^2(1+r(x)^2)^{\frac{\delta}{2}} \quad \text{on } M \tag{4.2}$$

for some $H > 0$ and $\delta \in \mathbb{R}$. The estimates we shall provide are quite sharp and naturally divide into two types: from below and from above. In both cases the method we use in establishing their validity resembles the old idea in Ahlfors' proof of the Schwarz lemma ([Ahl38]).

Some further estimates, which cannot be obtained with the previous method, are provided by direct comparison with the aid of the maximum principle (see section 4.3). The chapter ends with some nonexistence result for the Yamabe problem, which complement those described in Chapter 3.

4.1 Estimates from below

We begin with an estimate from below for positive supersolutions (see [RZ07], Theorem 2.1).

Theorem 4.1. *Let $(M, \langle\,,\,\rangle)$ be a complete manifold with Ricci tensor satisfying (4.2); let $a(x)$, $b(x) \in C^0(M)$ with $b(x) > 0$ on $M \backslash B_R(o)$ for some $R > 0$ and*

$$\liminf_{r(x) \to +\infty} \frac{a(x)}{r(x)^{\alpha}} > 0 \tag{4.3}$$

with $\alpha > \max\{-2, \delta\}$. Let $\psi(t)$ be a positive, nondecreasing function defined in a neighbourhood of infinity such that, for some $\varepsilon \in (0, 1)$,

$$\psi(t) = O\left(\psi\left(\frac{t}{1+\varepsilon}\right)\right) \qquad as \ t \to +\infty \tag{4.4}$$

and assume that

$$\liminf_{r(x)\to+\infty} \frac{a(x)}{b(x)} \psi(r(x)) > 0. \tag{4.5}$$

Then, any positive solution $u \in C^2(M)$ of

$$\Delta u + a(x)u - b(x)u^\sigma \le 0, \quad \sigma > 1, \quad on \ M \tag{4.6}$$

satisfies

$$u(x) \ge C\psi(r(x))^{-\frac{1}{\sigma-1}} \tag{4.7}$$

for $r(x) \gg 1$ and some constant $C > 0$.

Proof. We fix $q \in M$, with $r(q) \gg 1$ and we set $\rho(x) = dist(x, q)$. Fix $T > 0$ and consider on $B_T(q)$ the function

$$F(x) = \frac{u(x)}{[T^2 - \rho^2(x)]^\xi}$$

for some $\xi > 1$. Note that, as $x \to \partial B_T(q)$, $F(x) \to +\infty$, while $F(x) > 0$ on $B_T(q)$. Thus F attains a positive absolute minimum at $\bar{x} \in B_T(q)$. Using a trick of Calabi, [Cal57], which enables us to suppose that ρ is smooth near $\bar{x}$, we deduce

$$i) \ \nabla \log F(\bar{x}) = 0, \quad ii) \ \Delta \log F(\bar{x}) \ge 0. \tag{4.8}$$

Using (4.8) i) a computation yields

$$\frac{\nabla u}{u}(\bar{x}) = -2\xi \frac{\rho \nabla \rho}{T^2 - \rho^2}(\bar{x}), \tag{4.9}$$

while from (4.8) ii) and (4.9) we have at $\bar{x}$,

$$\frac{\Delta u}{u} - 4\xi(\xi - 1)\frac{\rho^2}{[T^2 - \rho^2]^2} + \xi \frac{\Delta \rho^2}{T^2 - \rho^2} \ge 0. \tag{4.10}$$

We want to estimate $\Delta \rho^2$; since M is complete, we have

$$\text{Ric} \ge -(m - 1)Z^2 \quad on \ \overline{B_T(q)} \tag{4.11}$$

for some constant $Z > 0$. Therefore, the Laplacian Comparison Theorem and a computation using Gauss' lemma imply

$$\Delta \rho^2 \le 2[m + (m - 1)Z\rho] \quad on \ \overline{B_T(q)}. \tag{4.12}$$

Inserting (4.6) and (4.12) into (4.10) we have at $\bar{x}$,

$$u \ge b^{-\frac{1}{\sigma-1}}\left\{a + 4\xi(\xi - 1)\frac{\rho^2}{[T^2 - \rho^2]^2} - 2\xi \frac{m + (m - 1)Z\rho}{T^2 - \rho^2}\right\}^{\frac{1}{\sigma-1}}.$$

Since $\bar{x}$ is the minimum of F on $B_T(q)$ we then deduce

$$[T^2 - \rho^2(\bar{x})]^\xi u(y) \geq [T^2 - \rho^2(y)]^\xi u(\bar{x})$$
$$\geq [T^2 - \rho^2(y)]^\xi b(\bar{x})^{-\frac{1}{\sigma-1}}$$
$$\times \left\{ a(\bar{x}) + 4\xi(\xi - 1)\frac{\rho^2(\bar{x})}{[T^2 - \rho^2(\bar{x})]^2} - 2\xi\frac{m + (m-1)Z\rho(\bar{x})}{T^2 - \rho^2(\bar{x})} \right\}^{\frac{1}{\sigma-1}}$$

for each $y \in B_T(q)$. In particular, for $y = q$,

$$u(q) \geq \left[\frac{a(\bar{x})}{b(\bar{x})}\right]^{\frac{1}{\sigma-1}} \left\{ 1 + \frac{4\xi(\xi-1)}{a(\bar{x})}\frac{\rho^2(\bar{x})}{[T^2 - \rho^2(\bar{x})]^2} - \frac{2\xi}{a(\bar{x})}\frac{m + (m-1)Z\rho(\bar{x})}{T^2 - \rho^2(\bar{x})} \right\}^{\frac{1}{\sigma-1}}. \tag{4.13}$$

We set

$$f(t) = \frac{1}{2} + \frac{2\xi(\xi-1)}{a(\bar{x})}\frac{t^2}{[T^2 - t^2]^2} - \frac{2\xi}{a(\bar{x})}\frac{(m-1)Zt}{T^2 - t^2}$$

and

$$g(t) = \frac{1}{2} + \frac{2\xi(\xi-1)}{a(\bar{x})}\frac{t^2}{[T^2 - t^2]^2} - \frac{2\xi}{a(\bar{x})}\frac{m}{T^2 - t^2}$$

on $[0, T)$ and, considering the parabola (note $\xi > 1$)

$$w = \frac{2\xi(\xi-1)}{a(\bar{x})}y^2 - \frac{2\xi(m-1)Z}{a(\bar{x})}y + \frac{1}{2}, \qquad y \in [0, +\infty),$$

we deduce that $f(t)$ attains on $[0, T)$ its minimum value

$$\bar{f} = \frac{1}{2} - \frac{1}{2}\frac{\xi}{\xi-1}\frac{(m-1)^2Z^2}{a(\bar{x})}.$$

As for $g(t)$, we have $g(0) = \frac{1}{2} - \frac{2\xi}{a(\bar{x})}\frac{m}{T^2}$ and $\lim_{t\to T^-} g(t) = +\infty$; furthermore

$$g'(t) = \frac{4\xi}{a(\bar{x})}\frac{t}{[T^2 - t^2]^3}\{(\xi - 1 - m)t^2 + (\xi - 1 + m)T^2\}$$
$$\geq \frac{8}{a(\bar{x})}\frac{t^3}{[T^2 - t^2]^3}\xi(\xi - 1) \geq 0$$

on $[0, T)$ because of our choice of ξ. It follows that $g(t)$ attains on $[0, T)$ its minimum value

$$\bar{g} = g(0) = \frac{1}{2} - \frac{2\xi}{a(\bar{x})}\frac{m}{T^2}.$$

Going back to (4.13) we obtain

$$u(q) \geq \left[\frac{a(\bar{x})}{b(\bar{x})}\right]^{\frac{1}{\sigma-1}} \left\{ 1 - \frac{\xi}{2(\xi-1)}\frac{(m-1)^2Z^2}{a(\bar{x})} - \frac{2\xi}{a(\bar{x})}\frac{m}{T^2} \right\}^{\frac{1}{\sigma-1}}. \tag{4.14}$$

We now choose
$$T = \varepsilon r(q)$$
and we observe that, since $\forall x \in B_T(q),\ r(q) - T \le r(x) \le r(q) + T$, with our choice of T, using (4.3) we have on $B_T(q)$,

$$\begin{cases} a(x) \ge C(\varepsilon)r(q)^\alpha, \\ \mathrm{Ric}\ \ge -(m-1)H^2\big[1 + W(\varepsilon)^2 r(q)^2\big]^{\frac{\delta}{2}} \end{cases} \tag{4.15}$$

for some constants $C(\varepsilon),\ W(\varepsilon) > 0$, depending only on $\varepsilon > 0$ and on the sign of α and δ respectively. Therefore,

$$\Lambda(\bar{x})^{\sigma-1} = 1 - \frac{\xi}{2(\xi-1)}\frac{(m-1)^2 Z^2}{a(\bar{x})} - \frac{2\xi}{a(\bar{x})}\frac{m}{T^2} \tag{4.16}$$

$$\ge 1 - \frac{\xi(m-1)^2 H^2}{2(\xi-1)C(\varepsilon)}\left[\frac{1}{r^2(q)} + W(\varepsilon)^2\right]^{\frac{\delta}{2}} r(q)^{\delta-\alpha} - \frac{2m\xi}{C(\varepsilon)}\frac{1}{\varepsilon^2}r(q)^{-2-\alpha}.$$

Thus, using the assumption $\alpha > \max\{-2, \delta\}$ we can suppose to have chosen $R > 0$ sufficiently large such that, $\forall q$ with $r(q) > R$, we have $\Lambda(\bar{x})^{\sigma-1} \ge \frac{1}{2}$. Hence

$$u(q) \ge \frac{1}{2}\left[\frac{a(\bar{x})}{b(\bar{x})}\right]^{\frac{1}{\sigma-1}}$$

and in turn

$$u(q)\psi(r(q))^{\frac{1}{\sigma-1}} \ge \frac{1}{2}\left[\psi(r(q)) \min_{B_T(q)} \frac{a(x)}{b(x)}\right]^{\frac{1}{\sigma-1}}. \tag{4.17}$$

We claim that

$$\liminf_{r(q)\to+\infty}\left[\psi(r(q)) \min_{B_T(q)} \frac{a(x)}{b(x)}\right] > 0 \tag{4.18}$$

so that (4.7) follows at once from (4.17). To prove the claim suppose that (4.18) is false, then there exists a sequence $\{y_n\}$ in M, with $r(y_n) \to +\infty$, such that

$$\lim_{r(y_n)\to+\infty}\left[\psi(r(y_n)) \min_{B_{T_n}(y_n)} \frac{a(x)}{b(x)}\right] = 0, \tag{4.19}$$

where $T_n = \varepsilon r(y_n)$. Let $\{z_n\} \in \overline{B_{T_n}(y_n)}$ realize the minimum, that is

$$\min_{\overline{B_{T_n}(y_n)}} \frac{a(x)}{b(x)} = \frac{a(z_n)}{b(z_n)}.$$

Fix $\eta > 0$ and choose n sufficiently large so that, from (4.19), we have

$$\frac{a(z_n)}{b(z_n)} < \frac{\eta}{\psi(r(y_n))}. \tag{4.20}$$

Since $z_n \in B_{T_n}(y_n)$,

$$r(y_n) \geq \frac{r(z_n)}{1 + \varepsilon}$$

and therefore, since ψ is nondecreasing and (4.4) holds we have

$$\frac{a(z_n)}{b(z_n)} < \frac{\eta}{\psi(\frac{r(z_n)}{1+\varepsilon})} \leq \frac{C\eta}{\psi(r(z_n))} \tag{4.21}$$

for some constant $C > 0$ independent of n. In other words

$$\frac{a(z_n)}{b(z_n)}\psi(r(z_n)) < C\eta$$

for $n \gg 1$. Since $\eta > 0$ was chosen arbitrarily this contradicts (4.5) and completes the proof of the theorem. $\qquad\square$

Remark. If we assume $\psi(t)$ nonincreasing, the conclusion of Theorem 4.1 holds substituting the requirement (4.4) with

$$\psi(t) = O\left(\psi\left(\frac{t}{1 - \varepsilon}\right)\right) \quad \text{as } t \to +\infty, \tag{4.22}$$

for some $\varepsilon \in (0,1)$. Thus Theorem 4.1 is valid in particular with $\psi(t) = Ct^\beta (\log t)^\gamma$ for $t \gg 1$, $\beta, \gamma \in \mathbb{R}$, $C > 0$ and $\varepsilon \in (0,1)$.

Remark. Clearly, (4.4) or (4.22) are not satisfied if $\psi(t)$ is of exponential type, for instance $\psi(t) = Ce^{\beta t}$, $C, \beta > 0$. In this case condition (4.4) has to be substituted with

$$\psi(t) = O(\psi(t - T_0)) \quad \text{as } t \to +\infty \tag{4.23}$$

for some $T_0 > 0$. The result continues to hold with the requirement

$$\alpha > \max\{0, \delta\} \tag{4.24}$$

but the proof has to be modified as follows. Following the argument of Theorem 4.1 we arrive at (4.14). Now we choose $T = T_0$, then (4.15) holds with $C(\varepsilon), W(\varepsilon)$ substituted with $C(T_0), W(T_0)$, positive constants depending only on T_0 and the sign of α and δ respectively. We then proceed, under the modified assumption (4.3) with $\alpha > \max\{0, \delta\}$, directly to

$$u(q)\psi(r(q))^{\frac{1}{\sigma-1}} \geq \frac{1}{2}\left[\psi(r(q)) \min_{B_{T_0}(q)} \frac{a(x)}{b(x)}\right]^{\frac{1}{\sigma-1}}. \tag{4.25}$$

The remainder of the proof is the same as in Theorem 4.1 using (4.23) instead of (4.4). Clearly if $\psi(t)$ is nonincreasing, then (4.23) has to be substituted with

$$\psi(t) = O(\psi(t + T_0)) \quad \text{as } t \to +\infty \tag{4.26}$$

for some $T_0 > 0$.

Remark. In case $\alpha = \delta$ the proof of Theorem 4.1 can be modified to obtain the same conclusion, provided a further condition on H is satisfied.

Indeed, if case (4.4) or (4.22) holds, assume that the Ricci tensor of M satisfies (4.2) with $\delta > -2$, and having set

$$A = \liminf_{r(x) \to +\infty} \frac{a(x)}{r(x)^\alpha} > 0,$$

suppose that H satisfies

$$0 < \frac{1}{2} H^2 < \frac{A}{(m-1)^2} \left(\frac{1-\varepsilon}{1+\varepsilon} \right)^{|\delta|}. \tag{4.27}$$

Under these assumptions, following the proof of Theorem 4.1, we arrive at (4.16) which takes the form

$$\Lambda(\bar{x})^{\sigma-1} \geq 1 - \frac{\xi}{2(\xi-1)} \frac{(m-1)^2 H^2}{C(\varepsilon)} \left[\frac{1}{r^2(q)} + W(\varepsilon)^2 \right]^{\frac{\delta}{2}} - \frac{2m\xi}{C(\varepsilon)} \frac{1}{\varepsilon^2} r(q)^{-2-\delta}$$

where $C(\varepsilon)$ and $W(\varepsilon)$ are given by

$$C(\varepsilon) = A(1 - (\mathrm{sgn}(\delta))\varepsilon)^\delta W(\varepsilon) = (1 + (\mathrm{sgn}(\delta))\varepsilon) \tag{4.28}$$

and $\mathrm{sgn}(\delta)$ is the signum of δ. Using (4.27) and $\delta > -2$ we can choose $\xi > 1$ and $R > 0$ both sufficiently large that, for some $\eta \in (0,1)$, $r(q) > R$ gives

$$\frac{\xi}{(\xi-1)} \frac{(m-1)^2 H^2}{C(\varepsilon)} \left[\frac{1}{r^2(q)} + W(\varepsilon)^2 \right]^{\frac{\delta}{2}} < \eta$$

and

$$1 - \frac{2m\xi}{C(\varepsilon)} \frac{1}{\varepsilon^2} r(q)^{-2-\delta} \geq \eta.$$

Thus,

$$\Lambda(\bar{x})^{\sigma-1} \geq \frac{\eta}{2}.$$

The remainder of the proof is as above.

If $\psi(t)$ satisfies (4.23) or (4.26), since $\delta > 0$ because of (4.24) and $\alpha = \delta$, we require

$$0 < H^2 < \frac{A}{(m-1)^2} \frac{(1 - \frac{T_0}{R_0})^\delta}{(1 + \frac{T_0}{R_0})^\delta}$$

for $R_0 > T_0$ and the theorem is still valid.

Remark. We underline that, unfortunately, in the geometric case of Yamabe's equation, for $m \geq 3$ say, under assumption (4.2) one has

$$a(x) = -c_m^{-1} S(x) \leq \frac{m(m-2)}{4} H^2 \left(1 + r(x)^2 \right)^{\delta/2},$$

hence (4.27) cannot be satisfied.

Remark. It is worth observing that the proof of Theorem 4.1 gives no uniform positive lower bounds on compact domains for the positive solutions of (4.1). As we shall see this contrasts with the case of the estimates from above.

4.2 Estimates from above

We now derive an estimate from above for nonnegative subsolutions in case $\delta \geq -2$. We recall that a *subsolution* (resp. a *supersolution*) of a Yamabe-type equation is a solution of the differential inequality

$$\Delta u + a(x)u - b(x)u^\sigma \geq 0 \quad (\text{resp. } \leq 0).$$

First we need the following (see [RRV97], [PRS05b], Lemma 2.6)

Lemma 4.2. *Let $a(x), b(x) \in C^0(M)$, $\sigma > 1$, $0 < \tilde{T} < T$ and $\Omega \subset\subset B_{\tilde{T}}(q) \subset M$. Assume that $b(x) > 0$ on $\overline{B_T(q)}$. Then, there exists a constant $C > 0$ such that any nonnegative C^2-solutions u on $\overline{B_T(q)}$ of*

$$\Delta u + a(x)u - b(x)u^\sigma \geq 0 \tag{4.29}$$

satisfies

$$\sup_\Omega u \leq C. \tag{4.30}$$

Proof. We let $\rho(x) = \mathrm{dist}(x, q)$ and, on the compact ball $\overline{B_T(q)}$ we consider the continuous function

$$F(x) = \left[T^2 - \rho(x)^2\right]^{\frac{2}{\sigma-1}} u(x) \tag{4.31}$$

where u is any nonnegative C^2-solution of (4.29). Note that $F(x)|_{\partial B_T(q)} \equiv 0$. If $u \equiv 0$ on $B_T(q)$, there is nothing to prove; otherwise F has a positive absolute maximum at some point $\bar{x} \in B_T(q)$. In particular, $u(\bar{x}) > 0$. Again, using Calabi's trick, [Cal57], we can assume that ρ is smooth near $\bar{x}$. Then

$$\text{i) } \nabla \log F(\bar{x}) = 0; \quad \text{ii) } \Delta \log F(\bar{x}) \leq 0. \tag{4.32}$$

From (4.32) i) we obtain

$$\frac{\nabla u}{u} = \frac{4}{\sigma - 1} \frac{\rho \nabla \rho}{T^2 - \rho^2} \quad \text{at } \bar{x}; \tag{4.33}$$

while from (4.32) ii) and (4.33) we deduce

$$0 \geq \frac{\Delta u}{u} - \frac{8}{\sigma - 1}\left(\frac{2}{\sigma - 1} + 1\right)\frac{\rho^2}{(T^2 - \rho^2)^2} - \frac{2}{\sigma - 1}\frac{\Delta \rho^2}{T^2 - \rho^2} \quad \text{at } \bar{x}. \tag{4.34}$$

Again,

$$\mathrm{Ric} \geq -(m-1)A^2 \quad \text{on } \overline{B_T(q)} \tag{4.35}$$

for some constant $A \geq 0$, so that, by the Laplacian comparison theorem,

$$\Delta \rho^2 \leq 2[m + (m-1)A\rho] \quad \text{on } \overline{B_T(q)}. \tag{4.36}$$

Inserting (4.29) and (4.36) into (4.34) we obtain that, at $\bar{x}$,

$$u \leq b^{-\frac{1}{\sigma-1}} \left\{ a_+ + \frac{8(\sigma-1)}{(\sigma-1)^2} \frac{\rho^2}{(T^2-\rho^2)^2} + \frac{4}{\sigma-1} \frac{m+(m-1)A\rho}{T^2-\rho^2} \right\}^{\frac{1}{\sigma-1}}, \tag{4.37}$$

where a_+ is the positive part of a. Next, we recall the elementary inequality

$$(v+w)^\delta \leq C^2(v^\delta + w^\delta), \quad v, w \geq 0 \tag{4.38}$$

valid for any fixed $\delta > 0$ provided $C = C(\delta) > 0$ is sufficiently large. We use (4.38) in (4.37) with $\delta = \frac{1}{\sigma-1}$ to get, at $\bar{x}$,

$$u \leq C^2 \left\{ \max_{B_T(q)} \frac{a_+}{b} \right\}^{\frac{1}{\sigma-1}} + C^2 \left\{ \min_{B_T(q)} b \right\}^{-\frac{1}{\sigma-1}} \left\{ \frac{\rho^2}{(T^2-\rho^2)^2} + \frac{m+(m-1)A\rho}{T^2-\rho^2} \right\}^{\frac{1}{\sigma-1}}$$

whence

$$F(\bar{x}) \leq C^2 \left\{ \left[\max_{B_T(q)} \frac{a_+}{b} \right] T^4 \right\}^{\frac{1}{\sigma-1}} + C^2 \left\{ \min_{B_T(q)} b \right\}^{-\frac{1}{\sigma-1}} \left\{ T^2 + AT^3 \right\}^{\frac{1}{\sigma-1}}.$$

On the other hand, $\forall y \in \overline{B_{\widetilde{T}(q)}}$, we have $F(y) \leq F(\bar{x})$ and therefore

$$u(y) \leq C^2 \left\{ \max_{B_T(q)} \frac{a_+}{b} \right\}^{\frac{1}{\sigma-1}} \left\{ \frac{T^4}{[T^2-\rho^2(y)]^2} \right\}^{\frac{1}{\sigma-1}} \tag{4.39}$$

$$+ C^2 \left\{ \min_{B_T(q)} b \right\}^{-\frac{1}{\sigma-1}} \left\{ \frac{T^2}{[T^2-\rho^2(y)]^2} + \frac{AT^3}{[T^2-\rho^2(y)]^2} \right\}^{\frac{1}{\sigma-1}}.$$

Therefore, using the fact that $\rho(y) \leq \widetilde{T} < T$, from (4.39) we immediately obtain (4.30). $\qquad \square$

We are now ready to prove the desired estimate (see [RRV97] and [PRS05b], Proposition 2.7).

Proposition 4.3. *Let $a(x), b(x) \in C^0(M)$ and let $\sigma > 1$. Suppose that the Ricci tensor satisfies*

$$\text{Ric} \geq -(m-1)B^2(1+r(x)^2)^{\frac{\delta}{2}} \tag{4.40}$$

for some $B > 0$, $\delta \geq -2$. Let ω, ψ be nondecreasing positive functions defined in a neighbourhood of $+\infty$ with the following properties: there exists $\varepsilon \in (0, 1)$ for which

$$\text{i) } \psi\left(\frac{t}{1-\varepsilon}\right) = O(\psi(t)); \quad \text{ii) } \omega\left(\frac{t}{1-\varepsilon}\right) = O(\omega(t)) \quad \text{as } t \to +\infty. \tag{4.41}$$

Assume that, for some $R_0 > 0$ and $\forall x : r(x) \geq R_0$, we have

$$\text{i) } \frac{a_+(x)}{b(x)} \leq \frac{1}{\omega(r(x))}; \quad \text{ii) } b(x) \geq \psi(r(x))r(x)^{\frac{\delta}{2}-1}. \tag{4.42}$$

If u is a nonnegative C^2-solution on M of

$$\Delta u + a(x)u - b(x)u^\sigma \geq 0, \tag{4.43}$$

then

$$u(x) = O\left(\omega(r(x))^{-\frac{1}{\sigma-1}} + \psi(r(x))^{-\frac{1}{\sigma-1}}\right) \quad \text{as } r(x) \to +\infty. \tag{4.44}$$

Proof. Let $q \in M \backslash \overline{B_{2R_0}(o)}$ and set $\rho(x) = \text{dist}(x, q)$. We fix a radius $T > 0$ so small as to insure that $B_T(q) \subset M \backslash \overline{B_{2R_0}(o)}$. Because of (4.40) we have

$$\text{Ric} \geq -(m-1)B^2\left[1 + (r(q) + \text{sgn}(\delta)T)^2\right]^{\frac{\delta}{2}} \quad \text{on } \overline{B_T(q)}. \tag{4.45}$$

In (4.39) of Lemma 4.2 we choose $y = q$ so that $\rho(q) = 0$ and

$$A = B\left[1 + (r(q) + \text{sgn}(\delta)T)^2\right]^{\frac{\delta}{4}}. \tag{4.46}$$

In this way, we get

$$u(q) \leq C^2\left\{\max_{B_T(q)} \frac{a_+}{b}\right\}^{\frac{1}{\sigma-1}} \tag{4.47}$$

$$+ C^2\left\{\min_{B_T(q)} b\right\}^{-\frac{1}{\sigma-1}}\left\{\frac{1}{T^2} + \frac{B}{T}\left[1 + (r(q) + \text{sgn}(\delta)T)^2\right]^{\frac{\delta}{4}}\right\}^{\frac{1}{\sigma-1}}.$$

Set $T = \varepsilon[r(q) - R_0]$ in (4.47), and note that $B_T(q) \subset M \backslash \overline{B_{R_0}(o)}$. Since $\delta \geq -2$ and $r(q) \geq 2R_0$, it is easy to see that there exists a constant $C > 0$ independent of q such that

$$\left\{\frac{1}{T^2} + \frac{B}{T}\left[1 + (r(q) + \text{sgn}(\delta)T)^2\right]^{\frac{\delta}{4}}\right\}^{\frac{1}{\sigma-1}} \leq C^2 r(q)^{(\frac{\delta}{2}-1)\frac{1}{\sigma-1}}. \tag{4.48}$$

Now, recalling that $b(x)$ satisfies (4.42) ii) and that ψ is nondecreasing, we obtain

$$\min_{B_T(q)} b \geq C^2 r(q)^{\frac{\delta}{2}-1} \psi(r(q) - T) \geq C^2 \psi((1-\varepsilon)r(q)) r(q)^{\frac{\delta}{2}-1}$$

for some constant $C = C(R_0) > 0$. On the other hand, (4.42) gives

$$\max_{B_T(q)} \frac{a_+}{b} \leq \frac{1}{\omega((1-\varepsilon)r(q) + \varepsilon R_0)} \leq \frac{1}{\omega((1-\varepsilon)r(q))}.$$

Inserting these inequalities into (4.47) and using (4.41) we obtain

$$u(q) \leq C^2 \Big\{ \omega((1-\varepsilon)r(q))^{-\frac{1}{\sigma-1}} + \psi((1-\varepsilon)r(q))^{-\frac{1}{\sigma-1}} \Big\}$$
$$\leq C^2 \Big\{ \omega(r(q))^{-\frac{1}{\sigma-1}} + \psi(r(q))^{-\frac{1}{\sigma-1}} \Big\}$$

as required. □

As a special case we have the following result (see [RZ07], Theorem 2.3)

Theorem 4.4. *Let* $(M, \langle\,,\,\rangle)$ *be a complete manifold with Ricci tensor satisfying*

$$\mathrm{Ric} \geq -(m-1)H^2(1+r^2)^{\frac{\delta}{2}}$$

for some constant $H > 0$ *and* $\delta \geq -2$. *Let* $a(x), b(x) \in C^0(M)$ *and satisfying*

$$a(x) \leq Ar(x)^\alpha, \quad \alpha \geq \frac{\delta}{2} - 1, \tag{4.49}$$

$$b(x) \geq Br(x)^\beta, \quad \beta \leq 1 - \frac{\delta}{2} + \alpha \tag{4.50}$$

for $r(x) \gg 1$ *and some constants* $A, B > 0$. *Then any nonnegative solution* $u \in C^2(M)$ *of*

$$\Delta u + a(x)u - b(x)u^\sigma \geq 0, \quad \sigma > 1 \quad on \ M \tag{4.51}$$

satisfies

$$u(x) \leq Cr(x)^{-\frac{\beta - \alpha}{\sigma - 1}} \tag{4.52}$$

for $r(x) \gg 1$ *and some constant* $C > 0$.

As in the case of Theorem 4.1, there are a number of versions of Proposition 4.3. We describe here (without proof) one of them that can be deduced from Proposition 4.1 of [RRV97].

Theorem 4.5. *Let* $(M, \langle\,,\,\rangle)$ *be a complete manifold with Ricci tensor satisfying*

$$\mathrm{Ric} \geq -(m-1)H^2(1+r^2)^{\frac{\delta}{2}}$$

for some constants $H > 0$ *and* $\delta \geq -2$. *Let* $a(x), b(x) \in C^0(M)$ *and let* $\psi(t)$ *be a nondecreasing positive function in a neighbourhood of* $+\infty$ *such that there exists* $T_0 > 0$ *for which*

$$\psi(t + T_0) = O(\psi(t)) \quad as \ t \to +\infty. \tag{4.53}$$

Assume, depending on the sign of δ, *that either*

$$\begin{cases} \text{i) for } -2 \leq \delta \leq 0, \ \limsup_{r(x) \to +\infty} a_+(x) < +\infty \ and \ \liminf_{r(x) \to +\infty} \dfrac{b(x)}{\psi(r(x))} = A > 0 \\ \ or \\ \text{ii) for } \delta > 0, \ \limsup_{r(x) \to +\infty} \dfrac{a_+(x)}{r(x)^\delta} < +\infty \ and \ \liminf_{r(x) \to +\infty} \dfrac{b(x)}{r(x)^\delta \psi(r(x))} = A > 0. \end{cases}$$

$$\tag{4.54}$$

If u is a nonnegative C^2-solution on M of

$$\Delta u + a(x)u - b(x)u^\sigma \geq 0, \quad \sigma > 1,$$

then

$$u(x) = O\left(\psi(r(x))^{-\frac{1}{\sigma-1}}\right) \quad as\ r(x) \to +\infty. \tag{4.55}$$

Furthermore, in case $A = +\infty$ in (4.54), the conclusion (4.55) improves to

$$u(x) = o\left(\psi(r(x))^{-\frac{1}{\sigma-1}}\right) \quad as\ r(x) \to +\infty. \tag{4.56}$$

4.3 Sharpness of the previous results

We now show by way of examples that the estimates we obtained are sharp. We begin with Theorem 4.1. Let $g(r) \in C^\infty([0,+\infty])$, $g(r) > 0$ for $r > 0$ be such that

$$g(r) = \begin{cases} r & \text{on } [0, \frac{1}{2}], \\ e^{\frac{1}{m-1} \int_0^r (1+s^2)^{\frac{\delta}{4}} ds} & \text{on } [1,+\infty) \end{cases}$$

for some $\delta \geq -2$. We define the *model manifold* $M_g = \mathbb{R}^m$ in the sense of Greene and Wu, [GW79], with metric on $M_g \setminus \{0\} = (0,+\infty) \times \mathbb{S}^{m-1}$ given by

$$\langle\,,\,\rangle = dr^2 + g(r)^2 d\theta^2$$

where $d\theta^2$ is the canonical metric on the unit sphere $\mathbb{S}^{m-1}$. Note that, since $g(r) = r$ on $[0, \frac{1}{2}]$, $\langle\,,\,\rangle$ can be smoothly extended to all of M_g. The Ricci curvature in the radial direction is given by

$$\text{Ric}(\nabla r, \nabla r) = -(m-1)\frac{g''(r)}{g(r)}$$

while, in the direction orthogonal to ∇r determined by the unit vector ν,

$$\text{Ric}(\nu, \nu) = -\frac{(m-2)}{g(r)^2}(1 - g'(r)^2) - \frac{g''(r)}{g(r)}.$$

Thus, a simple computation shows that there exists an appropriate $H > 0$ such that

$$\text{Ric} \geq -(m-1)H^2(1+r^2)^{\frac{\delta}{2}} \quad \text{on } M_g.$$

Next, we consider the function

$$w(r) = (\mu + P(r))^{\frac{\alpha-\beta}{\sigma-1}} > 0 \quad \text{on } [0,+\infty),$$

where $\mu > 0$, $\beta \geq \alpha > \delta$ and $P \in C^2([0,+\infty))$ satisfies

$$P(r) = r, \text{ for } r \gg 1, \quad P(r) \geq 0 \text{ on } [0,+\infty), \quad P'(0) = 0.$$

We define
$$H_w(r) = w^{-\sigma}\left\{w'' + (m-1)\frac{g'(r)}{g(r)}w' + A(1+r)^\alpha w\right\}$$

for some $A > 0$ constant. Computing we have,

$$(1+r)^{-\beta}H_w(r) = \frac{(\mu+P(r))^{\beta-\alpha}}{(1+r)^{\beta-\alpha}}\left\{\left[\frac{P''(r)}{(\mu+P(r))(1+r)^\alpha}\right.\right.$$
$$+\frac{\alpha-\beta-\sigma+1}{\sigma-1}\frac{P'(r)}{(\mu+P(r))^2(1+r)^\alpha}$$
$$\left.\left.+(m-1)\frac{g'(r)}{g(r)}\frac{P'(r)}{(\mu+P(r))(1+r)^\alpha}\right]+A\right\}.$$

Next, we note that
$$(m-1)\frac{g'(r)}{g(r)} = (1+r^2)^{\frac{\delta}{4}}$$

on $[1,+\infty)$, and therefore

$$(m-1)\frac{g'(r)}{g(r)}\frac{P'(r)}{(\mu+P(r))(1+r)^\alpha} = \frac{P'(r)}{(\mu+P(r))}(1+r^2)^{\frac{\delta}{4}}(1+r)^{-\alpha} \text{ on } [1,+\infty).$$
$$(4.57)$$

Furthermore, the left-hand side of (4.57) is bounded near zero since $P'(0) = 0$. It follows that, up to choosing $\mu \gg 1$, since $P(r) = r$ for $r \gg 1$, we can choose $B > 0$ such that
$$H_w(r) \le B(1+r)^\beta$$

on $[0,+\infty)$. Recalling the definition of $H_w(r)$, we then have

$$w'' + (m-1)\frac{g'(r)}{g(r)}w' + A(1+r)^\alpha w - B(1+r)^\beta w^\sigma \le 0. \qquad (4.58)$$

Setting $u(x) = w(r(x))$, $a(x) = A(1+r(x))^\alpha$ and $b(x) = B(1+r(x))^\beta > 0$ on M_g we then deduce
$$\Delta u + a(x)u - b(x)u^\sigma \le 0.$$

Moreover,
$$\frac{a(x)}{b(x)} = \frac{A}{B}(1+r(x))^{\alpha-\beta}$$

on M_g and we can therefore choose $\psi(t) = t^{\beta-\alpha}$. Since $\beta \ge \alpha$, ψ is nondecreasing, satisfies (4.4), (4.5) and according to Theorem 4.1,

$$u(x) \ge Cr(x)^{\frac{\alpha-\beta}{\sigma-1}} \qquad (4.59)$$

for $r(x) \gg 1$ and some constant $C > 0$. Since $u(x) = (\mu+r(x))^{\frac{\alpha-\beta}{\sigma-1}}$ for $r(x) \gg 1$, estimate (4.7) cannot be improved.

In the above example, let, as before, $\delta \geq -2$. Choose $\alpha > -2$ and $A > 0$. Then, for every $\beta \in \mathbb{R}$ we can find $\mu \gg 1$ so that

$$(1+r)^{-\beta} H_w(r) \geq B \quad \forall r > 0,$$

and for some $B > 0$. Therefore, $u(x) = w(r(x))$ satisfies the inequality

$$\Delta u + a(x)u - b(x)u^\sigma \geq 0 \quad \text{on } M_g. \tag{4.60}$$

Again, we observe that

$$u(x) = (\mu + r(x))^{\frac{\alpha - \beta}{\sigma - 1}} \quad \text{on } M_g \setminus B_1.$$

Thus, fixing $\delta > -2$ and choosing $\alpha \geq \delta/2 - 1$ and $\beta \leq 1 - \delta/2 + \alpha$, applying Theorem 4.4 to the differential inequality (4.60) we have

$$u(x) \leq Cr(x)^{\frac{\alpha - \beta}{\sigma - 1}}$$

for some constant $C > 0$ and $r(x) >> 1$, showing that the upper bound provided by Theorem 4.4 cannot be improved.

4.4 Some further estimates

We now complement the analysis of the previous section by considering the situation where

$$\liminf_{r(x) \to +\infty} \frac{a(x)}{r(x)^\alpha} > -\infty. \tag{4.61}$$

The technique used in the proof of Theorem 4.1 cannot be applied but it turns out that in this case an estimate similar to (4.7), depending directly on Ricci, can be obtained in an elementary way with the aid of the maximum principle. In fact we have (see [RZ07], Theorem 2.4)

Theorem 4.6. *Let* $(M, \langle , \rangle)$ *be a complete manifold with radial Ricci curvature satisfying*

$$\text{Ric}\,(\nabla r, \nabla r) \geq -(m-1)H^2 \big(1 + r(x)^2\big)^{\delta/2}$$

on M for some $H > 0$, $\delta > -2$ and let $a(x), b(x) \in C^0(M)$ and satisfy (4.61) and

$$\limsup_{r(x) \to +\infty} \frac{b(x)}{r(x)^\beta e^{(\sigma-1)\gamma \left(1 + r(x)^{\frac{\delta+2}{2}}\right)}} < +\infty \tag{4.62}$$

for some $\alpha < \delta$, $\beta \leq \delta$, $\gamma > \frac{2H}{2+\delta}(m-1)$, $\sigma > 1$. Then, any positive solution of

$$\Delta u + a(x)u - b(x)u^\sigma \leq 0 \quad \text{on } M \tag{4.63}$$

satisfies

$$\liminf_{r(x) \to +\infty} \frac{u(x)}{e^{-\gamma r(x)^{\frac{\delta+2}{2}}}} > 0. \tag{4.64}$$

Proof. First of all, using Proposition 1.15, the Laplacian comparison theorem and the lower bound on Ricci, we deduce

$$\Delta r \leq (m-1)Hr^{\delta/2}(1+o(1)) \quad \text{as } r \to +\infty. \tag{4.65}$$

Next, we choose $R > 0$ sufficiently large such that, using (4.61) and (4.62),

$$a(x) \geq -Ar(x)^{\alpha}, \ b(x) \leq Br(x)^{\beta}e^{(\sigma-1)\gamma\left(1+r(x)^{\frac{\delta+2}{2}}\right)} \tag{4.66}$$

on $M \setminus B_R$ for some $A, B > 0$. We let $0 < \xi < \min_{\overline{B_R}} u(x)$. To simplify the writing, set $\theta = \frac{\delta}{2} + 1$ and define

$$v(x) = \xi e^{-\gamma\left(1+r(x)^{\theta}\right)} - u(x).$$

Note that, since $\gamma > 0$, by our choice of ξ we have

$$v(x) \leq \xi - u(x) < 0 \quad \text{on } \overline{B_R}. \tag{4.67}$$

We now reason by contradiction and suppose that

$$\liminf_{r(x) \to +\infty} \frac{u(x)}{e^{-\gamma(1+r(x)^{\theta})}} = 0.$$

This means that there exists a sequence $\{x_n\} \subset M$, $r(x_n) \to +\infty$, such that

$$\frac{u(x_n)}{e^{-\gamma(1+r(x_n)^{\theta})}} \to 0 \quad \text{as } n \to +\infty. \tag{4.68}$$

Since $\gamma, \xi > 0$ and $u(x) > 0$, $v^* = \sup_M v(x) < +\infty$. Furthermore, because of (4.68),

$$v(x_n) = e^{-\gamma\left(1+r(x_n)^{\theta}\right)}\left\{\xi - \frac{u(x_n)}{e^{-\gamma(1+r(x_n)^{\theta})}}\right\} > 0 \tag{4.69}$$

for n sufficiently large. Thus $v^* > 0$. Finally, $\gamma, \theta > 0$ force that v^* has to be attained at some point $\bar{x} \in M$. Let Ω be the level set

$$\Omega = \{x \in M : v(x) > 0\};$$

because of (4.67), $\Omega \subset M \setminus \overline{B_R}$ and

$$u(x) < \xi e^{-\gamma\left(1+r(x)^{\theta}\right)} \quad \text{on } \Omega. \tag{4.70}$$

On Ω we compute

$$\Delta v = \xi\left[-\gamma\theta e^{-\gamma\left(1+r(x)^{\theta}\right)}r^{\theta-1}\Delta r + \gamma\theta\left(\gamma\theta r^{2\theta-2} + (1-\theta)r^{\theta-2}\right)e^{-\gamma\left(1+r(x)^{\theta}\right)}\right] - \Delta u$$

so that, using (4.70), (4.65), (4.66) and (4.63),

$$\Delta v \geq \xi r^{\frac{\delta}{2}-1+\theta} e^{-\gamma\left(1+r(x)^{\theta}\right)} \left\{ \gamma\theta \left[\gamma\theta r^{\theta_1 - \frac{\delta}{2}} + (1-\theta)r^{-1-\frac{\delta}{2}} - (m-1)H(1+o(1)) \right] \right.$$
$$\left. - A r^{\alpha-\theta+1-\frac{\delta}{2}} - \xi^{\sigma-1} B r^{\beta-\theta+1-\frac{\delta}{2}} \right\}$$
$$= \xi r^{\delta} e^{-\gamma\left(1+r(x)^{\frac{\delta+2}{2}}\right)} \left\{ \frac{\delta+2}{2}\gamma \left[\frac{\delta+2}{2}\gamma - \frac{\delta}{2}r^{-1-\frac{\delta}{2}} - (m-1)H(1+o(1)) \right] \right.$$
$$\left. - A r^{\alpha-\delta} - \xi^{\sigma-1} B r^{\beta-\delta} \right\}.$$

Note that, because of our assumptions on $\alpha, \beta, \gamma, \delta$ we can choose R sufficiently large and $\xi > 0$ sufficiently small that

$$\Delta v > 0 \quad \text{on } \Omega$$

contradicting the fact that $\bar{x} \in \Omega$. $\qquad\square$

The case $\delta = -2$ is, loosely speaking, border line between Euclidean and hyperbolic geometry. As seen in Proposition 1.15 the bound

$$\text{Ric}\,(\nabla r, \nabla r) \geq -(m-1)H^2\left(1+r^2\right)^{-1} \quad \text{on } M, \quad H > 0 \qquad (4.71)$$

implies the estimate

$$\Delta r \leq (m-1)r^{-1}\frac{1+\sqrt{1+4H^2}}{2}(1+o(1)) \quad \text{as } r \to +\infty. \qquad (4.72)$$

Proceeding in a way similar to that of the argument of Theorem 4.6 we can prove the following (see [RZ07], Theorem 2.5)

Theorem 4.7. *Let* $(M, \langle\,,\,\rangle)$ *be a complete manifold with radial Ricci curvature satisfying* (4.71). *Let* $a(x), b(x) \in C^0(M)$ *satisfy* (4.61) *and*

$$\limsup_{r(x)\to+\infty} \frac{b(x)}{r(x)^{\beta}} < +\infty \qquad (4.73)$$

with $\alpha < -2, \beta \leq \gamma(\sigma-1) - 2$ *and*

$$\gamma > (m-1)\frac{1+\sqrt{1+4H^2}}{2} - 1. \qquad (4.74)$$

Then any positive solution u *of*

$$\Delta u + a(x)u - b(x)u^{\sigma} \leq 0 \quad \text{on } M, \quad \sigma > 1$$

satisfies

$$\liminf_{r(x)\to+\infty} \frac{u(x)}{r(x)^{-\gamma}} > 0. \qquad (4.75)$$

Both estimates of Theorems 4.6 and 4.7 are quite sharp with respect to the range of the exponent γ. To simplify computations let us consider (4.75). Towards this aim, for a chosen $R_0 > 1$, let M_g be the model with $g(r) \in C^\infty(M)$ positive on $[0, +\infty)$ and such that

$$g(r) = \begin{cases} r & \text{on } \left[0, \tfrac{1}{2}\right], \\ r^{\frac{1}{2}\left(1+\sqrt{1+4\widetilde{H}^2}\right)} & \text{on } [R_0, +\infty) \end{cases} \tag{4.76}$$

for some $\widetilde{H} > 0$. Since $\mathrm{Ric}(\nabla r, \nabla r) = -(m-1)\frac{g''(r)}{g(r)}$ we have

$$\mathrm{Ric}(\nabla r, \nabla r) = -(m-1)\widetilde{H}^2 r^{-2} = -(m-1)\widetilde{H}^2(1+r^{-2})(1+r^2)^{-1}$$

on $M_g \setminus B_{R_0}$. We let $H^2 = \widetilde{H}^2(1 + R_0^{-2})$ so that

$$\mathrm{Ric}(\nabla r, \nabla r) \geq -(m-1)H^2(1+r^2)^{-1} \tag{4.77}$$

on $M_g \setminus B_{R_0}$. It is not hard to show that $g(r)$ can be defined on $\left[\tfrac{1}{2}, R_0\right]$ in such a way that (4.77) holds on all of M_g. Now we set

$$\gamma = \frac{m-1}{2}\left(1 + \sqrt{1+4H^2}\right)^{-1}$$

and we choose $A, B > 0$ sufficiently large that

$$\gamma(\gamma+2)R_0^2 - \gamma(m-1) \inf_{[0,R_0]} r\frac{g'(r)}{g(r)}(1+r^2)^{-\frac{\gamma}{2}-1} \tag{4.78}$$

$$\leq \gamma\left(1+R_0^2\right)^{-\frac{\gamma}{2}-1} + \left(1+R_0^2\right)^{-\frac{\gamma}{2}} A \inf_{[0,R_0]} (1+r)^\alpha$$

$$+ \beta\xi_0^{\sigma-1}\left(1+R_0^2\right)^{-\frac{\sigma\gamma}{2}} \inf_{[0,R_0]} (1+r)^\beta$$

for some fixed $\xi_0 > 0$ and $\alpha, \beta \in \mathbb{R}$. We set

$$a(x) = -A(1+r(x))^\alpha, \quad b(x) = B(1+r(x))^\beta$$

and we define, with $\xi \geq \xi_0$,

$$v(x) = \xi\left(1+r(x)^2\right)^{-\frac{\gamma}{2}}. \tag{4.79}$$

Then, a simple computation shows that on M_g,

$$\Delta v + a(x)v - b(x)v^\sigma = \xi\Big\{ \gamma(\gamma+2)r^2(x)\left(1+r^2(x)\right)^{-\frac{\gamma}{2}-2} - \gamma\left(1+r^2(x)\right)^{-\frac{\gamma}{2}-1}$$

$$- \gamma(m-1)\frac{g'(r(x))}{g(r(x))}r(x)\left(1+r^2(x)\right)^{-\frac{\gamma}{2}-1} - \tag{4.80}$$

$$- A(1+r(x)^\alpha)\left(1+r^2(x)\right)^{-\frac{\gamma}{2}} -$$

$$- B\xi^{\sigma-1}(1+r(x))^\beta\left(1+r^2(x)\right)^{-\frac{\sigma\gamma}{2}} \Big\}.$$

Thus, since $\xi \geq \xi_0$, (4.78) and inspection of (4.80) show that

$$\Delta v + a(x)v - b(x)v^\sigma \leq 0 \tag{4.81}$$

on B_{R_0}. Next, we note that on $M \setminus B_{R_0}$,

$$(m-1)\frac{g'(r)}{g(r)} = \frac{m-1}{2r}\left(1 + \sqrt{1+4H^2}\right) = (1 + \gamma - \varepsilon)\frac{1}{r} \tag{4.82}$$

for some $\varepsilon = \varepsilon(R_0)$ with $\varepsilon \to 0$ as $R_0 \to +\infty$. Rearranging the terms in (4.80) and using (4.82) we have

$$\Delta v + a(x)v - b(x)v^\sigma \tag{4.83}$$

$$= \xi r^2(x)\left(1+r^2(x)\right)^{-\frac{\gamma}{2}-2}\left\{\varepsilon\gamma - \gamma(\gamma + 2 - \varepsilon)r(x)^{-2}\right.$$

$$- Ar(x)^{\alpha+2}\left(1 + \frac{1}{r(x)}\right)^\alpha \left(1 + \frac{1}{r^2(x)}\right)^2$$

$$\left. - B\xi^{\sigma-1}r(x)^{\beta+2-\gamma(\sigma-1)}\left(1 + \frac{1}{r(x)}\right)^\beta \left(1 + \frac{1}{r^2(x)}\right)^{2-\frac{\gamma}{2}(\sigma-1)}\right\}.$$

If $\beta = (\sigma - 1)\gamma - 2$ we can choose $\xi \geq \xi_0$ sufficiently large that the above yields the validity of (4.81) on $M \setminus B_{R_0}$.

Note that, in this example, the range of α plays no role.

4.5 Nonexistence results for the Yamabe problem

In this section we apply the nonexistence results of Chapter 3 and the *a priori* estimates from above of Chapter 4 to give nonexistence results in the geometrical case of the Yamabe equation; similar results can, of course, be obtained for Yamabe-type equations on the complete, noncompact, manifold $(M, \langle \, , \, \rangle)$. We begin with the next result, that was first proved in [BRS98].

Theorem 4.8. *Let $(M, \langle \, , \, \rangle)$ be a complete Riemannian manifold with scalar curvature $S(x)$. Assume*

$$\mathrm{Ric} \geq -(m-1)B^2 \tag{4.84}$$

for some $B > 0$ and that there exists a $\varphi > 0$ solution of

$$L\varphi \leq -K\frac{|\nabla\varphi|^2}{\varphi} \tag{4.85}$$

for some $K > -1$, with $L = \Delta - \frac{m-1}{4m}S(x)$, $m = \dim M \geq 3$. Let $K(x) \in C^\infty(M)$ satisfy

$$\begin{cases} K(x) \leq 0 & \text{on } M, \\ K(x) \leq -Cr(x)^{\frac{2}{m-1}+\gamma}e^{2Br(x)} & \text{for } r(x) \gg 1 \end{cases} \tag{4.86}$$

and some constants $C, \gamma > 0$. Then the metric $\langle\,,\,\rangle$ cannot be pointwise conformally deformed to a metric $\widetilde{\langle\,,\,\rangle}$ of scalar curvature $K(x)$.

Remark. Condition (4.86) is sharp only in the term e^{2Br}. Indeed, for the Yamabe problem on hyperbolic space $\mathbb{H}^m_{-B^2}$, the second of (4.86) can be relaxed to

$$K(x) \le -C \frac{e^{2Br}}{r(x) \log r(x)} \quad \text{for } r(x) \gg 1$$

(see Theorem 7.3 and the related Remark 2; see also the Remark after Theorem 6.15).

Proof. First of all we recall that if $\widetilde{\langle\,,\,\rangle} = u^{\frac{4}{m-2}} \langle\,,\,\rangle$, then $u > 0$ has to satisfy

$$c_m \Delta u - S(x)u + K(x)u^{\frac{m+2}{m-2}} = 0, \quad c_m = \frac{4(m-1)}{m-2},$$

so that, with our notation for Yamabe-type equations,

$$a(x) = -\frac{m-2}{4(m-1)}S(x), \quad b(x) = -\frac{m-2}{4(m-1)}K(x).$$

We want to apply Theorem 3.11 with the choices

$$\sigma = \frac{m+2}{m-2}, \quad \alpha = \frac{m}{m-2}.$$

Indeed, in this case

$$L_{\sigma,\alpha} = \Delta + \frac{(m-1)^2}{m(m-2)}a(x) = \Delta - \frac{m-1}{4m}S(x) = L,$$

so that (4.85) gives $\lambda_1^{L_{\sigma,\alpha}}(M) \ge 0$. From (4.86) we have $b(x) \ge 0$ on M and $b(x) > 0$ for $r(x) \gg 1$. We now have to show the existence of $C > 0$ such that

$$a(x) \le Cb(x) \quad \text{for } r(x) \gg 1. \tag{4.87}$$

Now, (4.84) implies

$$\sup_M a(x) \le \frac{m-2}{4}mB^2 \tag{4.88}$$

so that, since $b(x) > 0$ for $r(x) \gg 1$, (4.87) is satisfied for C sufficiently large. This also shows that in order to verify (3.99) and (3.100) it is enough to show that

$$u^{2\frac{m-1}{m-2}} \in L^1(M). \tag{4.89}$$

Towards this aim we apply Theorem 4.5 with the choice $\delta = 0$ and

$$\psi(t) = t^{\frac{2}{m-1}+\gamma}e^{2Bt}$$

for some $\gamma > 0$ as in (4.86). Then assumptions (4.54) i) and ii) are satisfied because of (4.88) and (4.86). It follows that

$$u(x) \leq Cr(x)^{-\frac{1}{2}\frac{m-2}{m-1} - \frac{m-2}{4}\gamma} e^{-\frac{m-2}{2} Br(x)} \tag{4.90}$$

for some constant $C > 0$ and $r(x) \gg 1$. By Bishop's volume comparison theorem (see Chapter 1) and (4.90)

$$\int_{\partial B_r} u(x)^{2\frac{m-1}{m-2}} \leq Cr^{-1-\frac{m-1}{2}\gamma} e^{-(m-1)Br} e^{(m-1)Br} = Cr^{-1-\frac{m-1}{2}\gamma}$$

and since $\gamma > 0$ it follows that

$$\int_{\partial B_r} u(x)^{2\frac{m-1}{m-2}} \in L^1(+\infty),$$

that is, (4.89) is satisfied. □

Suppose now that we are able to solve Yamabe's equation

$$c_m \Delta u - S(x)u + K(x)u^{\frac{m+2}{m-2}} = 0 \quad \text{on } M, \ u > 0 \tag{4.91}$$

on the complete manifold $(M, \langle\,,\,\rangle)$; we then ask if the new metric $\widetilde{\langle\,,\,\rangle} = u^{\frac{4}{m-2}}\langle\,,\,\rangle$, $m = \dim M \geq 3$ is still complete. The *a priori* estimates of Chapter 4 give us an answer, in the negative, with the following

Theorem 4.9. *Let* $(M, \langle\,,\,\rangle)$ *be a complete Riemannian manifold of dimension* $m \geq 3$ *and scalar curvature* $S(x)$. *Suppose that*

$$\mathrm{Ric} \geq -(m-1)H^2(1+r(x))^{\frac{\delta}{2}} \tag{4.92}$$

for some $H > 0$, $\delta \geq -2$. *Let* $K(x) \in C^\infty(M)$ *and assume that*

$$K(x) \leq -K^2 r(x)^{\delta+2} [\log r(x)]^{2(1+\gamma)} \ \text{for } r(x) \gg 1 \tag{4.93}$$

and some constants $K, \gamma > 0$. *Then any metric conformal to* $\langle\,,\,\rangle$ *and with scalar curvature* $K(x)$ *is incomplete.*

Proof. As above we let $a(x) = -\frac{S(x)}{c_m}$, $b(x) = -\frac{K(x)}{c_m}$. We apply Proposition 4.3 with the choices

$$\omega(t) = Ct^2 (\log t)^{2(1+\gamma)}, \quad \psi(t) = Ct^{3+\frac{\delta}{2}} (\log t)^{2(1+\gamma)}$$

for $t \gg 1$ and some $C > 0$. We note that ω and ψ satisfy (4.41) i), ii) $\forall \varepsilon \in (0,1)$. Furthermore, due to (4.92) and (4.93), we respectively have

$$b(x) \geq \frac{K^2}{C_m} r(x)^{\delta+2} [\log r(x)]^{2(1+\gamma)} \geq \psi(r(x))r(x)^{\frac{\delta}{2}-1} \quad \text{for } r(x) \gg 1,$$

$$\frac{a_+(x)}{b(x)} \le \frac{1}{\omega(r(x))} \quad \text{for } r(x) \gg 1$$

provided $C > 0$ is chosen appropriately. An application of Proposition 4.3 yields

$$u(x) = O\left(\left[r(x)(\log r(x))^{1+\gamma} \right]^{-\frac{m-2}{2}} + \left[r(x)(\log r(x))^{1+\gamma} \right]^{-\frac{m-2}{2}} r(x)^{-\frac{m-2}{2}\left(\frac{\delta}{2}+1\right)} \right)$$

as $r(x) \to +\infty$. Since $\delta \ge -2$, from this latter we deduce that

$$u(x) \le C\left[r(x)(\log r(x))^{1+\gamma} \right]^{-\frac{m-2}{2}} \quad \text{for } r(x) \gg 1 \tag{4.94}$$

and some $C > 0$. To show that the metric $\widetilde{\langle\,,\,\rangle} = u^{\frac{4}{m-2}}\langle\,,\,\rangle$ is not complete, we reason as follows. Let $\{x_n\}$ be a sequence diverging in M, that is, $\{x_n\}$ is definitely outside any fixed compact set in M. Since $(M, \langle\,,\,\rangle)$ is complete, $r(x_n) \to +\infty$ as $n \to +\infty$; in particular we can suppose that $\{x_n\} \subset M \backslash B_R$ on which (4.94) holds. For n fixed, by the Hopf-Rinow theorem (see e.g. [Lee97]) there exists γ_n, a unit speed geodesic from o to x_n, realizing the distance $r(x_n)$. Thus if t is the arclength parameter of γ_n, then

$$r(\gamma_n(t)) = t.$$

For $\tilde{r}(x) = \text{dist}_{\widetilde{\langle\,,\,\rangle}}(o, x)$ we have

$$\tilde{r}(x_n) \le \int_0^{r(x_n)} u^{\frac{2}{m-2}}(\gamma_n(t))\, dt$$

$$\le C^{\frac{2}{m-2}} \int_0^{r(x_n)} \left[r(\gamma_n(t))(\log r(\gamma_n(t)))^{1+\gamma} \right]^{-1} dt$$

$$= C^{\frac{2}{m-2}} \int_0^{r(x_n)} \left[t(\log t)^{1+\gamma} \right]^{-1} dt \le C^{\frac{2}{m-2}} \int_0^{+\infty} \left[t(\log t)^{1+\gamma} \right]^{-1} dt \le \widehat{C}$$

with $\widehat{C}$ an absolute constant. Thus,

$$\tilde{r}(x_n)) \le \widehat{C} \quad \forall\, n \in \mathbb{N},$$

and the metric $\widetilde{\langle\,,\,\rangle}$ cannot be complete since $\{x_n\}$ is a divergent sequence in M. $\qquad\square$

In the next result we show that even a slow decay to 0 of $K(x)$ at infinity under some additional condition implies nonexistence. This will be an application of Theorems 3.12 and 4.4.

Theorem 4.10. *Let $(M, \langle\,,\,\rangle)$ be a complete Riemannian manifold of dimension $m \ge 3$ and scalar curvature $S(x)$. Suppose that*

$$\text{Ric} \ge -(m-1)H^2(1 + r(x)^2)^{\frac{\delta}{2}}$$

for some $H > 0$ and $-2 \le \delta < 2$. For $L = \Delta - \frac{1}{c_m} S(x)$ assume that

$$\lambda_1^L(\operatorname{supp} S_-) > 0, \tag{4.95}$$

where S_- is the negative part of S. Let $K(x) \in C^\infty(M)$ be such that $K(x) \le 0$ on M and

$$K(x) \le -\mathfrak{K}^2 r(x)^\beta \tag{4.96}$$

for $r(x) \gg 1$ and some $\mathfrak{K} > 0$, $\delta < \beta \le 1 + \frac{\delta}{2}$. Then the metric $\langle\,,\rangle$ cannot be pointwise conformally deformed to a metric $\widetilde{\langle\,,\rangle}$ of scalar curvature $K(x)$.

Proof. We reason by contradiction and suppose that $\widetilde{\langle\,,\rangle}$ exists. Then we have a solution $u > 0$ of the differential equation

$$\Delta u + a(x)u - b(x)u^{\frac{m+2}{m-2}} \quad \text{on } M$$

with, as we saw in Theorem 4.9,

$$a(x) = -\frac{S(x)}{c_m}, \quad b(x) = -\frac{K(x)}{c_m}.$$

The assumption on the Ricci tensor yields

$$a(x) \le A r(x)^\delta \quad \text{for } r(x) \gg 1$$

and an appropriate constant $A > 0$, while from (4.96) we deduce

$$b(x) \ge B r(x)^\beta.$$

Note that, due to the choice of the range of the parameters δ and β, the assumptions of Theorem 4.4 are satisfied and we conclude that

$$u(x) \le C r(x)^{-\frac{m}{4}(\beta - \delta)}.$$

Then $u(x) \to 0$ as $x \to \infty$. Using (4.95) and $K(x) \le 0$ on M we apply Theorem 3.12 to obtain a contradiction to the existence of u. $\qquad\square$

Chapter 5

Uniqueness

The aim of this chapter is to prove some uniqueness results for positive solutions of Yamabe-type equations. Our first theorem in this direction depends only on the sign of the coefficient $b(x)$ of the nonlinear term and, loosely speaking, on an L^2-type estimate of the distance, at infinity, of the two solutions under consideration. It is worth observing that this very general result is sharp and that the L^2-type condition cannot be substituted with a corresponding L^p condition with $p > 2$.

Our second result, Theorem 5.2, is obtained *via* a comparison result whose proof requires a version of the weak maximum principle (see Theorem 5.3) which is interesting in its own. The strength of the theorem lies both in the fact that it requires that the solutions approach one another at infinity in the weak sense of condition (5.13) below, and in the fact that the remaining assumptions are very general. Counterexamples are given after each result to prove sharpness.

We end the chapter with a geometric application to the group of conformal diffeomorphisms of a complete manifold and to uniqueness for the Yamabe problem.

5.1 A sharp integral condition

The first theorem gives a condition for two positive solutions of

$$\Delta u + a(x)u - b(x)u^\sigma = 0, \quad \sigma > 1, \quad \text{on } M \tag{5.1}$$

to coincide. The condition is expressed in integral form and, as we shall see, is quite sharp.

Theorem 5.1. *Let $a(x), b(x) \in C^0(M)$ and assume that*

$$\text{i) } b(x) \geq 0 \text{ on } M; \quad \text{ii) } b(x) \not\equiv 0 \text{ on } M. \tag{5.2}$$

Let $u, v \in C^2(M)$ be positive solutions of (5.1) *and suppose that*

$$\left\{\int_{\partial B_r} (u-v)^2\right\}^{-1} \notin L^1(+\infty). \tag{5.3}$$

Then $u \equiv v$ on M.

Note that condition (5.3) is implied by $u - v \in L^2(M)$ or even

$$\int_{B_R} (u-v)^2 = o(R^2) \quad \text{as } R \to +\infty.$$

Proof. Applying the divergence theorem on a geodesic ball B_R of radius R to the vector field

$$W = (v^2 - u^2)\nabla\left(\log \frac{v}{u}\right),$$

using (5.1) and rearranging, we obtain

$$\int_{\partial B_R} \langle W, \nabla r\rangle - \int_{B_R}\left|\nabla u - \frac{u}{v}\nabla v\right|^2 - \int_{B_R}\left|\nabla v - \frac{v}{u}\nabla u\right|^2 \tag{5.4}$$
$$= \int_{B_R} b(x)(v^2 - u^2)(v^{\sigma-1} - u^{\sigma-1}).$$

Since the right-hand side is a nonnegative and nondecreasing function of R it tends to a nonnegative limit $B \leq +\infty$ as $R \to +\infty$. Next by Gauss' lemma and Schwarz's inequality

$$|\langle W, \nabla r\rangle| \leq |v-u||v+u|\left|\frac{\nabla v}{v} - \frac{\nabla u}{u}\right| \leq |v-u|\left(\left|\nabla v - \frac{v}{u}\nabla u\right| + \left|\nabla u - \frac{u}{v}\nabla v\right|\right)$$

and therefore

$$\left\{\int_{\partial B_R}\langle W, \nabla r\rangle\right\}^2 \leq 2\left\{\int_{\partial B_R}(v-u)^2\right\}\left\{\int_{\partial B_R}\left|\nabla v - \frac{v}{u}\nabla u\right|^2 + \left|\nabla u - \frac{u}{v}\nabla v\right|^2\right\}. \tag{5.5}$$

We claim that

$$\left|\nabla v - \frac{v}{u}\nabla u\right|^2 + \left|\nabla u - \frac{u}{v}\nabla v\right|^2 \in L^1(M). \tag{5.6}$$

Indeed, assume by contradiction that

$$G(R) = \int_{B_R}\left|\nabla v - \frac{v}{u}\nabla u\right|^2 + \left|\nabla u - \frac{u}{v}\nabla v\right|^2 \to +\infty \quad \text{as } R \to +\infty.$$

Since

$$\int_{\partial B_R}\langle W, \nabla r\rangle - G(R) \to B \geq 0 \quad \text{as } R \to +\infty,$$

$\forall R \geq \bar{R}$ sufficiently large,

$$\int_{\partial B_R} \langle W, \nabla r \rangle \geq \frac{1}{2} G(R) \geq 0.$$

Therefore, using (5.5),

$$\left\{ \frac{1}{2} G(R) \right\}^2 \leq \left\{ \int_{\partial B_R} \langle W, \nabla r \rangle \right\}^2 \leq 2 G'(R) \left\{ \int_{\partial B_R} (v-u)^2 \right\}.$$

Arguing exactly as in Step 1 of the proof of Theorem 3.2 we contradict (5.3) and conclude that (5.6) holds. Now, because of (5.3) and (5.6) there exists an increasing sequence $\{r_k\} \uparrow +\infty$ such that

$$\lim_{k \to +\infty} \left\{ \int_{\partial B_{r_k}} (u-v)^2 \right\} \left\{ \int_{\partial B_{r_k}} \left| \nabla v - \frac{v}{u} \nabla u \right|^2 + \left| \nabla u - \frac{u}{v} \nabla v \right|^2 \right\} = 0$$

and then, using (5.5) we conclude that

$$\lim_{k \to +\infty} \int_{\partial B_{r_k}} \langle W, \nabla r \rangle = 0.$$

Next, we evaluate (5.4) along the sequence $\{r_k\}$ and let $k \to +\infty$ to obtain

$$\int_M b(x)(v^2 - u^2)(v^{\sigma-1} - u^{\sigma-1}) + \int_M \left[\left| \nabla v - \frac{v}{u} \nabla u \right|^2 + \left| \nabla u - \frac{u}{v} \nabla v \right|^2 \right] = 0. \quad (5.7)$$

Due to the assumptions on $b(x)$, $\left| \nabla u - \frac{u}{v} \nabla v \right| = 0$ on M, so that $u = Av$ for some constant $A > 0$; substituting into (5.7) yields

$$(1 - A^2)(1 - A^{\sigma-1}) \int_M b(x) v^{\sigma+1} = 0$$

and since $v > 0$, $b(x) \geq 0$, $b(x) \not\equiv 0$ this forces $A = 1$, that is, $u \equiv v$ on M. $\qquad \square$

Remark. Despite its generality, assumption (5.3) of Theorem 5.1 is quite sharp. Indeed suppose $(M, \langle \, , \, \rangle)$ is Euclidean space $\mathbb{R}^m$ with its canonical metric; let $a(x)$ and $b(x)$ be nonnegative continuous functions satisfying

$$a(x) \leq \frac{(m-2)^2}{4} |x|^{-2},$$

$$b(x) \leq \frac{|x|^{\frac{(m-2)(\sigma+1)}{2}}}{(\log |x|)^{\sigma+1} (\log \log |x|)(\log \log \log |x|)^2}$$

with equality holding for sufficiently large $|x|$. We shall prove in the next section that the equation

$$\Delta u + a(x)u - b(x)u^\sigma = 0, \quad \sigma > 1 \quad \text{on } \mathbb{R}^m \qquad (5.8)$$

has a family of positive solutions u_α ($\alpha > 0$) satisfying

$$u_\alpha(0) = \alpha \quad \text{and} \quad u_\alpha(x) \sim |x|^{-\frac{m-2}{2}} \log|x| \quad \text{as } |x| \to +\infty.$$

If $\alpha_1 \neq \alpha_2$, then u_{α_1} and u_{α_2} are distinct solutions for which

$$|u_{\alpha_1} - u_{\alpha_2}|^p r^{m-1} \leq C r^{m-1-\frac{m-2}{2}p}(\log r)^p$$

for $r = |x| \gg 1$. Thus

$$\int_{\partial B_r} |u_{\alpha_1} - u_{\alpha_2}|^p \leq C r^{m-1-\frac{m-2}{2}p}(\log r)^p.$$

Hence, for $p > 2$,

$$\left\{ \int_{\partial B_r} |u_{\alpha_1} - u_{\alpha_2}|^p \right\}^{-1} \notin L^1(+\infty),$$

while, for $p = 2$, (5.3) just falls short of being met. Note that this same example shows that the case $p = 2$ is special for the validity of the theorem.

5.2   A remark on the asymptotic behaviour of solutions: examples in $\mathbb{R}^m$ and $\mathbb{H}^m$

Let $b(x) \geq 0$ and let u and v be positive solutions of the equation

$$\Delta u - b(x)u^\sigma = 0, \quad \sigma > 1 \text{ on } (M, \langle\,,\,\rangle) \tag{5.9}$$

such that

$$\lim_{x \to \infty} (u(x) - v(x)) = 0. \tag{5.10}$$

Then, a simple application of the maximum principle shows that u and v coincide. Indeed, setting $w = u - v$, we have

$$\Delta(w^2) = 2w\Delta w + 2|\nabla w|^2 \geq 2w\Delta w$$
$$= 2(u - v)b(x)(u^\sigma - v^\sigma) \geq 0$$

so that w^2 is a nonnegative subharmonic function on M which tends to zero at infinity. Thus it attains a nonnegative maximum and it is therefore identically zero by the maximum principle (see, [CN92]).

This suggests that two positive solutions u and v of

$$\Delta u + a(x)u - b(x)u^\sigma = 0, \quad \sigma > 1 \text{ on } (M, \langle\,,\,\rangle)$$

may coincide if they have the "same" behaviour at infinity. The following example shows that, in general, it may be that a simple limit condition like (5.10) is not enough to conclude. Let B be the unit ball in $\mathbb{R}^m$ and, for $a > 1$, define

$$\beta_a(r) = \frac{1}{m(m-2)a^2}(a^2 - r^2)^{-\frac{m-2}{2}} \quad r \in [0, 1), \ m \geq 3.$$

A simple computation shows that β satisfies

$$\begin{cases} \beta_a'' + \frac{m-1}{r}\beta_a' = \beta_a^{\frac{m+2}{m-2}}, & r \in [0,1), \\ \beta_a(0) = \frac{1}{m(m-2)a^m}, \ \beta_a'(0) = 0, \end{cases}$$

so that $\beta_a(|x|)$ is a family of positive radial solutions of (5.9) on B with $b(x) \equiv 1$, $\sigma = \frac{m+2}{m-2}$. Operating the change of variable

$$r(t) = \frac{t}{(1+t^{m-2})^{1/(m-2)}}$$

we obtain a solution $\alpha_a(t) = \beta_a(r(t))$ of

$$\begin{cases} \alpha_a'' + \frac{m-1}{t}\alpha_a' = \tilde{b}(t)\alpha_a^{\frac{m+2}{m-2}}, & t \in [0,+\infty) \\ \alpha_a(0) = \frac{1}{m(m-2)a^m}, \ \alpha_a'(0) = 0, \end{cases}$$

with

$$\tilde{b}(t) = \left(1+t^{m-2}\right)^{-2\frac{m-1}{m-2}}.$$

Thus α_a gives rise to a family of positive radial solutions $u_a(x)$ of

$$\Delta u_a - \left(1+|x|^{m-2}\right)^{-2\frac{m-1}{m-2}} u_a^{\frac{m+2}{m-2}} = 0 \quad \text{on } \mathbb{R}^m$$

for which

$$\lim_{|x|\to+\infty} u_a(x) = \frac{1}{m(m-2)a^2}(a^2-1)^{-\frac{m-2}{2}}.$$

It is clear that two solutions u_{a_1} and u_{a_2} of the family coincide if and only if

$$\lim_{|x|\to+\infty} u_{a_1} = \lim_{|x|\to+\infty} u_{a_2},$$

that is, $a_1 = a_2$. Note that the geometrical meaning of these solutions is obvious: $\mathbb{R}^m$ can be conformally deformed to a complete Riemannian manifold of negative scalar curvature sufficiently close to zero. Thus, in this case, uniqueness is expressed by (5.10). However, we may consider the same family of solutions on the hyperbolic space $\mathbb{H}^m$ of constant negative curvature -1 by identifying it with $\mathbb{R}^m$ with metric, in polar coordinates, $ds^2 = dr^2 + (\sinh r)^2 d\theta^2$. In this case the situation changes completely: indeed, let

$$\gamma_a(t) = \beta_a\left(\tanh\frac{t}{2}\right)\left[\frac{2}{1-\tanh^2\frac{t}{2}}\right]^{-\frac{m-2}{2}}.$$

Then,

$$\begin{cases} \gamma_a'' + (m-1)\coth(t)\gamma_a' + \frac{m(m-2)}{4}\gamma_a = \gamma_a^{\frac{m+2}{m-2}} & \text{on } [0,+\infty), \\ \gamma_a'(0) = 0, \ \gamma_a(0) = \frac{1}{2^{\frac{(m-2)}{2}}m(m-2)a^m} \end{cases}$$

and

$$w_a(x) = \frac{1}{m(m-2)a^2}\left(a^2 - \tanh^2 \frac{\rho(x)}{2}\right)^{-\frac{m-2}{2}}\left[2\cosh^2 \frac{\rho(x)}{2}\right]^{-\frac{m-2}{2}},$$

with $\rho(x) = \mathrm{dist}_{\mathbb{H}^m}(x, o)$, is a family of positive solutions of

$$\Delta w_a + \frac{m(m-2)}{4}w_a - w_a^{\frac{m+2}{m-2}} = 0 \quad \text{on } \mathbb{H}^m.$$

We observe that

$$w_a(x) \sim \frac{2^{-\frac{m-2}{2}}}{m(m-2)a^2}(a^2 - 1)^{-\frac{m-2}{2}}e^{-\frac{m-2}{2}\rho(x)} \quad \text{as } \rho(x) \to +\infty.$$

Thus,

$$w_a(r) \to 0 \quad \text{as } r(x) \to +\infty$$

independently of $a > 1$ and with the same speed.

In the next section we show that uniqueness of solutions converging to zero is indeed special.

5.3   Uniqueness via the weak maximum principle

In this section we prove a number of uniqueness results for nonnegative solutions of equation (5.1). We begin with the following (see [RZ07]).

Theorem 5.2. *Let* $(M, \langle\,,\,\rangle)$ *be a complete manifold,* $a(x), b(x) \in C^0(M)$, $\sigma > 1$, $\tau \geq 0$, $\beta + \tau(\sigma - 1) > -2$ *and suppose that* $b(x) > 0$ *on* M *and*

$$\text{i) } b(x) \geq Br(x)^\beta \text{ for } r(x) \gg 1; \quad \text{ii) } \sup_M \frac{a_-(x)}{b(x)}r(x)^{\tau(1-\sigma)} < +\infty. \tag{5.11}$$

Suppose that $u, v \in C^2(M)$ *are two nonnegative solutions of*

$$\Delta u + a(x)u - b(x)u^\sigma = 0 \quad \text{on } M \tag{5.12}$$

satisfying

$$C^{-1}r(x)^\tau \leq u(x), v(x) \leq Cr(x)^\tau \tag{5.13}$$

for $r(x) \gg 1$ *and some constant* $C > 0$. *If*

$$\liminf_{r \to +\infty} \frac{\log \mathrm{vol}\, B_r}{r^{2+\beta+\tau(\sigma-1)}} < +\infty, \tag{5.14}$$

then $u \equiv v$ *on* M.

Theorem 5.2 will follow immediately from a corresponding comparison result whose proof is based on the next weak maximum principle at infinity.

5.3.1 A useful form of the weak maximum principle

The next result, which is a version of what is known in literature as the *weak maximum principle at infinity* (see [RSV05], [PRS05b] and [MRS10] for a more general version), turns out to be quite useful in many contexts, both geometrical and analytical. It is worth observing that the function u under consideration is not necessarily bounded but is only required to satisfy some growth control (expressed in (5.17)). The other main assumption of the theorem is a control on the volume of geodesic balls (see (5.16)).

Theorem 5.3. *Let $(M, \langle \, , \rangle)$ be a complete manifold, and let $\sigma, \mu \in \mathbb{R}$, be such that*

$$\text{i) } \sigma \geq 0; \quad \text{ii) } 2 - \sigma - \mu > 0. \tag{5.15}$$

Assume that

$$\liminf_{r \to +\infty} \frac{\log \operatorname{vol} B_r}{r^{2-\sigma-\mu}} = d_0 < +\infty. \tag{5.16}$$

Let $u \in C^2(M)$ and suppose that

$$\widehat{u} = \limsup_{r(x) \to +\infty} \frac{u(x)}{r(x)^\sigma} < +\infty. \tag{5.17}$$

Then, given $\gamma \in \mathbb{R}$ such that

$$\Omega_\gamma = \{x \in M : u(x) > \gamma\} \neq \emptyset,$$

we have

$$\inf_{\Omega_\gamma} [1 + r(x)]^\mu \Delta u \leq d_0 \max\{\widehat{u}, 0\} \, C(\sigma, \eta) \tag{5.18}$$

with

$$C(\sigma, \mu) = \begin{cases} 0 & \text{if } \sigma = 0; \\ (2 - \sigma - \mu)^2 & \text{if } \sigma > 0, \mu + 2(\sigma - 1) < 0; \\ \sigma(2 - \sigma - \mu) & \text{if } \sigma > 0, \mu + 2(\sigma - 1) \geq 0. \end{cases} \tag{5.19}$$

Proof. We begin by observing that if ν is any constant and we set $u_\nu = u + \nu$, then $\Delta u_\nu = \Delta u$ and

$$\{x \in M : u(x) > \gamma\} = \Omega_\gamma = \{x \in M : u_\nu(x) > \gamma + \nu\}.$$

Furthermore, if $\sigma > 0$, $\widehat{u} = \widehat{u}_\nu$ so, in order to estimate

$$\inf_{\Omega_\gamma} [1 + r]^\mu \Delta u$$

we may replace u by a suitable translate. Next, fix $b > \max\{\widehat{u}, 0\}$; then, there exists a constant ν such that

$$\frac{u_\nu}{(1 + r)^\sigma} < b \quad \text{on } M \quad \text{and} \quad u_\nu(x_0) > 0 \quad \text{for some } x_0 \in M. \tag{5.20}$$

We will assume that a constant ν has been selected in such a way that (5.20) holds. In accordance with the observation made at the beginning of the proof, we are going to replace u with u_ν, and for ease of notation we will even suppress the subscript ν. We want to show (5.18) and we may suppose, without loss of generality, that $\gamma \geq 0$. Next, let

$$K = \inf_{\Omega_\gamma} [1+r]^\mu \Delta u.$$

Clearly, the theorem amounts to showing that K is bounded above by the right-hand side of (5.19). If $K \leq 0$ there is nothing to prove; so let us assume $K > 0$, in which case u is nonconstant on any connected component of Ω_γ, and

$$[1+r]^\mu \Delta u \geq K > 0 \quad \text{on } \Omega_\gamma. \tag{5.21}$$

We fix $\theta \in \left(\frac{1}{2},1\right)$ and we choose $R_0 > 0$ large enough that $|\nabla u| \not\equiv 0$ on the nonempty open set $B_{R_0} \cap \Omega_\gamma$. Given $R > R_0$, let $\varphi \in C^\infty(M)$ be a cut-off function such that

$$0 \leq \varphi \leq 1, \quad \varphi \equiv 1 \text{ on } B_{\theta R}, \ \varphi \equiv 0 \text{ on } M \backslash B_R, \ |\nabla \varphi| \leq \frac{c}{R(1-\theta)} \tag{5.22}$$

for some absolute constant $c > 0$. Let also $\lambda \in C^1(\mathbb{R})$ and $F(v,r) \in C^1(\mathbb{R}^2)$ be such that

$$0 \leq \lambda \leq 1; \ \lambda = 0 \text{ on } (-\infty,\gamma]; \ \lambda > 0, \ \lambda' \geq 0 \text{ on } (\gamma,+\infty) \tag{5.23}$$

and

$$F(v,r) > 0; \quad \frac{\partial F}{\partial v}(v,r) < 0 \ \text{ on } [0,+\infty) \times [0,+\infty). \tag{5.24}$$

Finally, let W be the C^1 vector field defined on Ω_γ by

$$W = \varphi^2 \lambda(u) F(v,r) \nabla u \tag{5.25}$$

where v is given by

$$v = \alpha(1+r)^\sigma - u \tag{5.26}$$

and $\alpha > b$ is a constant. Note that, according to (5.20), $v > 0$ on Ω_γ, and, in fact, since $u > \gamma > 0$ on Ω_γ, we have

$$(\alpha - b)(1+r)^\sigma \leq v \leq \alpha(1+r)^\sigma \quad \text{on } \Omega_\gamma. \tag{5.27}$$

We note that W vanishes on $\partial(\Omega_\gamma \cap B_R)$ and it extends to a C^1 vector field on the whole of M by setting it equal to 0 in the complement of $\Omega_\gamma \cap B_R$. A computation that uses the properties of φ, λ and F, the definition (5.26) of v, inequality (5.21) and Cauchy-Schwarz inequality yields

$$\text{div } W = \varphi^2 \lambda(u) F(v,r) \Delta u + 2\varphi \lambda(u) F(v,r) \langle \nabla \varphi, \nabla u \rangle$$

$$+ \varphi^2 \lambda'(u) F(v,r) |\nabla u|^2 + \varphi^2 \lambda(u) \frac{\partial F}{\partial v}(v,r) \langle \nabla v, \nabla u \rangle$$

$$+ \varphi^2 \lambda(u) \frac{\partial F}{\partial r}(v,r) \langle \nabla u, \nabla r \rangle$$

$$\geq \varphi^2 \lambda(u) F(v,r)(1+r)^{-\mu} K - 2\varphi \lambda(u) F(v,r) |\nabla \varphi| |\nabla u|$$

$$+ \varphi^2 \lambda(u) \frac{\partial F}{\partial v}(v,r) \langle \nabla u, \alpha\sigma(1+r)^{\sigma-1} \nabla r - \nabla u \rangle$$

$$+ \varphi^2 \lambda(u) \frac{\partial F}{\partial r}(v,r) \langle \nabla u, \nabla r \rangle$$

$$\geq -2\varphi \lambda(u) F(v,r) |\nabla \varphi| |\nabla u| +$$

$$+ \varphi^2 \lambda(u) \left| \frac{\partial F}{\partial v}(v,r) \right|$$

$$\times \left\{ \frac{F(v,r)}{\left| \frac{\partial F}{\partial v} \right|} K(1+r)^{-\mu} + |\nabla u|^2 + \left[\frac{\frac{\partial F}{\partial r}}{\left| \frac{\partial F}{\partial v} \right|} - \alpha\sigma(1+r)^{\sigma-1} \right] \langle \nabla u, \nabla r \rangle \right\},$$

that is,

$$\text{div } W \geq -2\varphi \lambda(u) F(v,r) |\nabla \varphi| |\nabla u| + \varphi^2 \lambda(u) \left| \frac{\partial F}{\partial v}(v,r) \right| B(\nabla u, r) \qquad (5.28)$$

with

$$B(\nabla u, r) = |\nabla u|^2 + \frac{F(v,r)}{\left| \frac{\partial F}{\partial v} \right|} K(1+r)^{-\mu} + \left[\frac{\frac{\partial F}{\partial r}}{\left| \frac{\partial F}{\partial v} \right|} - \alpha\sigma(1+r)^{\sigma-1} \right] \langle \nabla u, \nabla r \rangle.$$

$$(5.29)$$

For notational convenience we set $\eta = \mu + 2(\sigma - 1)$ so that by (5.15), $\sigma - \eta > 0$, and we divide the proof into three cases.

Case I. $\sigma > 0, \eta < 0$.

We let

$$F(v,r) = e^{-qv(1+r)^{-\eta}}$$

where $q > 0$ is a constant that will be specified later. An elementary computation which uses the estimate for v given in (5.27) shows that

$$0 \geq \frac{\frac{\partial F}{\partial r}(v,r)}{\left| \frac{\partial F}{\partial v}(v,r) \right|} - \alpha\sigma(1+r)^{\sigma-1} = \eta v(1+r)^{-1} - \alpha\sigma(1+r)^{\sigma-1} \geq -\alpha(\sigma-\eta)(1+r)^{\sigma-1}$$

$$(5.30)$$

and

$$\frac{F(v,r)}{\left| \frac{\partial F}{\partial v} \right|(v,r)} = \frac{1}{q}(1+r)^{\eta}. \qquad (5.31)$$

Inserting (5.30) and (5.31) into (5.29) and using Cauchy-Schwarz inequality we deduce

$$B(\nabla u, r) \geq |\nabla u|^2 + \frac{K}{q}(1+r)^{2(\sigma-1)} - \alpha(\sigma-\eta)(1+r)^{\sigma-1}|\nabla u|. \qquad (5.32)$$

At this point we need to estimate the right-hand side of (5.32) so as to have

$$B(\nabla u, r) \geq \Lambda|\nabla u|^2, \qquad (5.33)$$

with a constant $\Lambda > 0$ independent of ∇u and r. A simple computation shows that if we choose

$$q < \frac{4K}{[\alpha(\sigma-\eta)]^2}, \qquad (5.34)$$

then (5.33) holds for every

$$0 < \Lambda \leq 1 - \frac{q[\alpha(\sigma-\eta)]^2}{4K}. \qquad (5.35)$$

With this choice of Λ we insert (5.33) and the expression of $\frac{\partial F}{\partial v}$ into (5.28) to deduce

$$\operatorname{div} W \geq -2\varphi\lambda(u)F(v,r)|\nabla\varphi||\nabla u| + q\varphi^2(1+r)^{-\eta}F(v,r)\Lambda|\nabla u|^2.$$

We integrate this inequality once in $\Omega_\gamma \cap B_R$, apply the divergence theorem, and recall that W vanishes on $\partial(\Omega_\gamma \cap B_R)$ to get

$$\frac{q\Lambda}{2}\int_{\Omega_\gamma \cap B_R} \varphi^2\lambda(u)F(v,r)(1+r)^{-\eta}|\nabla u|^2 \leq \int_{\Omega_\gamma \cap B_R} \varphi\lambda(u)F(v,r)|\nabla\varphi||\nabla u|.$$

Applying Hölder's inequality to the integral on the right-hand side and simplifying we finally obtain

$$\frac{q^2\Lambda^2}{4}\int_{\Omega_\gamma \cap B_R} \varphi^1\lambda(u)F(v,r)(1+r)^{-\eta}|\nabla u|^2 \leq \lambda(u)F(v,r)(1+r)^{\eta}|\nabla\varphi|^2. \quad (5.36)$$

By the volume growth assumption (5.16), $\forall d > d_0$ there exists a divergent sequence $R_k \uparrow +\infty$ with $R_1 > 2R_0$ such that

$$\log \operatorname{vol}(B_{R_k}) \leq dR_k^{\sigma-\eta}. \qquad (5.37)$$

Noting that $\theta R_k > \frac{1}{2}R_k > R_0$, we apply (5.36) with $R = R_k$, use the bound for $|\nabla\varphi|$ and the fact that $\lambda \leq 1$, to get

$$E = \left(\frac{q\lambda}{2}\right)^2 \int_{\Omega_\gamma \cap B_{R_0}} \lambda(u)F(v,r)|\nabla u|^2 \qquad (5.38)$$

$$\leq (1+\theta R_k)^\eta[(1-\theta)R_k]^{-2}\int_{\Omega_\gamma \cap (B_{r_k}\setminus B_{\theta R_k})} F(v,r). \qquad (5.39)$$

Now we observe that $E > 0$, since $\nabla u \not\equiv 0$ on $\Omega_\gamma \cap B_{R_0}$. On the other hand, using the bound (5.27) for v and the expression of F we get

$$F(v, r) \le e^{-q(\alpha-b)(1+\theta R_k)^{\sigma-\eta}} \quad \text{on } \Omega_\gamma \cap (B_{R_k} \setminus B_{\theta R_k}).$$

Inserting this into the right-hand side of (5.38), using (5.37) and rearranging we obtain the inequality

$$0 < E \le CR_k^{\eta-2} e^{dR_k^{\sigma-\eta} - q(\alpha-b)(1+\theta R_k)^{\sigma-\eta}} \tag{5.40}$$

where $C > 0$ is a constant independent of k. In order for this inequality to hold for every k we must have

$$d \ge (\alpha - b)q\theta^{\sigma-\eta},$$

whence, letting θ tend to 1,

$$d \ge (\alpha - b)q.$$

We set $\alpha = tb$, insert the definition (5.34) of q in the above inequality, solve with respect to K, and then let $\tau \to 1$ to obtain

$$K \le d\frac{b(\sigma-\eta)^2}{4}\frac{t^2}{t-1},$$

whence minimizing with respect to $t > 1$ and letting $d \to d_0$ and $b \to \max\{\hat{u}, 0\}$, we obtain

$$K \le d_0 \max \hat{u}, 0(\sigma-\eta)^2. \tag{5.41}$$

Case II. $\sigma = 0$ (and therefore $\eta < 0$ by (5.15)).

We proceed exactly as in the previous case and conclude that for any C^2 function u bounded above (5.41) holds. But, as noted earlier, $\forall \nu$,

$$\liminf_{\Omega_\gamma}(1+r)^\mu \Delta u = \inf_{\{x : u_\nu(x) > \gamma+\nu\}}(1+r)^\mu \Delta u_n u. \tag{5.42}$$

Thus we may choose ν in such a way that $\hat{u}_\nu \le 0$, apply (5.41) to u_ν to conclude that the right-hand side of (5.42) is nonpositive and then deduce that $K \le 0$ as required.

Case III. $\sigma > 0$, $\eta \ge 0$.

We choose

$$F(v, r) = F(v) = e^{-qv^{\frac{\sigma-\eta}{\sigma}}}$$

where $q > 0$ is a constant to be specified later. Since $\sigma - \eta > 0$ by assumption, a computation gives

$$\frac{\partial F}{\partial v}(v, r) = -q\frac{\sigma-\eta}{\sigma}v^{-\frac{\eta}{\sigma}}F(v) < 0,$$

while clearly

$$\frac{\partial F}{\partial r} \equiv 0.$$

Using the above expression, recalling that $v \geq (\alpha - b)(1+r)^\sigma$ on Ω_γ, and proceeding as in Case I, we estimate

$$B(\nabla u, r) \geq |\nabla u|^2 - \alpha\sigma(1+r)^{\sigma-1}|\nabla u| + \frac{K\sigma(\alpha - b^{4/\sigma}}{q(\sigma - \eta)}(1+r)^{2(\sigma-1)}.$$

It is now a simple matter to observe that, $\forall\, r \geq 0$ fixed, the right-hand side of the above inequality is bounded from below by $\Lambda|n\nabla u|^2$ provided

$$\Lambda \leq 1 - \frac{q\alpha^2\sigma^2(\sigma - \eta)}{4K\sigma(\alpha - b)^{\eta/\sigma}}. \tag{5.43}$$

Since the right-hand side of the above inequality is independent of r, for every such Λ we have

$$B(\nabla u, r) \geq \Lambda|\nabla u|^2.$$

In particular, if $\tau \in (0,1)$ and we choose

$$q = \frac{4\tau K\sigma(\alpha - b)^{\eta/\sigma}}{\alpha^2\sigma^2(\sigma - \eta)} \quad \text{and} \quad \Lambda = 1 - \tau, \tag{5.44}$$

then Λ is positive and satisfies (5.43). Substituting into (5.28) and using the expression of $\frac{\partial F}{\partial v}$ we deduce that

$$\operatorname{div} W \geq -2\varphi^2\lambda(u)F(v)|\nabla\varphi||\nabla u|\Lambda q\frac{\sigma - \eta}{\sigma}\varphi^2\lambda(u)v^{\eta/\sigma}|\nabla u|^2.$$

We now proceed as in Case I, repeating, with minor adaptations, the arguments that lead to (5.40), to conclude that

$$0 < E = \int_{\Omega_\gamma \cap B_{R_0}} \lambda(u)F(v)|\nabla u|^2 \leq C(1 + R_k)^\eta[(1 - \theta)R_k]^{-2}$$

$$\times \int_{\Omega_\gamma \cap (B_{R_k} \setminus B_{\theta R_k})} F(v, r)$$

where $C > 0$ is a constant independent of k and θ. Using the inequality

$$F(v) \leq e^{-q(\alpha-b)\frac{\sigma-\eta}{\sigma}(1+\theta R_k)^{\sigma-\eta}}$$

valid on $\Omega_\gamma \cap (B_{R_k} \setminus B_{\theta R_k})$ and the volume growth estimate (5.38), we deduce that

$$0 < E \leq C(\alpha - b)^{-q}(1 + \theta R_k)^{-q\sigma}(1 + R_k)^\sigma R_k^{d-2}$$

with $C > 0$ independent of k. This forces $\sigma + d - 2 \geq q\sigma$, and letting $d \downarrow d_0$,

$$\sigma + d_0 - 2 \geq q\sigma.$$

Thus, if $\sigma + d_0 - 2 < 2$ we get a contradiction, on $K \leq 0$. Otherwise, if $\sigma + d_0 - 2 \geq 0$, substituting the value of q, solving for K, letting $\alpha = tb$, $t > 1$, with $\tau \uparrow 1$ and $b \downarrow \max\{\widehat{u}, 0\}$, we obtain

$$K \leq \frac{1}{4} \max\{\widehat{u}, 0\} \sigma(\sigma + d_0 - 2) \frac{t^2}{t-1};$$

whence, minimizing over $t > 1$,

$$K \leq \max\{\widehat{u}, 0\} \sigma(\sigma + d_0 - 2).$$

This completes the proof if $\sigma > 0$. The case $\sigma = 0$ is dealt with as in Case II. $\square$

We note, for future use, the following corollary (see [PRS05b], Corollary 4.4):

Corollary 5.4. *Let* $(M, \langle\,,\,\rangle)$ *be a complete manifold, and let* $b(x) \in C^0(M)$ *satisfy*

$$b(x) \geq \frac{C}{(1+r(x))^\mu}$$

for some $\mu \in \mathbb{R}$. *Let* $u \in C^2(M)$ *be a solution of the differential inequality*

$$\Delta u \geq b(x) f(u) \tag{5.45}$$

and suppose that, for some $0 \leq \sigma < 2 - \mu$, *either* $\sigma > 0$, $\liminf_{t \to \infty} f(t) > 0$ *and*

$$u(x) = o\left(r(x)^\sigma\right) \quad \text{as } r(x) \to +\infty,$$

or $\sigma = 0$ *and* u *is bounded above. If*

$$\liminf_{r \to \infty} \frac{\operatorname{vol} B_r}{r^{2-\sigma-\mu}} < +\infty,$$

then $u^* = \sup u < +\infty$ *and* $f(u^*) \leq 0$.

Proof. Observe first of all that under the stated assumptions it follows from Theorem 5.3 that, for every $\gamma < u^*$,

$$\inf_{\Omega_\gamma} b(x)^{-1} \Delta u \leq C^{-1} \inf_{\Omega_\gamma} (1 + r(x))^\mu \Delta u \leq 0.$$

Now, if u were unbounded, then Ω_γ would be nonempty for every γ and it would follow from (5.45) that

$$\inf_{t > \gamma} f(t) \leq 0, \tag{5.46}$$

contradicting the assumption that f is bounded away from zero at infinity. Thus u is necessarily bounded above and $u^* < +\infty$. Arguing as above shows that (5.45) holds for every $\gamma < u*$, and therefore $f(u^*) \leq 0$. $\square$

5.3.2   A comparison result

We are now ready to prove the desired *comparison result* (see [RZ07] and also [MR10] for the case of a diffusion-type operator), from which the proof of Theorem 5.2 follows at once.

Theorem 5.5. *Let* $a(x), b(x) \in C^0(M)$, $\sigma > 1$, $\tau \geq 0$, $\beta + \tau(\sigma - 1) > -2$ *and suppose that* $b(x) > 0$ *on* M *and*

$$\text{i) } b(x) \geq Br(x)^\beta, \ B > 0 \text{ for } r(x) \gg 1; \quad \text{ii) } \sup_M \frac{a_-(x)}{b(x)} r(x)^{\tau(1-\sigma)} < +\infty. \quad (5.47)$$

Let $u, v \in C^2(M)$ *be nonnegative solutions of*

$$\Delta u + a(x)u - b(x)u^\sigma \geq 0 \geq \Delta v + a(x)v - b(x)v^\sigma \quad (5.48)$$

on M, *satisfying*

$$\text{i) } v(x) \geq C_1 r(x)^\tau; \quad \text{ii) } u(x) \leq C_2 r(x)^\tau \quad (5.49)$$

for $r(x) \gg 1$ *and some positive constants* C_1, C_2. *If*

$$\liminf_{r \to +\infty} \frac{\log \operatorname{vol} B_r}{r^{2+\beta+\tau(\sigma-1)}} < +\infty, \quad (5.50)$$

then $u(x) \leq v(x)$ *on* M.

Proof. First of all, let $u(x) \not\equiv 0$, otherwise there is nothing to prove. Next, we observe by (5.48), (5.49) and the strong maximum principle that $v(x) > 0$ on M. This fact, $u(x) \not\equiv 0$, and (5.49) i), ii) imply that

$$\xi = \sup_M \frac{u(x)}{v(x)} \quad (5.51)$$

satisfies $0 < \xi < +\infty$. If $\xi \leq 1$, then $u(x) \leq v(x)$ on M. Let us assume, by contradiction, $\xi > 1$ and define

$$\varphi = u - \xi v.$$

Note that $\varphi \leq 0$ on M. We claim

$$\sup_M r(x)^{-\tau} \varphi(x) = 0. \quad (5.52)$$

Indeed, let $\{x_n\} \subset M$ be a sequence realizing ξ; then

$$r(x_n)^{-\tau} \varphi(x_n) = r(x_n)^{-\tau} v(x_n) \left\{ \frac{u(x_n)}{v(x_n)} - \xi \right\}. \quad (5.53)$$

Now observe that
$$r(x_n)^{-\tau}v(x_n)$$
is bounded because otherwise (5.49) ii) would imply $\xi = 0$. Then, it follows from (5.53) that $r(x_n)^{-\tau}\varphi(x_n) \to 0$ as $n \to +\infty$ proving (5.52). We now use (5.48) to obtain
$$\Delta\varphi \geq -a(x)\varphi + b(x)(u^\sigma - (\xi v)^\sigma) + b(x)v^\sigma\xi(\xi^{\sigma-1} - 1). \qquad (5.54)$$

We write
$$(u^\sigma - (\xi v)^\sigma)(x) = h(x)\varphi(x) \qquad (5.55)$$

where
$$h(x) = \begin{cases} \sigma u(x)^{\sigma-1} & \text{if } u(x) = \xi v(x), \\ \dfrac{\sigma}{u(x)-\xi v(x)} \int_{\xi v(x)}^{u(x)} t^{\sigma-1}\, dt & \text{if } u(x) \neq \xi v(x), \end{cases}$$

is continuous and nonnegative on M. Fix the level set
$$\Omega_1 = \{x \in M : \varphi(x) > -1\},$$

and note that $(1 + r(x))^{-\tau}v(x)$ is bounded above on Ω_1. Indeed, using (5.49) ii), for every $y \in \Omega_1$ we have
$$\xi\frac{v(y)}{(1+r(y))^\tau} = \frac{(u-\varphi)(y)}{(1+r(y))^\tau} \leq \tilde{C}_2 + \frac{1}{(1+r(y))^\tau} \leq \tilde{C}_2 + 1.$$

Thus there exists a constant $C_3 > 0$ such that
$$v(x) \leq C_3(1 + r(x))^\tau \quad \text{on } \Omega_1. \qquad (5.56)$$

Now, from the mean value theorem for integrals, for some $y \in (\xi v(x), u(x))$ or $y \in (u(x), \xi v(x))$,
$$h(x) = \frac{\sigma}{u(x)-\xi v(x)}(u(x) - \xi v(x))y^{\sigma-1} \qquad (5.57)$$
$$\leq \sigma\left[u^{\sigma-1}(x) + (\xi v(x))^{\sigma-1}\right]$$
$$\leq C(1 + r(x))^{\tau(\sigma-1)} \qquad \text{on } \Omega_1$$

because of (5.49) ii) and (5.56). Now we note that, since $b(x) > 0$ on M we can rewrite (5.47) i) as
$$b(x) \geq \tilde{B}(1 + r(x))^\beta \quad \text{on } M, \qquad (5.58)$$

for some appropriate $\tilde{B} > 0$. Using $b(x) > 0$ and (5.55), from (5.54) and $\varphi \leq 0$ we deduce
$$\frac{1}{b(x)}\Delta\varphi \geq \left(\frac{a_-(x)}{b(x)} + h(x)\right)\varphi(x) + v^\sigma(x)\xi(\xi^{\sigma-1} - 1)$$

and therefore, from $\xi > 1$, (5.47) ii), (5.57), (5.49) i) and $\varphi \leq 0$,
$$(1 + r(x))^{-\sigma\tau}\frac{1}{b(x)}\Delta\varphi \geq C(1 + r(x))^{-\tau}\varphi(x) + D\xi(\xi^{\sigma-1} - 1) \quad \text{on } \Omega_1$$

for some appropriate constants $C, D > 0$. Next, we choose $0 < \varepsilon < 1$ sufficiently small so that

$$C(1 + r(x))^{-\tau} \varphi(x) \geq -\frac{1}{2} D\xi(\xi^{\sigma-1} - 1) \tag{5.59}$$

on

$$\Omega_\varepsilon = \{x \in M : \varphi(x) > -\varepsilon\} \subset \Omega_1.$$

Note that this is possible since $\tau \geq 0$. Then, on Ω_ε, $\Delta\varphi \geq 0$ so that (5.58) implies

$$(1 + r(x))^{-\beta-\sigma\tau} \Delta\varphi \geq (1 + r(x))^{-\sigma\tau} \frac{1}{b(x)} \Delta\varphi$$

and thus

$$\inf_{\Omega_\varepsilon} (1 + r)^{-\beta-\sigma\tau} \Delta\varphi \geq \frac{1}{2} D\xi(\xi^{\sigma-1} - 1) > 0 \tag{5.60}$$

since $\xi > 1$. This fact, together with (5.50), contradicts Theorem 5.3.   □

Combining the *a priori* estimates given in Theorems 4.1 and 4.4, the volume growth estimates of Chapter 1 and Theorem 5.2, we obtain

Theorem 5.6. *Let* $(M, \langle\,,\,\rangle)$ *be a complete manifold with Ricci tensor satisfying*

$$\mathrm{Ric} \geq -(m-1)H^2(1 + r^2)^{\delta/2} \tag{5.61}$$

for some $H > 0, \delta \geq -2$. *Let* $a(x), b(x) \in C^0(M)$ *with* $b(x) > 0$ *on* M *and assume that, for some* $A, B > 0$,

$$A^{-1}r(x)^\alpha \leq a(x) \leq Ar(x)^\alpha, \tag{5.62}$$
$$B^{-1}r(x)^\beta \leq b(x) \leq Br(x)^\beta \tag{5.63}$$

for $r(x) \gg 1$ *and with* $\alpha > \delta, \alpha \geq \max\left\{\beta, \beta + \frac{\delta}{2} - 1\right\}$. *Then, there exists at most one nonnegative, nontrivial solution* $u \in C^2(M)$ *of*

$$\Delta u + a(x)u - b(x)u^\sigma = 0, \quad \sigma > 1, \quad on\ M. \tag{5.64}$$

Remark. The geometric case of Yamabe's equation for which $\alpha \geq \delta$ will be dealt with in Section 5.3.

Proof. First observe, by the maximum principle, that if $u \geq 0$ and $u \not\equiv 0$ is a solution of (5.64), then $u > 0$ on M. Thus if $u \not\equiv 0$, according to the *a priori* estimates of Theorems 4.1 and 4.4 we have (note that $\alpha \geq \frac{\delta}{2} - 1$ because $\alpha > \delta \geq -2$, while $\beta \leq 1 - \frac{\delta}{2} + \alpha$ because of our assumptions)

$$C^{-1}r(x)^{-\frac{\beta-\alpha}{\sigma-1}} \leq u(x) \leq Cr(x)^{-\frac{\beta-\alpha}{\sigma-1}} \tag{5.65}$$

for some constant $C > 0$ and $r(x) \gg 1$. To apply Theorem 5.2, since $a_-(x) \equiv 0$ for $r(x) \gg 1$ because of (5.62) we only need to have $\alpha \geq \beta$ and $\alpha > -2$. This latter is guaranteed by $\alpha > \delta$, while the first is satisfied by assumption. It remains to check that (5.14) holds, which in this case becomes

$$\liminf_{r \to +\infty} \frac{\log \mathrm{vol}\, B_r}{r^{2+\alpha}} < +\infty. \tag{5.66}$$

Towards this aim, we need to estimate $\mathrm{vol}\, B_r$ from above using assumption (5.61). Using Proposition 1.15, we see that if $\delta \geq -2$, then

$$\log \mathrm{vol}\, B_r \leq Cr^{1+\frac{\delta}{2}} \quad \text{as } r \to +\infty$$

for some constant $C > 0$, and (5.66) is satisfied provided

$$\alpha \geq \frac{\delta}{2} - 1. \tag{5.67}$$

If $\delta = -2$, then

$$\mathrm{vol}\, B_r \leq C \int_0^r s^{\frac{1+\sqrt{1+4H^2}}{2}(m-1)} \, ds \asymp r^{\frac{1+\sqrt{1+4H^2}}{2}(m-1)+1},$$

and (5.66) is satisfied provided
$$\alpha > -2. \tag{5.68}$$

Note that $\alpha > \delta$ implies (5.67) and (5.68). □

5.3.3 Uniqueness of ground states

Our aim is now to prove uniqueness of positive solutions converging to zero at infinity, or *ground states* (see also Chapter 3, section 3.2). As the above examples show we have to impose some extra condition. Surprisingly, this condition is again related to $\lambda_1^L(M) \geq 0$, as we shall see below. We begin with the following general result proved in [RRV97]; we denote with $W^{1,2}(M)$ the Sobolev space of functions which are in $L^2(M)$ and have L^2 weak gradient.

Theorem 5.7. Let $(M, \langle\,,\,\rangle)$ be a complete manifold and let $a(x), b(x) \in C^0(M)$, $b(x) \geq 0$, $b(x) \not\equiv 0$. Let u_1 and u_2 be positive solutions of

$$\Delta u + a(x)u - b(x)u^\sigma = 0. \tag{5.69}$$

If there exists $\varphi \in C^0(M) \cap W^{1,2}_{loc}(M)$, $\varphi > 0$ *and such that*

$$\Delta \varphi + a(x)\varphi - \sigma b(x) u_i^{\sigma-1}\varphi \leq 0 \tag{5.70}$$

in the weak sense on M *for* $i = 1, 2$, *and*

$$u_1(x) - u_2(x) = o(\varphi(x)) \quad \text{as } r(x) \to +\infty, \tag{5.71}$$

then $u_1 \equiv u_2$.

Proof. We divide the proof in two steps.

Step I. For $\varepsilon > 0$ we set $v = v_\varepsilon = u_1 + \varepsilon\varphi$. We claim that

$$H(x) = \Delta v + a(x)v - b(x)v^\sigma \leq 0 \tag{5.72}$$

in the weak sense on M. Indeed we find

$$H(x) = \Delta u_1 + a(x)u_1 - b(x)(u_1 + \varepsilon\varphi)^\sigma + \varepsilon\Delta\varphi + \varepsilon a(x)\varphi. \tag{5.73}$$

From (5.69) we get

$$H(x) = \varepsilon\left[\Delta\varphi + a(x)\varphi - \frac{1}{\varepsilon}b(x)((u_1 + \varepsilon\varphi)^\sigma - u_1^\sigma)\right]. \tag{5.74}$$

Since $(u_1 + \varepsilon\varphi)^\sigma - u_1^\sigma \geq \varepsilon\sigma u_1^{\sigma-1}\varphi$, by convexity we have

$$H(x) \leq \varepsilon\left[\Delta\varphi + a(x)\varphi - \sigma b(x)u_1^{\sigma-1}\varphi\right] \tag{5.75}$$

which implies (5.72) from (5.70).

Step II. We claim that

$$u_1 \geq u_2. \tag{5.76}$$

From (5.69) and (5.72) we have

$$\frac{\Delta u_2}{u_2} - \frac{\Delta v}{v} - b(x)\left[u_2^{\sigma-1} - v^{\sigma-1}\right] \geq 0 \tag{5.77}$$

weakly on M. Since $(u_2^2 - v^2)_+ \in C^0(M) \cap W^{1,2}(M)$ and compactly supported, by (5.71)), it is an admissible test function and we have

$$\int_M \left(\frac{\Delta u_2}{u_2} - \frac{\Delta v}{v}\right)(u_2^2 - v^2)_+ - \int_M b(x)\left[u_2^{\sigma-1} - v^{\sigma-1}\right](u_2^2 - v^2)_+ \geq 0. \tag{5.78}$$

Suppose now that

$$\Omega = \{x \in M : u_2(x) > v(x)\} \neq \emptyset.$$

Then, (5.78) yields

$$\int_\Omega \left|\nabla u_2 - \frac{u_2}{v}\nabla v\right|^2 + \left|\nabla v - \frac{v}{u_2}\nabla u_2\right|^2 + \int_\Omega b(x)\left[u_2^{\sigma-1} - v^{\sigma-1}\right](u_2^2 - v^2) \leq 0 \tag{5.79}$$

and since $b(x) \geq 0$, $b(x) \not\equiv 0$ we conclude, as in Theorem 5.1, that

$$u_2 \leq u_1 + \varepsilon\varphi \qquad \forall \varepsilon > 0,$$

that is, (5.76). Similarly, $u_1 \leq u_2$ and therefore $u_1 \equiv u_2$. $\qquad\square$

Corollary 5.8. *Let* $(M, \langle \, , \, \rangle)$ *be a complete manifold and let* $a(x), b(x) \in C^0(M)$, $b(x) \geq 0$, $b(x) \not\equiv 0$. *Suppose* ψ *is a positive solution of*

$$\Delta \psi + a(x)\psi \leq 0 \quad on \ M. \tag{5.80}$$

If u_1 *and* u_2 *are positive solutions of* (5.69) *on* M *satisfying*

$$u_1(x) - u_2(x) = o(\psi(x)) \quad as \ r(x) \to +\infty, \tag{5.81}$$

then $u_1 \equiv u_2$.

Proof. It is enough to show that ψ satisfies (5.70) of Theorem 5.7. But this is obvious, since $b(x) \geq 0$, $u_i > 0$ and (5.80) imply

$$0 \geq \Delta \psi + a(x)\psi \geq \Delta \psi + a(x)\psi - \sigma b(x)u_i^{\sigma-1}\psi, \quad i = 1, 2. \qquad \square$$

Remark. By a result of Fisher-Colbrie and Schoen, [FCS80], condition (5.80) is equivalent to

$$\lambda_1^L(M) \geq 0$$

with $L = \Delta + a(x)$ (see the discussion at the beginning of Chapter 3). We observe that if

$$\rho(t) = C\left(\cosh \frac{t}{2}\right)^{2-m} > 0 \quad on \ [0, +\infty),$$

then the function

$$\psi(x) = \rho(r(x))$$

is a radial solution of

$$\Delta_{\mathbb{H}^m}\psi + \frac{m(m-2)}{4}\psi = 0$$

on the m-dimensional hyperbolic space $\mathbb{H}^m$, of constant negative curvature -1 ($m \geq 3$). In this case condition (5.81) becomes

$$(u_1(x) - u_2(x))e^{\frac{m-2}{2}r(x)} \to 0 \quad as \ r(x) \to +\infty.$$

In the case of the example in section 5.2 we have

$$(w_{a_1}(x) - w_{a_2}(x))e^{\frac{m-2}{2}r(x)} \sim \frac{2^{-\frac{m-2}{2}}}{m(m-2)}\left[\frac{(a_1^2 - 1)^{-\frac{m-2}{2}}}{a_1^2} - \frac{(a_2^2 - 1)^{-\frac{m-2}{2}}}{a_2^2}\right]$$

and this converges to zero if and only if $a_1 = a_2$. This shows that (5.81) cannot be relaxed to $O(\psi(x))$ as $r(x) \to +\infty$.

It is clear that the applicability of Corollary 5.8 depends on the knowledge of the asymptotic behaviour of the solution ψ of (5.80). We shall come back to this later, in section 7.3.2.

5.4 Some geometric applications and further uniqueness

We apply the uniqueness results obtained so far to the problem of characterizing isometries into the group of conformal diffeomorphisms of the complete manifold M. We then go back to uniqueness for the Yamabe problem; this will be achieved *via* an L^∞ *a priori* estimate of independent interest.

5.4.1 Conformal diffeomorphisms

Let $(M, \langle \, , \, \rangle)$ be a Riemannian manifold with scalar curvature $S(x)$. To simplify the writing we consider the case $m = \dim M \geq 3$, but one could treat the case $m = 2$ with few changes.

Suppose that $\varphi : M \to M$ is a *conformal diffeomorphism*, that is, φ is a diffeomorphism such that, for some $u \in C^\infty(M), u > 0$,

$$\varphi^* \langle \, , \, \rangle = u^{\frac{4}{m-2}} \langle \, , \, \rangle \tag{5.82}$$

(u is usually called the *stretching factor* of the diffeomorphism).

We shall express this fact by writing $\varphi \in \mathrm{Conf}(M)$, where $\mathrm{Conf}(M)$ denotes the group of conformal diffeomorphisms. If $u \equiv 1$, then φ is an isometry. If we denote by $\mathrm{Iso}(M)$ the group of isometries, then, clearly, $\mathrm{Iso}(M) \subseteq \mathrm{Conf}(M)$. Note that if φ is an isometry, then it preserves the scalar curvature (in fact the sectional curvature). It is thus meaningful to investigate *when a conformal diffeomorphism preserving the scalar curvature is an isometry*.

If φ preserves scalar curvature, then its stretching factor u is a positive solution of the (special) Yamabe equation

$$\Delta u = -\frac{m-2}{4(m-1)} S(x) u \left\{ u^{\frac{4}{m-2}} - 1 \right\}. \tag{5.83}$$

Since the inverse diffeomorphism φ^{-1} also preserves the scalar curvature and

$$(\varphi^{-1})^* \langle \, , \, \rangle = u^{-\frac{4}{m-2}} \langle \, , \, \rangle,$$

we conclude that proving that φ is an isometry amounts to showing that $u \leq 1$.

After this preparation we prove the next (see [RRV94a])

Theorem 5.9. *Let $(M, \langle \, , \, \rangle)$ be a complete manifold with Ricci tensor satisfying*

$$\mathrm{Ric} \geq -(m-1)H^2(1 + r(x)^2)^{\delta/2} \quad \text{on } M \tag{5.84}$$

for some $H > 0$, $\delta \geq -2$. Suppose furthermore that the scalar curvature $S(x)$ satisfies $S(x) \leq 0$ on M and

$$S(x) \leq -\frac{d^2}{(1 + r(x))^\mu} \quad \text{for } r(x) \gg 1 \tag{5.85}$$

and some $d > 0$ and $\mu < 1 - \frac{\delta}{2}$. Then any conformal transformation of $(M, \langle\,,\,\rangle)$ which preserves the scalar curvature is an isometry.

Proof. Let $a(x) = b(x) = -\frac{m-2}{4(m-1)}S(x)$. In Proposition 4.3 we chose $\omega(t) = C_1 > 0$ constant and

$$\psi(t) = C_2 t^{1-\frac{\delta}{2}-\mu}, \quad \mu \leq 1 - \frac{\delta}{2}.$$

Then, it follows from (5.83) and (5.85) that the solution $u > 0$ of (5.83) satisfies

$$u^* = \sup_M u < +\infty.$$

According to the discussion above, in order to show that φ is an isometry it is enough to prove that $u^* \leq 1$. We observe that from Theorem 3.10 and the Remark thereafter we can suppose $S(x) < 0$ on M without loss of generality. Now we reason by contradiction and we suppose $u^* > 1$. We fix $n \in \mathbb{N}$ sufficiently large such that $1 < u^* - \frac{1}{n} < u^*$ and we set

$$\Omega_n = \left\{ x \in M : u(x) > u^* - \frac{1}{n} \right\}.$$

Fix $R > 0$ sufficiently large that (5.85) holds on $M \backslash \overline{B}_R$. Then, using (5.85) and the fact that $u^* - \frac{1}{n} > 1$ we have

$$(1 + r(x))^\mu \Delta u \geq \frac{m-2}{4(m-1)} d^2 u \left(u^{\frac{4}{m-2}} - 1 \right) \geq C > 0 \quad \text{on } \Omega_n \cap (M \backslash \overline{B}_R)$$

for some constant C while

$$(1 + r(x))^\mu \Delta u \geq C > 0 \quad \text{on } \Omega_n \cap \overline{B}_R$$

since $S(x) < 0$ on M. Now, if

$$\liminf_{r \to +\infty} \frac{\log \operatorname{vol} B_r}{r^{2-\mu}} < +\infty \tag{5.86}$$

we can apply Theorem 5.3 to obtain the desired contradiction. Proceeding as in the proof of Theorem 5.6 we see that (5.84) implies (5.86) in the assumptions that $\mu < 1 - \frac{\delta}{2}$. $\qquad\square$

Remark. In case $\delta > -2$ the conclusion in the above theorem is reached under the milder assumption $\mu \leq 1 - \frac{\delta}{2}$. Indeed, as shown in [RRV94a], Corollary 1, the results holds for $\nu \leq 1 - \frac{\delta}{2}$ in the whole range $\delta \geq -2$.

Similarly to what we did above we can use the uniqueness result contained in Theorem 5.1 to obtain the following version:

Theorem 5.10. *Let* $(M, \langle\,,\,\rangle)$ *be a complete manifold of dimension* $m \geq 3$ *and scalar curvature* $S(x)$ *satisfying*

i) $S(x) \leq 0$ *on* M;   ii) $S(x) \not\equiv 0$ *on* M.

Let $\varphi : M \to M$ *be a conformal diffeomorphism whose stretching factor* u *satisfies*

$$\left\{ \int_{\partial B_r} (u-1)^2 \right\}^{-1} \notin L^1(+\infty).$$

Then φ *is an isometry.*

5.4.2   Uniqueness for the Yamabe problem

As we pointed out in Section 4.1 , see the remark after the proof of Theorem 4.1, the lower estimates on u obtained there do not apply to solutions of the Yamabe problem since condition (4.27) is not satisfied under the usual assumption (4.2). This implies that Theorem 5.6 cannot be applied to deduce uniqueness for the Yamabe problem. However, the next result where no assumptions on the Ricci tensor are required does hold (see [RZ07], Theorem 3.4).

Theorem 5.11. *Let* $(M, \langle\,,\,\rangle)$ *be a complete manifold of dimension* $m \geq 3$ *and scalar curvature* $S(x)$ *satisfying*

$$\sup_M S(x) < +\infty. \tag{5.87}$$

Let $K(x) \in C^\infty(M), K(x) < 0$ *on* M *and suppose that*

$$K(x) \leq -\frac{B}{r(x)^\beta} \quad \text{for } r(x) \gg 1, \tag{5.88}$$

some constant $B > 0$, $\beta < 2$ *and*

$$\inf_M \frac{S(x)}{K(x)} > -\infty. \tag{5.89}$$

Assume

$$\liminf_{r \to +\infty} \frac{\log \operatorname{vol} B_r}{r^{2-\beta}} < +\infty. \tag{5.90}$$

Then, there exists at most one conformal deformation to a new metric $g = u^{\frac{4}{m-2}}\langle\,,\,\rangle$ *with scalar curvature* $K(x)$ *and such that* $u_* = \inf_M u > 0$.

Proof. Suppose such a metric exist. Thus $u > 0$ satisfies

$$c_m \Delta u - S(x)u + K(x)u^{\frac{m+2}{m-2}} = 0$$

on M. We claim that (5.87), (5.88), (5.89) and (5.90) imply

$$u^* = \sup_M u < +\infty. \tag{5.91}$$

Postponing for the moment the proof of the claim, we have

$$0 < u_* \leq u \leq u^* < +\infty.$$

A direct application of Theorem 5.5 with $\tau = 0$ gives uniqueness of u. □

It remains to show that u is bounded above under the sole assumption of the volume growth condition (5.90): this is precisely the content of the next remarkable result.

5.4.3 An L^∞ a priori estimate

We prove here the following result of very general interest, which was obtained in [PRS03b].

Theorem 5.12. *Let $(M, \langle\,,\,\rangle)$ be a complete manifold and $a(x), b(x) \in C^0(M)$. Assume*

$$\sup_M a_+(x) < +\infty \tag{5.92}$$

and that $b(x)$ satisfy

$$b(x) > 0 \quad on\ M, \quad b(x) \geq \frac{B}{r(x)^\beta} \tag{5.93}$$

for $r(x) \gg 1$, $B > 0$ and some $0 \leq \beta < 2$. Assume further that

$$\frac{a_+(x)}{b(x)} \leq E \quad on\ M. \tag{5.94}$$

Let $u \in C^1(M)$ be a nonnegative solution of

$$\Delta u + a(x)u - b(x)u^\sigma \geq 0 \quad on\ M \tag{5.95}$$

with $\sigma > 1$. If

$$\liminf_{r \to +\infty} \frac{\log \mathrm{vol}\, B_r}{r^{2-\beta}} < +\infty, \tag{5.96}$$

then

$$u(x) \leq E^{\frac{1}{\sigma-1}} \quad on\ M. \tag{5.97}$$

To prove Theorem 5.12, we first need two technical results.

Lemma 5.13. *Let* $a(x), b(x) \in C^0(M)$, $a(x) = a_+(x) - a_-(x)$, *with* $a_\pm \geq 0$, $b(x) > 0$, *and assume that, for some* $E > 0$ *we have*

$$\frac{a_+(x)}{b(x)} \leq E \quad on \ M. \tag{5.98}$$

Assume that $u \in C^2(M)$ *and* $\gamma > 0$ *are such that*

$$\Omega_\gamma = \{x \in M \ : \ u(x) > \gamma\} \neq \emptyset \tag{5.99}$$

and that u *satisfies*

$$\Delta u \geq b(x)u^\sigma - a(x)u + \frac{D}{u}|\nabla u|^2, \tag{5.100}$$

on $\overline{\Omega_\gamma}$, *for some constants* $D \in \mathbb{R}$ *and* $\sigma > 1$. *Let* $\lambda : \mathbb{R} \to [0, +\infty)$ *be a* C^1, *nondecreasing function such that* $l(t) = 0$ *for* $t \leq \gamma$. *Then, there exists* $R > 0$ *sufficiently large, and a constant* $C > 0$ *such that, for every* $r > R$, *and for every* $\alpha > \max 1 - D, 1$,

$$\int_{B_r} b(x)\lambda(u) \leq \left\{\frac{E}{\gamma^{\sigma-1}} + \frac{1}{\gamma^{\sigma-1}}\frac{C}{r^2}\frac{1}{\inf_{B_{2r}} b}\frac{(\alpha+\sigma-1)^2}{D+\alpha-1}\right\}^{\frac{\alpha+\sigma-1}{\sigma-1}}\int_{B_{2r}} b(x)\lambda(u). \tag{5.101}$$

Proof. Let $R > 0$ large enough that $B_R \cap \Omega_\gamma \neq \emptyset$, and fix $\zeta > 1$ such that

$$2 + \frac{1+\delta}{\sigma-1}\left(\frac{1}{\zeta} - 1\right) > 0. \tag{5.102}$$

We choose a C^∞ cut-off function $\psi : M \to [0,1]$ such that, for every $r \geq R$,

$$\text{i)} \ \psi \equiv 1 \text{ on } B_r; \quad \text{ii)} \ \psi \equiv 0 \text{ on } M \setminus B_{2r}; \quad \text{iii)} \ |\nabla\psi| \leq \frac{C_0}{r}\psi^{1/\zeta}, \tag{5.103}$$

for some constant $C_0 = C_0(\zeta) > 0$. Note that this is possible since $\zeta > 1$. Finally, we fix α and consider the vector field W defined by

$$W = \psi^{2(\alpha+\sigma-1)}\lambda(u)u^{\alpha-1}\nabla u.$$

Note that the properties of λ and ψ imply that u vanishes off $B_{2r} \cap \Omega_\gamma$.

A computation that uses (5.100), $\lambda' \geq 0$, $\alpha \geq 1-D$ and the Cauchy–Schwarz inequality, yields

$$\text{div } W \geq -2(\alpha+\sigma-1)\psi^{2(\alpha+\sigma-1)-1}\lambda(u)u^{\alpha-1}|\nabla u||\nabla\psi|$$
$$+ \lambda(u)\psi^{2(\alpha+\sigma-1)}\left[b(x)u^{\alpha+\sigma-1} - a(x)u^\alpha + ((D+\alpha-1))u^{\alpha-2}|\nabla u|^2\right].$$

Recalling that W is compactly supported, the divergence theorem yields

$$\int \psi^{2(\alpha+\sigma-1)}\lambda(u)b(x)u^{\alpha+\sigma-1} - \int \psi^{2(\alpha+\sigma-1)}\lambda(u)a(x)u^{\alpha}$$

$$\leq -(D+\alpha-1)\int \psi^{2(\alpha+\sigma-1)}\lambda(u)u^{\alpha-2}|\nabla u|^2$$

$$+ 2(\alpha+\sigma-1)\int \psi^{2(\alpha+\sigma-1)-1}\lambda(u)u^{\alpha-1}|\nabla u||\nabla\psi|. \qquad (5.104)$$

Now we apply the inequality $ab \leq \frac{\varepsilon^p a^p}{p} + \frac{b^q}{\varepsilon^q q}$, valid for $a, b \geq 0$, $\varepsilon > 0$, with p, q conjugate exponents, with the choices

$$p = q = 2, \quad \varepsilon = (2(D+\alpha-1))^{1/2}$$

to the second integral on the right-hand side of (5.104) to obtain

$$2(\alpha+\sigma-1)\int \psi^{2(\alpha+\sigma-1)-1}\lambda(u)u^{\alpha-1}|\nabla u||\nabla\psi|$$

$$\leq (D+\alpha-1)\int \psi^{2(\alpha+\sigma-1)}\lambda(u)u^{\alpha-2}|\nabla u|^2$$

$$+ \frac{(\alpha+\sigma-1)^2}{D+\alpha-1}\int \psi^{2(\alpha+\sigma-1)-2}\lambda(u)u^{\alpha}|\nabla\psi|^2.$$

Inserting this into (5.104) and using $\lambda \geq 0$ gives

$$\int \psi^{2(\alpha+\sigma-1)}\lambda(u)b(x)u^{\alpha+\sigma-1} \leq \int \psi^{2(\alpha+\sigma-1)}\lambda(u)a_+(x)u^{\alpha} \qquad (5.105)$$

$$+ \frac{(\alpha+\sigma-1)^2}{D+\alpha-1}\int \psi^{2(\alpha+\sigma-1)-2(1-1/\varsigma)}\lambda(u)u^{\alpha}\left(\psi^{-1/\varsigma}|\nabla\psi|\right)^2.$$

For ease of notation, we denote by I and II the integrals on the right-hand side.
    We use (5.98), multiply and divide by $b(x)$, and use Hölder inequality with conjugate exponents p and q to be chosen later, to estimate

$$I \leq E\left(\int \psi^{2(\alpha+\sigma-1)}\lambda(u)b(x)u^{\alpha p}\right)^{1/p}\left(\int \psi^{2(\alpha+\sigma-1)}\lambda(u)b(x)\right)^{1/q}. \qquad (5.106)$$

Similarly, using (5.103) iii), we see that

$$II \leq \frac{C_0^2}{r^2}\left(\int \psi^{2(\alpha+\sigma-1)}\lambda(u)b(x)u^{\alpha p}\right)^{1/p}$$

$$\times \left(\int \psi^{2(\alpha+\sigma-1)+2q(1/\varsigma-1)}\lambda(u)b(x)^{1-q}\right)^{1/q}, \qquad (5.107)$$

where $C_0 = C_0(\zeta)$ is the constant appearing in (5.103) iii). In the above inequalities, it is assumed that p and q are chosen in such a way that $2(\alpha + \sigma - 1) + 2q(1/\zeta - 1) > 0$ is positive.

To proceed we need to estimate the powers of u in the above integrals. We choose p in such a way that αp is equal to the exponent of u in the integral on the left-hand side of (5.105), namely $\alpha + \sigma - 1$. This implies

$$p = \frac{\alpha + \sigma - 1}{\alpha}, \quad q = \frac{\alpha + \sigma - 1}{\sigma - 1}.$$

Note that, since $\sigma > 1$, we have $p, q > 1$, and that, by our choice of ζ,

$$2(\alpha + \sigma - 1) + 2q\left(\frac{1}{\zeta - 1}\right) = (\alpha + \sigma - 1)\left[2 + \frac{2}{\sigma - 1}\left(\frac{1}{\zeta - 1}\right)\right] > 0.$$

Since $\psi \le 1$, and $\psi = 0$ off B_{2r}, we have

$$I \le E\left(\int \psi^{2(\alpha+\sigma-1)}\lambda(u)b(x)u^{\alpha+\sigma-1}\right)^{\frac{\alpha}{\alpha+\sigma-1}} \times \left(\int_{B_{2r}} \lambda(u)b(x)\right)^{\frac{\sigma-1}{\alpha+\sigma-1}}, \quad (5.108)$$

and

$$II \le \frac{C_0^2}{r^2} \frac{\gamma^{\delta-1}}{\inf_{B_{2r}} b}\left(\int \psi^{2(\alpha+\sigma-1)}\lambda(u)b(x)u^{(\alpha+\sigma-1)}\right)^{\frac{\alpha}{\alpha+\sigma-1}}$$

$$\times \left(\int_{B_{2r}} \lambda(u)b(x)\right)^{\frac{\sigma-1}{\alpha+\sigma-1}}. \quad (5.109)$$

Substituting (5.108) and (5.109) into (5.105), and rearranging yield

$$\int \psi^{2(\alpha+\sigma-1)}\lambda(u)b(x)u^{\alpha+\sigma-1}$$

$$\le \left(E + \frac{C}{r^2}\frac{1}{\inf_{B_{2r}} b}\frac{(\alpha+\sigma-1)^2}{D+\alpha-1}\right)^{\frac{\alpha+\sigma-1}{\sigma-1}}\int_{B_{2r}} \lambda(u)b(x)$$

with $C = C_0^2 C_1$. We estimate from below the integral on the left-hand side using $\psi = 1$ on B_r, and $\lambda(u) = 0$ if $u \le \gamma$, and arrive at

$$\int_{B_r} \lambda(u)b(x) \le \left(\frac{E}{\gamma^{\sigma-1}} + \frac{C}{r^2}\frac{\gamma^{1-\sigma}}{\inf_{B_{2r}} b}\frac{(\alpha+\sigma-1)^2}{D+\alpha-1}\right)^{\frac{\alpha+\sigma-1}{\sigma-1}}\int_{B_{2r}} \lambda(u)b(x),$$

which is the required conclusion. $\square$

Lemma 5.14. *Let $G, Q : [R, +\infty) \to [0, +\infty)$ be nondecreasing functions such that, for some constants $0 < \Lambda < 1$ and $B, \theta > 0$,*

$$G(r) \le \Lambda^{Br^\theta/Q(2r)}G(2r) \quad \forall r \ge R. \quad (5.110)$$

Then there exists a constant $S = S(\theta) > 0$ such that, for every $r \geq 2R$,

$$\frac{Q(r)}{r^\theta} \log G(r) \geq \frac{Q(r)}{r^\theta} \log G(R) + SB \log(1/\Lambda). \tag{5.111}$$

Proof. Let $r_0 = R$ and $r_k = 2^k r_0$. Then, for every $r \geq 2r_0$, there exists k such that $r_k \leq r \leq r_{k+1}$. Applying inequality (5.110) k-times, we obtain

$$G(r_0) \leq \Lambda^{B \sum_{j=0}^{k-1} r_j^\theta / Q(2r_j)} G(r_k). \tag{5.112}$$

Using the definition of r_k and the fact that Q is nondecreasing we estimate

$$\sum_{j=0}^{k-1} r_j^\theta / Q(2r_j) \geq \frac{r_0^\theta}{Q(2r_{k-1})} \sum_{j=0}^{k-1} 2^{j\theta} = \frac{r_{k+1}^\theta}{Q(2r_{k-1})} \frac{1 - 2^{-k\theta}}{2^\theta - 1} 2^{-\theta} \geq S \frac{r^\theta}{Q(r)}$$

with $S = 2^{-\theta}/(2^\theta - 1)$. Substituting into (5.112), and recalling that $0 < \Lambda < 1$ and that G is nondecreasing, we conclude that

$$G(r_0) \leq \Lambda^{BSr^\theta / Q(r)} G(r),$$

whence (5.111) follows by taking logarithms. □

We are now ready for the

Proof of Theorem 5.12. First of all, we note that since $\|a_+\|_\infty < +\infty$, we may assume that $\|b\|_\infty < +\infty$.

We let R, λ and γ be as in Lemma 5.13, and set, for $r \geq R$,

$$G(r) = \int_{B_r} \lambda(u) b(x).$$

Applying Lemma 5.13 with $D = 0$, the lower bound $b(x) \geq B/r(x)^\beta$, and the inequality

$$\frac{(\alpha + \sigma - 1)^2}{\alpha - 1} \leq 2^3 \alpha$$

valid for $\alpha \geq \max\{\sigma - 1, 2\}$, we deduce that there exists a constant C_3 depending only on σ such that, for every $r \geq R$, $\gamma < u^*$ and $\alpha > \max\{\sigma - 1, 2\}$,

$$G(r) \leq \left(\frac{E}{\gamma^{\sigma-1}} + \frac{C_3}{\gamma^{\sigma-1}} \frac{\alpha}{r^{2-\beta}} \right)^{\frac{\alpha+\sigma-1}{\sigma-1}} G(2r). \tag{5.113}$$

Now, assume by contradiction that $u^* = +\infty$, so that $\Omega_\gamma \neq \emptyset$ for every $\gamma > 0$. Fix $0 < \rho < 1$ and let $\gamma_0 > 0$ be such that, for every $\gamma \geq \gamma_0$,

$$\frac{E}{\gamma^{\sigma-1}} \leq 1 - 2\rho. \tag{5.114}$$

Next, we choose α of the form

$$\alpha = \alpha(r) = \rho \frac{\gamma^{\sigma-1}}{C_3} r^{2-\beta},$$

and observe that, since $\beta < 2$, there exists $R_1 \geq R$ such that $\alpha(r)$ satisfies the condition stated before (5.113) for every $r \geq R_1$ and

$$\left(\frac{E}{\gamma^{\sigma-1}} + \frac{C_3}{\gamma^{\sigma-1}} \frac{\alpha}{r^{2-\beta}} \alpha \right) \leq 1 - \rho.$$

Setting $\Lambda = 1 - \rho$, (5.113) gives

$$G(r) \leq \Lambda^{B\gamma^{\sigma-1}r^{2-\beta}} G(2r) \quad \forall r \geq R_1, \tag{5.115}$$

where $B > 0$ is a constant depending only on ρ and σ. We apply Lemma 5.14 to deduce that there exists a constant S such that, for every $r \geq R_1$,

$$\frac{1}{r^{2-\beta}} \log \int_{B_r} b(x)\lambda(u) \geq \frac{1}{r^{2-\beta}} \log \int_{B_{R_1}} b(x)\lambda(u) + SB\gamma^{\sigma-1} \log (1/\Lambda). \tag{5.116}$$

To reach the required contradiction, we choose λ satisfying $\sup \lambda = 1/\sup_M b$, so that $b(x)\lambda(u) \leq 1$, and let $r \to +\infty$ in (5.116). Since $2 - \beta > 0$, we conclude that

$$\liminf_{r \to +\infty} \frac{1}{r^{2-\beta}} \log \operatorname{vol} B_r \geq \liminf_{r \to +\infty} \frac{1}{r^{2-\beta}} \log \int_{B_r} b(x)\lambda(u) \geq SB\gamma^{\sigma-1} \log (1/\Lambda).$$

Since $\sigma > 1$, by taking γ sufficiently large, this contradicts (5.96). To complete the proof of the theorem we observe that the differential equation (5.95) yields

$$\Delta u \geq b(x)u(u^{\sigma-1} - E),$$

whence the required conclusion follows applying Corollary 5.4 with $f(u) = u \cdot (u^{\sigma-1} - E)$. $\qquad\qquad\square$

Remark. One might wonder whether a corresponding result on u_* holds. This is the case under further restrictions on the nonlinearity of the equation, see for instance [PRS05a]. However, this does not apply to Yamabe's equation, so that the assumption that $u_* > 0$ in the statement of Theorem 5.11 is necessary. See also the final remark of Section 4.1.

We conclude this section with a last application of Theorem 5.12. Towards this goal, let (M, g) be a complete Riemannian manifold with scalar curvature $S(x)$, and, as usual, let $r(x)$ be the Riemannian distance function from a fixed reference point o. Assume that $\varphi : (M, g) \to (N, h)$ is a conformal immersion, so that, assuming it will simplify matters, $m = \dim(M) \geq 3$,

$$\varphi^* h = u^{\frac{4}{m-2}} g \tag{5.117}$$

for some positive function $u \in C^\infty(M)$. Let also $K(x)$ be the scalar curvature of the conformal metric $\varphi^* h$.

Note that if $u \leq 1$ the length of curves in M with respect to the conformal metric $\varphi^* h$ is less than or equal to their lengths with respect to the original metric g, and therefore the Riemannian distance induced by $\varphi^* h$ is smaller than or equal to the original distance function. In this situation we say that φ is *weakly distance decreasing* .

Theorem 5.15. *Assume that*

$$\inf_M S(x) > -\infty$$

and that

$$K(x) < 0, \ K(x) \leq S(x) \ on \ M \quad and \quad K(x) \leq -\frac{A}{r(x)^\beta}$$

for $r(x) \gg 1$, and for some constants $A > 0$ and $\beta < 2$. If

$$\liminf_{r \to \infty} \frac{\log\left(\operatorname{vol} B_r\right)}{r^{2-\beta}} < +\infty,$$

then φ is weakly distance decreasing.

Proof. Recall that the stretching factor u of $\varphi^* h$ with respect to g satisfies the Yamabe equation

$$c_m \Delta u - S(x)u + K(x)u^{\frac{m+2}{m-2}} = 0 \quad \text{on } M.$$

Setting $S(x) = -\frac{S(x)}{c_m}$ and $b(x) == -\frac{K(x)}{c_m}$, the conditions imposed on $S(x)$ and $K(x)$ imply that all the assumptions of Theorem 5.12 are satisfied with $E = 1$. Hence $u \leq 1$ and φ is weakly distance decreasing. □

Remark. Theorem 5.15 has its roots in the work started, in the compact case, by A. Lichnerowicz, [Lic58], Obata, [Oba62b] (see Section 2.1.1), and K. Yano and T. Nagano, [YN59], and extended to the noncompact case by Yau, [Yau73] where it was assumed that $K(x) = S(x) \leq -\varepsilon < 0$ and that the sectional curvature of (M, g) was bounded from below.

We also remark that some negativity is necessary for the conclusion of Theorem 5.15 to hold, as the following elementary example shows: let (M, g) be a manifold of dimension $m \geq 3$ with nonnegative scalar curvature $S(x)$, and for every fixed $a > 1$ let $(N, h) = (M, a^2 g)$. Then $K(x) = a^{-\frac{2}{m-2}} S(x) \leq S(x)$ since $a > 1$ and the identity map is a conformal diffeomorphism of (M, g) onto (N, h) which is not weakly distance decreasing.

Chapter 6

Existence

In this chapter we provide existence results for nonnegative solutions of the Yamabe-type equation

$$\Delta u + a(x)u - b(x)u^{\sigma} = 0, \ \ \sigma > 1 \tag{6.1}$$

on the complete, noncompact, manifold M. Existence is obtained by various versions of the monotone iteration scheme (see the appendix at the end of the chapter) and we consequently concentrate on the construction of (global and local) super- and subsolutions for the problem. The first are obtained with some assumptions on the sign of $b(x)$ and on that of the first eigenvalue of $L = \Delta + a(x)$ on appropriate domains. In the situation at hand, subsolutions are generally harder to find and in what follows we give a number of sufficient conditions that allow their construction. We mention in this respect Theorem 6.15, in which existence is guaranteed under a very weak growth condition on $b(x)$, and also Theorem 6.16, where a further weakening the condition on the sign of $b(x)$ is balanced by the necessity of imposing a constant negative lower bound on the Ricci curvature.

We note that the assumptions of our existence theorems match those of the nonexistence results in the previous chapters.

Existence can be in particular guaranteed for the Yamabe problem, but we limit ourselves to explicitly state only one of the possible consequences at the end of the chapter, in Theorem 6.18, leaving the derivation of further results to the interested reader.

The chapter begins with some introductory material, in particular a useful comparison result and some facts from basic spectral theory; then we prove a result of Li, Tam and Yang (see [LTY98]) concerning the nonnegativity of the first eigenvalue of Schrödinger operators. These steps are used to outline a general procedure that reduces the existence problem to that of the existence of a positive subsolution; next, we provide different sufficient conditions for this latter. We then deal with a case where the existence of a supersolution is no longer a consequence of a spectral assumption and depends on a rather delicate construction. Finally, we consider specifically the case of the geometric Yamabe equation. The chapter ends with an appendix in which we describe the method of sub- and supersolutions, and where we also prove that the maximum of two subsolutions (or the minimum of two supersolutions) is again a subsolution (resp. supersolution).

6.1   A general procedure

The aim of this section is to describe the general procedure mentioned above. The various steps towards this goal, achieved in Theorem 6.16, are explained in the next subsections. The next proposition is a useful comparison result that shall be used throughout the chapter.

6.1.1   Another comparison result

Proposition 6.1. *Let $D \subset M$ be a bounded open set with boundary ∂D. Assume $a(x), b(x) \in C^0(\bar{D})$, and that $b(x)$ is nonnegative. Let $u, v \in C^0(\bar{D}) \cap C^2(D)$ be solutions on D of*

$$\Delta u + a(x)u - b(x)u^\sigma \geq 0 \tag{6.2}$$

and

$$\Delta v + a(x)v - b(x)v^\sigma \leq 0 \tag{6.3}$$

with $u \geq 0$, $v > 0$ and $\sigma \geq 1$. If $u \leq v$ on ∂D, then $u \leq v$ on D.

Proof. The proof is modeled on that of the generalized maximum principle (see for example [PW67]). Set $w = u/v$. A computation shows that

$$\Delta w \geq b(x)\big(u^{\sigma-1} - v^{\sigma-1}\big)w - 2\langle \nabla w, \nabla \log v \rangle.$$

Assume by contradiction that $u > v$ somewhere in D, so that for a sufficiently small $\varepsilon > 0$,

$$\Omega_\varepsilon = \{x \in D : w(x) > 1 + \varepsilon\} \neq \emptyset.$$

Then $\overline{\Omega}_\varepsilon \subset D$, and since $u > v$ on Ω_ε, $\sigma \geq 1$ and $b(x)$ is nonnegative,

$$\Delta w + 2\langle \nabla w, \nabla \log v \rangle \geq 0 \quad \text{on } \Omega_\varepsilon.$$

Since $w = 1 + \varepsilon$ on $\partial \Omega_\varepsilon$, by the maximum principle $w \leq 1 + \varepsilon$ on Ω_ε, and this yields the required contradiction. $\square$

6.1.2   More basic spectral theory and a result of Li, Tam and Yang

In this section we collect some other simple results (see also Chapter 3) from the spectral theory of Schrödinger operators . For more details, we refer to [PRS08] and references therein.

Let $a(x) \in C^0(M)$ and $L = \Delta + a(x)$. As we already know, if Ω is a nonempty open set, the first Dirichlet eigenvalue $\lambda_1^L(\Omega)$ is variationally defined by means of the formula

$$\lambda_1^L(\Omega) = \inf \left\{ \int_\Omega |\nabla \varphi|^2 - a(x)\varphi^2 : \varphi \in W_0^{1,2}(\Omega), \int_\Omega \varphi^2 = 1 \right\},$$

and, if Ω is bounded and both Ω and a are sufficiently regular, the infimum is attained by a unique normalized eigenfunction v defined on Ω and satisfying

$$\begin{cases} \Delta v + a(x)v + \lambda_1^L(\Omega)v = 0, & \text{on } \Omega; \\ v > 0 \quad \text{on } \Omega, \quad v \equiv 0 \quad \text{on } \partial\Omega. \end{cases}$$

We extend the above definition to an arbitrary bounded subset Υ of M by setting

$$\lambda_1^L(\Upsilon) = \sup \lambda_1^L(\Omega), \tag{6.4}$$

where the supremum is taken over all open bounded sets with smooth boundary Ω such that $\Upsilon \subset \Omega$ (see also Chapter 3). Note that, by definition, if $\Upsilon = \emptyset$, then $\lambda_1^L(S) = +\infty$. Finally, if Υ is an unbounded subset of M, we define

$$\lambda_1^L(\Upsilon) = \inf \lambda_1^L(D \cap \Upsilon),$$

where the infimum is taken over all bounded open sets D with smooth boundary. Note that if $\{D_n\}$ is an increasing sequence of open sets with smooth boundaries which exhausts M, then, by domain monotonicity,

$$\lambda_1^L(\Upsilon) = \lim_{n \to +\infty} \lambda_1^L(D_n \cap \Upsilon).$$

Since the first Dirichlet eigenvalue of the Laplacian of a ball B_r grows like r^{-2} as $r \to 0$ (see for instance Chavel [Cha84]), $\lambda_1^L(B_r) > 0$ provided r is sufficiently small, and one may think that the condition $\lambda_1^L(\Upsilon) > 0$ expresses the fact that Υ is small in a spectral sense. This notion of smallness is appropriate for our purposes. Indeed, P. Li, L. F. Tam and D. Yang, [LTY98], have established the following relationship between the first eigenvalue of

$$B_0 = \{x \in M : b(x) = 0\} \tag{6.5}$$

and the existence of a nontrivial supersolution of (6.1). We provide here a mildly improved proof of the original result.

Theorem 6.2. *Let $a(x), b(x) \in C^{0,\alpha}(M)$, $0 < \alpha \le 1$, and suppose that $b(x) \ge 0$ on M and that the set B_0 defined in (6.5) is bounded. Let Ω be a bounded open domain containing B_0. If*

$$\Delta u + a(x)u - b(x)u^\sigma = 0, \quad \sigma > 1 \tag{6.6}$$

has a positive supersolution on Ω, then $\lambda_1^L(B_0) \ge 0$. Conversely, if $\lambda_1^L(B_0) > 0$, then (6.6) has a positive supersolution on Ω.

Proof. Suppose (6.6) has a positive supersolution u on Ω and, by contradiction, assume that

$$\lambda_1^L(B_0) = -a$$

for some constant $a > 0$. By the definition of $\lambda_1^L(B_0)$ we can find a sequence of open sets $\Omega_i, i \in \mathbb{N}$, with smooth boundaries such that, $\forall i \in \mathbb{N}$,

$$\Omega_i \subset \Omega, \quad \overline{\Omega}_{i+1} \subset \Omega_i, \quad \bigcap_{i=0}^{+\infty} \Omega_i = B_0$$

and $\lambda_i = \lambda_1^L(\Omega_i) \to -a$ as $i \to +\infty$ monotonically. Corresponding to the eigenvalues λ_i there is a sequence of positive eigenfunctions v_i such that

$$\begin{cases} \Delta v_i + a(x)v_i + \lambda_i v_i = 0 & \text{on } \Omega_i, \\ v_i > 0 & \text{on } \Omega_i, \quad v_i = 0 \quad \text{on } \partial\Omega_i. \end{cases} \tag{6.7}$$

Note that, since $v_i > 0$ on Ω_i,

$$\langle \nabla v_i, \nu \rangle \le 0 \quad \text{on } \partial\Omega_i, \tag{6.8}$$

where ν is the outward unit normal to $\partial\Omega_i$. By assumption, $u > 0$ satisfies

$$\Delta u + a(x)u \le b(x)u^\sigma \quad \text{on } \Omega; \tag{6.9}$$

using the second Green's formula and (6.7), (6.8), (6.9) we have

$$0 \ge \int_{\partial\Omega_i} u \langle \nabla v_i, \nu \rangle = \int_{\partial\Omega_i} u \langle \nabla v_i, \nu \rangle - v_i \langle \nabla u, \nu \rangle = \int_{\Omega_i} u\Delta v_i - v_i \Delta u$$

$$\ge \int_{\Omega_i} -a(x)uv_i - \lambda_i uv_i + a(x)uv_i - b(x)v_i u^\sigma = -\int_{\Omega_i} v_i u \left(\lambda_i + b(x)u^{\sigma-1} \right),$$

that is,

$$\int_{\Omega_i} v_i u (\lambda_i + b(x)u^{\sigma-1}) \ge 0. \tag{6.10}$$

Now,

$$\lambda_i + b(x)u^{\sigma-1} \le -a + b(x)u^{\sigma-1},$$

but $b(x) \equiv 0$ on $\bigcap_i \Omega_i = B_0$, hence, for i sufficiently large,

$$\lambda_i + b(x)u^{\sigma-1} < 0 \quad \text{on } \Omega_1$$

contradicting (6.10).

Conversely, suppose $\lambda_1^L(B_0) > 0$. Let D and D' be bounded open domains such that

$$B_0 \subset\subset D' \subset\subset D \subset\subset \Omega$$

and $\lambda_1^L(D) > 0$. Let u_1 be a positive solution of

$$\begin{cases} \Delta u_1 + a(x)u_1 + \lambda_1^L(D)u_1 = 0, & \text{on } D; \\ u_1 = 0, & \text{on } \partial D. \end{cases} \tag{6.11}$$

Since $b(x) > 0$ on $M \backslash B_0$, and $\overline{\Omega} \backslash D' \subset\subset M \backslash B_0$,

$$\beta = \inf_{\Omega \backslash D'} b > 0.$$

We claim that a sufficiently large positive constant u_2 is a supersolution of (6.6) on $\Omega \backslash \overline{D'}$. Towards this end let

$$\alpha = \sup_{\Omega \backslash D'} a(x),$$

and note that $\alpha < +\infty$ since Ω is bounded. Then we have

$$\Delta u_2 + a(x)u_2 - b(x)u_2^\sigma = u_2(a(x) - b(x)u_2^{\sigma-1}) \le u_2(\alpha - \beta u_2^{\sigma-1}) \le 0$$

on $\Omega \backslash \overline{D'}$ provided

$$u_2 \ge \left(\frac{\alpha}{\beta}\right)^{\frac{1}{\sigma-1}}.$$

Now let $\psi \in C^\infty(M)$ be a cut-off function such that $0 \le \psi \le 1$, $\psi \equiv 1$ on D', $\text{supp}\,\psi \subset D$. Fix any constant $\gamma > 0$ and define

$$u = \gamma(\psi u_1 + (1 - \psi)u_2). \tag{6.12}$$

Since $b(x) \ge 0$ and $\lambda_1^L(D) > 0$, on $\overline{D'}$, we have

$$\begin{aligned} \Delta u + a(x)u - b(x)u^\sigma &= \Delta \gamma u_1 + a(x)\gamma u_1 - b(x)(\gamma u_1)^\sigma \\ &= -\left(\lambda_1^L(D) + b(x)(\gamma u_1)^{\sigma-1}\right)\gamma u_1 < 0 \end{aligned}$$

independently of the value of $\gamma > 0$. Moreover, in $\Omega \backslash D$,

$$\Delta u + a(x)u - b(x)u^\sigma = \Delta \gamma u_2 + a(x)\gamma u_2 - b(x)(\gamma u_2)^\sigma = \gamma\left(a(x)u_2 - b(x)\gamma^{\sigma-1}u_2^\sigma\right).$$

Now, for $\gamma \ge 1$,

$$b(x)\gamma^{\sigma-1} \ge b(x),$$

thus, on $\Omega \backslash D$,

$$\Delta u + a(x)u - b(x)u^\sigma \le \gamma(\Delta u_2 + a(x)u_2 - b(x)u_2^\sigma) \le 0$$

because of our choice of u_2. It remains to analyze the situation on $D \backslash \overline{D'}$. On this set

$$(\Delta + a(x))(\psi u_1 + (1 - \psi)u_2) \le C \tag{6.13}$$

for any constant $C > 0$ sufficiently large. Now

$$\inf_{D\backslash\overline{D'}} b(x)(\psi u_1 + (1 - \psi)u_2)^\sigma > 0$$

and we can therefore choose $C > 0$ sufficiently large that (6.13) holds together with

$$b(x)(\psi u_1 + (1 - \psi)u_2)^\sigma > C^{-1} \tag{6.14}$$

on $D\backslash\overline{D'}$. On this set we then have

$$\begin{aligned} \Delta u + a(x)u - b(x)u^\sigma &= \gamma(\Delta + a(x))(\psi u_1 + (1 - \psi)u_2) \\ &\quad - b(x)\gamma^\sigma(\psi u_1 + (1 - \psi)u_2)^\sigma \\ &\leq \gamma C - \gamma^\sigma C^{-1} = \gamma\big(C - \gamma^{\sigma-1}C^{-1}\big) \leq 0 \end{aligned}$$

provided $\gamma \geq C^{\frac{2}{\sigma-1}}$. Thus u is a supersolution of (6.6) on Ω. $\square$

6.1.3   Two useful lemmas

In the following lemmas we proceed to construct a solution of (6.6).

Lemma 6.3. *Let $a(x), b(x) \in C^{0,\alpha}(M)$, $0 < \alpha < 1$, and suppose that $b(x) \geq 0$ on M and that B_0 is bounded. Let $\Omega \supset B_0$ be a bounded open domain. Having set $L = \Delta + a(x)$, assume that*

$$\lambda_1^L(B_0) > 0. \tag{6.15}$$

Then, for every $n \in (0, +\infty)$, there exists a solution of the problem

$$\begin{cases} \Delta u + a(x)u - b(x)u^\sigma = 0, \ u > 0 & on \ \Omega, \ \sigma > 1, \\ u = n & on \ \partial\Omega. \end{cases} \tag{6.16}$$

Proof. By the definition of $\lambda_1^L(B_0)$ and (6.15) there exists an open domain D with smooth boundary such that $B_0 \subset D \subset\subset \Omega$ and $\lambda_1^L(D) > 0$. Let $\psi \in C^\infty(M)$ be a cut-off function such that $0 \leq \psi \leq 1$ and

$$\psi \equiv 1 \quad \text{on } D, \quad \psi \equiv 0 \quad \text{on } M\backslash\overline{\Omega}.$$

Fix $N \geq \max\big\{\sup_{\overline{\Omega}} |a(x)| + 1, \lambda_1^\Delta(M\backslash\overline{\Omega}) + 1\big\}$, where $\lambda_1^\Delta(M\backslash\overline{\Omega}) \geq 0$ is the bottom of the spectrum of the Laplacian on $M \setminus \overline{\Omega}$. Define

$$\bar{a}(x) = \psi(x)a(x) + N(1 - \psi(x)) \tag{6.17}$$

and consider the operator $\bar{L} = \Delta + \bar{a}(x)$. Since $\bar{a}(x) \equiv a(x)$ on D,

$$\lambda_1^{\bar{L}}(B_0) = \lambda_1^L(B_0) > 0.$$

Furthermore, since $N \geq \lambda_1(M \setminus \overline{\Omega}) + 1$, we have $\lambda_1^{\bar{L}}(M \setminus \overline{\Omega}) \leq -1$ and it follows that there exists $R > 0$ sufficiently large such that

$$\overline{\Omega} \subset B_R(o) \quad \text{and} \quad \lambda_1^{\bar{L}}(B_R(o)) < 0.$$

Let φ be the normalized eigenfunction of $\bar{L}$ on $B_R(o)$ relative to the eigenvalue $\lambda_1^{\bar{L}}(B_R(o))$, so that

$$\begin{cases} \bar{L}\varphi + \lambda_1^{\bar{L}}(B_R(o))\varphi = 0 & \text{on } B_R(o), \\ \varphi = 0 & \text{on } \partial B_R(o) \end{cases}$$

and $\|\varphi\|_{L^2(B_R(o))} = 1$. We fix $\gamma > 0$ sufficiently small that

$$\int_{B_R} |\nabla \varphi|^2 - \bar{a}(x)\varphi^2 + \gamma b(x)\varphi^2 = \lambda_1^{\bar{L}}(B_R(o)) + \gamma \int_{B_R} b(x)\varphi^2 < 0,$$

showing that the operator $\tilde{L} = \Delta + \bar{a}(x) - \gamma b(x)$ satisfies $\lambda_1^{\tilde{L}}(B_R(o)) < 0$. Let ψ be a positive eigenfunction corresponding to $\lambda_1^{\tilde{L}}(B_R(o))$. Then ψ satisfies

$$\begin{cases} \bar{L}\psi - \gamma b(x)\psi = -\lambda_1^{\tilde{L}}(B_R(o))\psi \geq 0 & \text{on } B_R(o), \\ \psi \equiv 0 & \text{on } \partial B_R(o). \end{cases}$$

If we choose

$$0 < \mu \leq \gamma^{\frac{1}{\sigma-1}} \left(\sup_{B_R(o)} \psi \right)^{-1},$$

then the function $v_- = \mu\psi$ satisfies

$$\begin{cases} \Delta v_- + \bar{a}(x)v_- - b(x)v_-^{\sigma} \geq 0, \; v_- > 0 & \text{on } B_R(o), \\ v_- \equiv 0 & \text{on } \partial B_R(o). \end{cases}$$

On the other hand, since $\lambda_1^{\bar{L}}(B_0) > 0$, by Theorem 6.2 there exists $v_+ > 0$ on $\overline{B_R(o)}$ satisfying

$$\begin{cases} \Delta v_+ + \bar{a}(x)v_+ - b(x)v_+^{\sigma} \leq 0, \; v_+ > 0 & \text{on } B_R(o), \\ v_+ \geq 0 & \text{on } \partial B_R(o). \end{cases}$$

Thus, by the monotone iteration scheme (see section 6.5), we find a solution w of the problem

$$\begin{cases} \Delta w + \bar{a}(x)w + -b(x)w^{\sigma} = 0, \; w > 0 & \text{on } B_R(o), \\ w \equiv 0 & \text{on } \partial B_R(o). \end{cases}$$

Note that $\inf_{\partial\Omega} w > 0$ since $\overline{\Omega} \subset B_R(o)$. Thus, recalling that $\bar{a}(x) \geq a(x)$ on $\overline{\Omega}$, it is easily verified that if $\alpha >$ is sufficiently large, then the function $w_+ = \alpha w$ satisfies

$$\begin{cases} \Delta w_+ + a(x)w_+ - b(x)w_+^\sigma \leq 0, \ w_+ > 0 & \text{on } \Omega, \\ w_+ \geq n & \text{on } \partial\Omega. \end{cases}$$

Finally, since $w_- \equiv 0$ is a subsolution of the problem, by the monotone iteration scheme (see section 6.5) we deduce the existence of a solution $u \geq 0$ of (6.16). To conclude, note that since u satisfies $\Delta u + \big(a(x) - b(x)u^{\sigma-1}\big)u = 0$ it follows from the strong maximum principle on page 35 of [GT01] that $u > 0$. □

Now we produce a solution blowing up at the boundary of Ω.

Lemma 6.4. *In the assumption of Lemma 6.3 there exists a solution of the problem*

$$\begin{cases} \Delta u + a(x)u - b(x)u^\sigma = 0, & u > 0 \quad \text{on } \Omega, \ \sigma > 1, \\ u = +\infty & \text{on } \partial\Omega. \end{cases} \tag{6.18}$$

Proof. By standard regularity theory (see [GT01]) it is enough to show that the sequence $\{u_n\}$, with u_n solution of (6.16), is bounded on any compact subset K of Ω. If $K \subset \Omega \backslash B_o$, then we can find a finite covering of balls B_i for K, $i = 1, \ldots, t$ such that $b(x) > 0$ on each B_i. Applying Lemma 4.2 we deduce the existence of a constant $C_1 > 0$ such that

$$u_n(x) \leq C_1 \quad \forall x \in K, \ \forall n \in \mathbb{N}. \tag{6.19}$$

It remains to find an upper bound of u_n on a neighbourhood of B_o. Towards this aim, for $\eta > 0$ we let

$$N_\eta = \{x \in M : d(x, B_o) < \eta\}$$

where $\eta > 0$ is small enough that $\overline{N_\eta} \subset \Omega$. Furthermore, by the definition of $\lambda_1^L(B_o)$ and the fact that $\lambda_1^L(B_o) > 0$, we can also suppose to have chosen η so small that

$$\lambda_1^L(N_\eta) > 0.$$

Now $\partial N_{\eta/2}$ is closed and bounded, hence compact by the completeness of M, therefore (6.19) holds on $\partial N_{\eta/2}$ for some constant $C_2 > 0$. Let φ be a positive eigenfunction corresponding to $\lambda_1^L(N_\eta)$. Then, there exists a positive constant μ such that $\mu\varphi > C_2$ on $\partial N_{\eta/2}$. On $N_{\eta/2}$ we have

$$\Delta(\mu\varphi) + a(x)(\mu\varphi) = -\lambda_1^L(N_\eta)(\mu\varphi) < 0$$

while, by (6.16),

$$\Delta u_n + a(x)u_n = b(x)u_n^\sigma \geq 0.$$

It follows from the generalized maximum principle, [PW67] Theorem 2.10, that the function $\frac{u_n}{\mu\varphi}$ cannot attain an interior positive maximum unless it is constant, and since,

$$u_n \leq C_2 < \mu\varphi \quad \text{on } \partial N_{\eta/2},$$

we conclude that

$$u_n \leq \mu\varphi \leq C_3$$

on $N_{\eta/2}$ with C_3 independent of n. This proves the lemma. $\square$

6.1.4   Existence of a maximal solution

With the aid of the previous lemma we are ready to prove the following general result; we denote with $C^{0,\alpha}(M)$ the space of locally Hölder continuous functions on M with exponent α.

Theorem 6.5. *Let* $a(x), b(x) \in C^{0,\alpha}_{loc}(M)$ *for some* $0 < \alpha \leq 1$. *Assume that* $b(x) \geq 0$ *and it is strictly positive outside a compact set, and that*

$$B_0 = \{x \in M : b(x) = 0\}$$

satisfies $\lambda_1^L(B_0) > 0$ *with* $L = \Delta + a(x)$. *If* $u_- \in C^0(M) \cap H^{1,2}_{loc}(M)$, $u_- \geq 0$, $u_- \not\equiv 0$, *is a global subsolution of*

$$\Delta u + a(x)u - b(x)u^\sigma = 0, \quad \sigma > 1, \tag{6.20}$$

on M, *then* (6.20) *has a maximal positive* C^2*-solution* .

Proof. We fix an exhausting sequence $\{D_k\} \subset M$ by open domains with smooth boundaries such that

$$B_0 \subset D_k \subset \overline{D_k} \subset D_{k+1} \quad \forall \, k,$$

and for every k we denote by u_k^∞ the solution of the problem

$$\begin{cases} \Delta u + a(x)u - b(x)u^\sigma = 0 & \text{on } D_k, \\ u = +\infty & \text{on } \partial D_k \end{cases}$$

obtained by applying Lemma 6.4. It follows from Proposition 6.1 that

$$u_- \leq u_{k+1}^\infty \leq u_k^\infty \quad \text{on } \bar{D}_k. \tag{6.21}$$

Thus $\{u_k^\infty\}$ converges monotonically to a function u which solves (6.20), and because of (6.21) $u \geq u_- \geq 0$, $u_- \not\equiv 0$. Thus $u \geq 0$ and $u \not\equiv 0$. By the maximum principle (see [GT01], page 35) it follows that $u > 0$ on M. Let now $u_1 > 0$ be a second solution of (6.20) on M. Again by Proposition 6.1, $u_1 \leq u_k^\infty$ on D_k $\forall k$, and therefore $u_1 \leq u$, as required to show that u is a maximal positive solution. $\square$

The next result is an immediate consequence of the above reasoning and it will be used in the proof of Theorem 6.8.

Theorem 6.6. *Let* $a(x), b(x) \in C^\infty(M)$ *and assume* $b(x) > 0$. *If* $u_- \geq 0$, $u_- \not\equiv 0$ *is a subsolution of*

$$\Delta u + a(x)u - b(x)u^\sigma = 0, \quad \sigma > 1 \quad \text{on } M \setminus \overline{B_{R_0}}, \tag{6.22}$$

then (6.22) *has a maximal positive smooth solution* $u \geq u_-$ *on* $M \setminus \overline{B_{R_0}}$.

6.2   Subsolutions and existence

In this section we give sufficient conditions to guarantee the existence of a positive
subsolution of equation (6.20). An application of Theorem 6.5 then yields the
existence of a maximal solution.

6.2.1   Existence with $\lambda_1^L(M) < 0$

The first assumption we consider to provide an *entire* (i.e., global), nonnegative
and nontrivial subsolution of (6.20) is the spectral condition (6.23) below.

Theorem 6.7. *Let* $a(x), b(x) \in C_{loc}^{0,\alpha}(M)$ *for some* $0 < \alpha \leq 1$. *Assume that*
$b(x) \geq 0$, $b(x) > 0$ *outside a compact set, and that* $\lambda_1^L(B_0) > 0$ *where* $B_0 =$
$\{x \in M : b(x) = 0\}$ *and* $L = \Delta + a(x)$. *Furthermore assume*

$$\lambda_1^L(M) < 0. \tag{6.23}$$

Then the equation

$$\Delta u + a(x)u - b(x)u^\sigma = 0, \quad \sigma > 1 \tag{6.24}$$

possesses a minimal and a maximal (possibly coinciding) positive solutions.

Proof. According to Theorem 6.5 it is enough to prove that there exists a minimal
solution. Since $\lambda_1^L(M) < 0$, we can find a relatively compact domain with smooth
boundary $\Omega_0 \supset B_0$ such that $\lambda_1^L(\Omega_0) < 0$. Arguing exactly as in the proof of
Lemma 6.3, we see that if $\gamma > 0$ is sufficiently small, then

$$\lambda_1^{\widetilde{L}}(\Omega_0) < 0,$$

where $\widetilde{L} = \Delta + a(x) - \gamma b(x)$ so that, if ψ is a positive eigenfunction corresponding
to $\lambda_1^{\widetilde{L}}(\Omega_0)$, then

$$\begin{cases} \Delta\psi + a(x)\psi \geq \gamma b(x)\psi & \text{on } \Omega_0, \\ \psi \equiv 0 & \text{on } \partial\Omega_0 \end{cases}$$

and, choosing $\mu \leq \gamma^{\frac{1}{\sigma-1}} \left(\sup_{\Omega_0} \psi\right)^{-1}$, the function v_- defined by $v_- = \mu\psi$ satisfies

$$\begin{cases} \Delta v_- + a(x)v_- \geq b(x)v_-^\sigma & \text{on } \Omega_0, \\ v_- \equiv 0 & \text{on } \partial\Omega_0. \end{cases} \tag{6.25}$$

Since $\lambda_1(B_0) > 0$, Theorem 6.2 guarantees the existence of a supersolution $v_+ > 0$
to (6.25), and possibly multiplying v_+ by a large positive constant we can also
suppose that $v_+ \geq v_-$ on Ω_0. By the monotone iteration scheme (see section 6.5),
we deduce the existence of a positive solution $u_0 \in C^2(\Omega_0)$ of

$$\begin{cases} \Delta u_0 + a(x)u_0 - b(x)u_0^\sigma & \text{on } \Omega_0, \\ u_0 = 0 & \text{on } \partial\Omega_0. \end{cases} \tag{6.26}$$

Now let $\Omega_1 \supset \overline{\Omega}_0 \supset \Omega_0$. By domain monotonicity

$$\lambda_1^L(\Omega_1) \le \lambda_1^L(\Omega_0) < 0$$

and the above procedure yields the existence of a positive solution u_1 of (6.26) on Ω_1. Since $\partial\Omega_0 \subset \Omega_1$, $u_0 = 0 < u_1$ on $\partial\Omega_0$, it follows by Proposition 6.1 that $u_0 \le u_1$ on Ω_0. Choosing an increasing exhaustion of M by relatively compact domains with smooth boundaries $\{\Omega_i\}_{i=0}^{+\infty}$, the above procedure produces a sequence $\{u_i\}$ of solutions of (6.26) on Ω_i satisfying

$$u_i \le u_{i+1} \quad \text{on } \Omega_i.$$

Using the procedure of Lemma 6.4 we see that $\{u_i\}$ is uniformly bounded on $\overline{\Omega}_k$ for $i \ge k+1$; therefore u_i converges to $u > 0$, solution of (6.24). The argument used to prove the monotonicity of $\{u_i\}$ shows that if $\tilde{u}$ is any other solution of (6.24), then $\tilde{u} \ge u_i$ on $\Omega_i \; \forall i \ge 0$. Thus $\tilde{u} \ge u$ and u is the required positive minimal solution of (6.24). □

In the same vein we prove the following theorem (see [BRS98]).

Theorem 6.8. *Let* $(M, \langle\,,\,\rangle)$ *be a complete manifold of dimension* $m \ge 3$ *satisfying*

$$\operatorname{Ric}\langle \nabla r, \nabla r\rangle \ge -(m-1)H^2(1+r^2)^{\delta/2} \quad \text{on } M, \tag{6.27}$$

for some constants $H > 0$ *and* $\delta \ge -2$. *Let* $a(x), b(x) \in C^\infty(M)$, $b(x) > 0$ *on* M *and suppose that, for some constants* $\gamma \le 0$, $\mu < 1 - \frac{\delta}{2}$,

$$\liminf_{r(x) \to +\infty} \frac{a(x)}{r(x)^{-\mu}} > 0 \tag{6.28}$$

and

$$\limsup_{r(x) \to +\infty} \frac{b(x)}{r(x)^{-\mu - \frac{4\gamma}{m-2}}} < +\infty. \tag{6.29}$$

Finally, given $1 < \sigma \le \frac{m+2}{m-2}$, *let* $L_\sigma = \Delta + \frac{(m-2)(\sigma-1)}{4} a(x)$ *and assume that*

$$\lambda_1^{L_\sigma}(M) < 0. \tag{6.30}$$

Then equation

$$\Delta u + a(x)u - b(x)u^\sigma = 0 \tag{6.31}$$

has a positive solution $u \in C^\infty(M)$ *on* M *satisfying*

$$\liminf_{r(x) \to +\infty} \frac{u(x)}{r(x)^{\frac{4\gamma}{(m-2)(\sigma-1)}}} > 0. \tag{6.32}$$

Remark. Two comments are in order. First of all, the theorem establishes the existence of a solution to (6.31) satisfying an explicit lower bound at infinity. In this respect it is worth noticing that, if $\delta > -2$ and $\mu < -\delta$, then any positive solution u of (6.31) satisfies the estimate (6.32) by the *a priori* estimates of Theorem 4.1. However in case $\delta \le \mu < 1 - \frac{\delta}{2}$, which is allowed since $\delta > -2$, the lower bound (6.32) does not follow from Theorem 4.1, and in this case one has to construct the appropriate subsolution as in the proof of Theorem 6.8.

The second observation concerns the validity of the spectral assumption (6.30). Note that in general, if $0 \le \sigma_1 \le \sigma_2$, then, by the variational characterization of the bottom of the spectrum, $\lambda_1^{L_{\sigma_1}}(M) \ge \frac{\sigma_1}{\sigma_2}\lambda_1^{L_{\sigma_2}}(M)$ (see the proof of Theorem 2 in [FCS80]). Indeed, for every $\phi \in C_0^\infty(M)$,

$$\int_M |\nabla \psi|^2 - \sigma_1 a(x)\psi^2 = \int_M |\nabla \psi|^2 - \frac{\sigma_1}{\sigma_2}\sigma_2 a(x)\psi^2 \ge \frac{\sigma_1}{\sigma_2}\int_M |\nabla \psi|^2 - \sigma_2 a(x)\psi^2$$

and the claim follows taking the infimum over all such ψ. Thus assumption (6.30) implies the condition $\lambda_1^L(M) \ge 0$ considered in Theorem 6.7.

Proof. Again according to Theorem 6.5 it suffices to find a nonnegative subsolution of (6.31) satisfying (6.32). We divide the proof into several steps. For the sake of simplicity we assume that the M has a *pole* o, i.e., a point such that $\exp : T_oM \to M$ is a diffeomorphism (see for instance [dC92], [GW79]). The general case is dealt with noting that, applying Lemma 1.12, the radial functions which are shown to be pointwise super-, respectively subsolutions within the cut locus, are in fact global super-, respectively subsolutions in the weak sense, and this is all that is needed to carry out the argument.

Step 1: reduction to the case where $\sigma = \frac{m+2}{m-2}$.

It suffices to show that there is a positive function $v_- \in C^0(M) \cap H_{loc}^{1,2}(M)$ which satisfies weakly on M,

$$L_\sigma v_- - \frac{(m-2)(\sigma-1)}{4}b(x)v_-^{\frac{m+2}{m-2}} \ge 0 \tag{6.33}$$

and

$$v_-(x) \ge Cr(x)^\gamma \quad \text{for } r(x) \gg 1 \tag{6.34}$$

and some constant $C > 0$. Indeed, since $1 < \sigma \le \frac{m+2}{m-2}$, a straightforward computation shows that if v_- satisfies (6.33), then the function u_- defined by

$$u_- = v_-^{-\frac{4}{(m-2)(\sigma-1)}}$$

satisfies

$$\Delta u_- + a(x)u_- - b(x)u_-^\sigma$$

$$\ge \frac{4}{(m-2)(\sigma-1)}\left(\frac{4}{(m-2)(\sigma-1)} - 1\right)v_-^{-\frac{4}{(m-2)(\sigma-1)}-2}|\nabla v_-|^2 \ge 0$$

weakly on M, and therefore u_- is a subsolution of (6.31) satisfying (6.32).

Step 2: construction of a subsolution outside a compact set.

Consider $w(x) = r(x)^\gamma = \beta(r(x))$ for $r(x) \geq 1$. Clearly $w(x)$ satisfies the lower bound (6.34). We claim that there exist $\delta_0 > 0$ and R_0 sufficiently large such that, $\forall 0 < \delta \leq \delta_0$, the function

$$\widehat{v} = \delta w$$

satisfies (6.33) weakly on $M \setminus \overline{B_{R_0}}$. Indeed, from (6.27), the Laplacian comparison theorem and Proposition 1.15, having fixed $\widetilde{H} > H'$ of Proposition 1.15, there exists $R_0 = R_0(\widetilde{H})$ sufficiently large such that

$$\Delta r(x) \leq (m-1)\widetilde{H}r(x)^{\delta/2} \quad \text{on } M \setminus \overline{B_{R_0}}.$$

Using (6.28) and (6.29) we can also suppose to have chosen R_0 so large that

$$a(x) \geq A_1 r(x)^{-\mu} \quad \text{and} \quad b(x) \leq A_2 r(x)^{-\mu - \frac{4\gamma}{m-2}} \tag{6.35}$$

on $M \setminus \overline{B_{R_0}}$ for some constants $A_1, A_2 > 0$. Since $\beta' \leq 0$ it follows that, on $M \setminus \overline{B_{R_0}}$,

$$
\begin{aligned}
L_\sigma w &= \Delta w + \frac{(m-2)(\sigma-1)}{4} a(x) w \\
&= \beta'' + (\Delta r)\beta' + \frac{(m-2)(\sigma-1)}{4} a(x)\beta \\
&\geq \gamma(\gamma-1)r^{\gamma-2} + (m-1)\widetilde{H}\gamma r^{\gamma-1+\delta/2} + \frac{(m-2)(\sigma-1)}{4} A_1 r^{\gamma-\mu}.
\end{aligned}
$$

Thus, in case $\mu < 1 - \frac{\delta}{2}$ we deduce, up to choosing R_0 sufficiently large, that there exists a constant $C > 0$ such that

$$L_\sigma w \geq C r^{\gamma-\mu} \quad \text{on } M \setminus \overline{B_{R_0}}.$$

To conclude note that, having set

$$b_w = w^{-\frac{m+2}{m-2}} L_\sigma w, \tag{6.36}$$

we have

$$b_w \geq C r^{-\mu - \frac{4\gamma}{m-2}} \quad \text{on } M \setminus \overline{B_{R_0}}.$$

Since

$$b_{\delta w} = \delta^{-\frac{4}{m-2}} b_w,$$

it follows from (6.35) that, if δ is sufficiently small,

$$b_{\delta w} \geq b(x) \frac{(m-2)(\sigma-1)}{4} \quad \text{on } M \setminus \overline{B_{R_0}}$$

as required to show that there exists δ_o sufficiently small that for every $\delta < \delta_o$, $\widehat{v}$ solves (6.33) weakly on $M \setminus \overline{B_{R_0}}$, and satisfies (6.34).

Step 3: construction of a weak subsolution v_- on M.

Since by assumption $\lambda_1^{L_\sigma}(M) < 0$, we can find a relatively compact set with smooth boundary $\Omega \supset \overline{B_{R_0}}$ such that $\lambda_1^{L_\sigma}(\Omega) < 0$. Let φ be a corresponding eigenfunction, so that

$$\begin{cases} \Delta\varphi + \frac{(m+2)(\sigma-1)}{4} a(x)\varphi = -\lambda\varphi, & \varphi > 0 \text{ in } \Omega, \\ \varphi = 0 \quad \text{on } \partial\Omega. \end{cases}$$

Arguing as above, it follows that if $\eta > 0$ is sufficiently small, then $\varphi_\eta = \eta\varphi$ is a subsolution of (6.33) on Ω. Finally let $\delta < \delta_0$ be small enough that the function $\widehat{v} < \varphi_\eta$ in $B_{R_0+\epsilon} \setminus B_{R_0}$. Note that since $\widehat{v} > 0$ on $\partial\Omega$ there exists an open set D with $\overline{D} \subset \Omega$ such that $\widehat{v} > \varphi_\eta$ in $\Omega \setminus \overline{D}$.

We claim that the function v_- defined by

$$v_- = \begin{cases} \varphi_\eta & \text{in } B_{R_0+\epsilon}, \\ \max\{\widehat{v}, \varphi_\eta\} & \text{in } \Omega \setminus \overline{B_{R_0}}, \\ \widehat{v} & \text{in } M \setminus \overline{D} \end{cases}$$

is the required subsolution. Note first of all that v_- is well defined by construction, and it is clearly a subsolution in $B_{R_0+\epsilon}$ and in $M \setminus \overline{D}$. On the other hand, v_- is a subsolution in $\Omega \setminus \overline{B_{R_0}}$ as maximum of two subsolutions, by Theorem 6.20. $\square$

6.2.2 $\lambda_1^L(M) < 0$: some sufficient conditions

The previous Theorems 6.7 and 6.8 point out the relevance of condition (6.23) in guaranteeing the existence of positive solutions to (6.24). It is therefore interesting to find sufficient conditions in order that (6.23) holds. We begin with the following elementary result.

Proposition 6.9. *Let* $(M, \langle\,,\,\rangle)$ *be a complete manifold,* $a(x) \in C^0(M)$, *and* $L = \Delta + a(x)$. *Set* $v(r) = \mathrm{vol}(\partial B_r)$ *and denote by*

$$\bar{a}(r) = \frac{1}{\mathrm{vol}(\partial B_r)} \int_{\partial B_r} a(x) \tag{6.37}$$

the spherical mean of $a(x)$. *Let* $A(r) \leq \bar{a}(r)$ *and suppose that* α *is a solution of the problem*

$$\begin{cases} (v(r)\alpha')' + v(r)A(r)\alpha \geq 0, \\ \alpha(0) = \alpha_0 > 0, \quad \alpha'(0) = 0, \end{cases} \tag{6.38}$$

and has a first zero at $r = T > 0$. *Then*

$$\lambda_1^L(M) < 0. \tag{6.39}$$

Proof. We consider the geodesic ball $B_T(o)$ and define

$$\varphi(x) = \alpha(r(x)).$$

Then, using the co-area formula (1.87) and (6.38) and integrating by parts we compute

$$
\begin{aligned}
\int_{B_T} |\nabla\varphi|^2 - a(x)\varphi^2 &= \int_{B_T} |\nabla\varphi|^2 - \bar{a}(r)\varphi^2 \\
&\leq \int_{B_T} |\nabla\varphi|^2 - A(r)\varphi^2 \\
&= \int_0^T [(\alpha')^2 v(r)\, dr - A(r)\alpha^2 v(r)]\, dr \\
&= -\int_0^T [(v(r)\alpha')' + A(r)\alpha v(r)]\alpha\, dr \leq 0.
\end{aligned}
$$

By the Rayleigh characterization $\lambda_1^L(B_T) \leq 0$ and by domain monotonicity we get (6.39). $\qquad\square$

Recalling that the solution of an ordinary differential equation is said to be *oscillating* if is has an infinite number of zeroes, we observe that, in order to show that a solution α of (6.38) has a first zero at $T > 0$, any oscillation criterion would do. For instance:

Proposition 6.10. *Let α be any solution of (6.38) with equality on $[0, +\infty)$ and assume that $\frac{1}{v} \in L^1(+\infty)$. Then α is oscillating provided*

$$\liminf_{r \to +\infty} A(r) \left[v(r) \int_r^{+\infty} \frac{ds}{v(s)} \right]^2 > \frac{1}{4}. \tag{6.40}$$

Proof. We perform the standard change of variables

$$t = K(r) = \left(\int_r^{+\infty} \frac{ds}{v(s)} \right)^{-1}. \tag{6.41}$$

Then, $K : (0, +\infty) \to (0, +\infty)$ is strictly increasing and setting

$$\gamma(t) = t\alpha(K^{-1}(t))$$

a direct computation shows that γ satisfies

$$\gamma'' + p(t)\gamma = 0 \tag{6.42}$$

with

$$p(t) = \frac{v^2(K^{-1}(t))A(K^{-1}(t))}{t^4}.$$

An application of Cauchy's theorem, together with the definition of K and (6.40) shows that

$$\liminf_{t \to +\infty} t \int_t^{+\infty} p(s) \, ds > \frac{1}{4}.$$

It follows from the Hille-Nehari oscillation theorem, see [Swa68] Chapter 2, that α is oscillating. $\square$

Remark. We note that in the above oscillating result $A(r)$ has necessarily to be positive in a neighborhood of $+\infty$, but neither here nor in Proposition 6.9 is $A(r)$ required to have a definite sign.

Suppose now we have an estimate of the type

$$\Delta r \le (m-1)\frac{h'}{h}(r), \quad h > 0 \text{ on } (0, +\infty)$$

which would typically follow from an appropriate lower bound for the Ricci curvature. Consider a solution of

$$\begin{cases} \left(h^{m-1}(r)\beta' \right)' + h^{m-1}(r)A(r)\beta = 0, \\ \beta'(0) = 0, \, \beta(0) = \beta_0 > 0 \end{cases}$$

and let T be the first zero of β. We set

$$\psi(x) = \beta(r(x));$$

then, if $A(r) \le \bar{a}(r)$,

$$\Delta\psi + \bar{a}(r(x))\psi = \beta'' + \Delta r\beta' + \bar{a}(r)\beta \ge \beta'' + \Delta r\beta' + A(r)\beta.$$

Therefore, if $\beta' \le 0$ (which is necessarily the case if $A(r) \ge 0$ on $[0, T)$), then ψ satisfies

$$\Delta\psi + \bar{a}(r(x))\psi \ge 0.$$

Since ψ is radial, using again the co-area formula,

$$\int_{B_T} |\nabla\psi|^2 - a(x)\psi^2 = \int_{B_T} |\nabla\psi|^2 - \bar{a}\psi^2 = -\int_{B_T} \psi(\Delta\psi + \bar{a}\psi) \le 0$$

and we conclude that $\lambda_1^L(B_T) \le 0$. Note that in this case

$$v(r) \le Ch^{m-1}(r).$$

We conclude this short account by quoting some new results in oscillation theory obtained in [BMR09] and [MRV], which we refer to for a complete treatment. With the notation of Proposition 6.9, let $v(r) = \mathrm{vol}(\partial B_r)$ and let $\bar{a}(r)$ be the spherical mean of the potential a. Suppose that there exists an oscillating solution $z \in \mathrm{Lip}_{\mathrm{loc}}$ of the problem

$$\begin{cases} (v(t)z'(t))' + \bar{a}(t)v(t)z(t) = 0 \text{ on } (0, +\infty), \\ z'(t) = O(1) \text{ as } t \searrow 0^+, \quad z(0^+) = z_0 > 0, \end{cases} \tag{6.43}$$

and let $0 < t_2 < t_3$ be two consecutive zeros, that is, $z(t_2) = z(t_3) = 0$ and $z \neq 0$ in (t_2, t_3). Similarly to what we did in Proposition 6.9, we consider the function $\varphi_z : M \to \mathbb{R}$ defined as

$$\varphi_z(x) = \begin{cases} z(r(x)) & r(x) \in [t_2, t_3] \\ 0 & r(x) \in [0, +\infty) \setminus [t_2, t_3]. \end{cases}$$

Then,

$$\lambda_1^L(B_{t_3} \setminus \bar{B}_{t_2}) \leq \frac{\int_{B_{t_3} \setminus \bar{B}_{t_2}} |\nabla \varphi_z|^2 - \int_{B_{t_3} \setminus \bar{B}_{t_2}} a\varphi_z^2}{\int_{B_{t_3} \setminus \bar{B}_{t_2}} \varphi_z^2} = \frac{\int_{t_2}^{t_3} v(z')^2 - \int_{t_2}^{t_3} \bar{a}vz^2}{\int_{t_2}^{t_3} vz^2}$$

$$= -\frac{\int_{t_2}^{t_3} \left[(vz')' + \bar{a}vz\right]z}{\int_{t_2}^{t_3} vz^2} = 0.$$

Therefore, by the domain monotonicity of eigenvalues, the oscillation of the solution z implies

$$\lambda_1^L(M \setminus B_R) < 0, \quad \text{for all } R \geq 0. \tag{6.44}$$

Recall now that the *index* of L is defined as the number of negative eigenvalues of $-L$. Hence condition (6.44), together with a result by D. Fisher–Colbrie [FC85] shows that L has infinite index, that is

$$\mathrm{ind}_L(M) = \sup_{\Omega \subset \subset M} \{\mathrm{ind}_L(\Omega)\} = +\infty.$$

In fact the conclusion follows without appealing to [FC85] by noting that taking pairs of consecutive zeros of z the corresponding functions φ_z constructed above are L^2-orthogonal and this is enough to conclude that $\mathrm{ind}_L(M) = +\infty$. Now fix a constant $B \geq 0$ and define, for every $t_1, t_2 \in [0, +\infty]$,

$$V(t_1, t_2) = e^{2B \int_{t_1}^{t_2} \frac{ds}{v(s)}}. \tag{6.45}$$

Then we have the following ([MRV], Theorem 14)

Theorem 6.11. *Let V and B as in definition (6.45) and suppose that the spherical mean $\bar{a}(r)$ of $a(x)$ satisfies*

$$\bar{a}(r)v^2(r) \geq -B^2$$

for all $r > 0$.

i) *If there exist $0 < a < b$ such that*

$$\int_{B_b \setminus B_a} a(x)dx > \begin{cases} 2B & \text{if } v^{-1} \notin L^1(+\infty), \\ 2B \frac{V(b,+\infty)}{V(b,+\infty)-1} & \text{if } v^{-1} \in L^1(+\infty), \end{cases}$$

then $\lambda_1^L(M) < 0$.

ii) *L is unstable at infinity, i.e., $\lambda_1^L(M \setminus B_R) < 0$ for every $R > 0$, provided either $v \in L^1(+\infty)$ and, for some $R > 0$,*

$$\limsup_{t \to \infty} \left\{ \int_{B_t \setminus B_R} a(x)dx \int_t^\infty \frac{ds}{v(s)} \right\} > 1,$$

or $v \notin L^1(+\infty)$ and

$$\lim_{t \to \infty} \left\{ \sup_{t \le q_1 < q_2 \le \infty} \int_{B_{q_2} \setminus B_{q_1}} a(x)dx \right\} > 2B.$$

In particular, in the assumption ii) *L has infinite index.*

Theorem 6.11 is a consequence of Theorems 6.12 and 6.14 below which are proved in [MRV] (see also Proposition 1.2 and Theorem 1.4 in [BMR09]).

Theorem 6.12. *Let $v(t)$ and $\bar{a}(t) \in L^\infty_{loc}([0, +\infty))$ be such that*

$$v(t) \ge 0, \quad \frac{1}{v(t)} \in L^\infty_{loc}((0, +\infty)), \quad \frac{1}{v} \notin L^1(0^+), \quad \lim_{t \to 0^+} v(t) = 0 \qquad (6.46)$$

and

$$\bar{a}(t) \ge -\frac{B^2}{v(t)^2} \qquad (6.47)$$

for some real constant $B \ge 0$. Let $z(t) \in Lip_{loc}([0, +\infty))$ be a solution of problem (6.43). If $z(t) \ne 0$ for all $t \in (0, +\infty)$, then for every $0 \le a < b$,

$$\int_a^b \bar{a}(s)v(s)ds \le \begin{cases} 2B & \text{if } \frac{1}{v} \notin L^1(+\infty), \\ 2B \frac{V(b,+\infty)}{V(b,+\infty)-1} & \text{if } \frac{1}{v} \in L^1(+\infty). \end{cases}$$

In order to prove Theorem 6.12 let us recall the following Riccati comparison result ([MRV], Lemma 18; compare also with Lemma 1.10 in Chapter 1):

Lemma 6.13. *Let G, $v \in C^0([0, +\infty))$, $v > 0$, and $q_i \in AC(T, T_i)$, $i = 1, 2$, be solutions of the Riccati differential inequalities*

$$q_1'(t) \ge G(t) + \frac{1}{v(t)}q_1^2(t), \quad q_2'(t) \le G(t) + \frac{1}{v(t)}q_2^2(t) \qquad (6.48)$$

a.e. in (T, T_i) satisfying

$$q_1(T) = q_2(T). \tag{6.49}$$

Then $T_1 \leq T_2$ and $q_1(t) \geq q_2(t)$ in $[T, T_1)$.

Conversely, if $q_i \in AC(T_i, \bar{t})$, $i = 1, 2$, are solutions of (6.48) a.e. in (T_i, T) satisfying $q_1(T) = q_2(T)$, then $T_1 \geq T_2$ and $q_1(t) \leq q_2(t)$ in $(T_1, T]$.

A proof of the lemma is a minor modification of that of Corollary 2.2 in [PRS08].

We are now ready for the

Proof of Theorem 6.12. Since, by assumption, $z(t) > 0$ on $(0, +\infty)$, the function

$$y(t) = -v(t)\frac{z'(t)}{z(t)}$$

is well defined on $(0, +\infty)$ and it is locally Lipschitz, as can be immediately seen by integrating (6.43) once. Furthermore, y satisfies

$$y' = \frac{y^2}{v(t)} + \bar{a}(t)v(t) \tag{6.50}$$

so that, according to (6.47),

$$y' \geq \frac{y^2 - B^2}{v(t)} \quad \text{on } (0, +\infty). \tag{6.51}$$

First we consider the case $\frac{1}{v} \notin L^1(+\infty)$. Since, by (6.46) we also have $\frac{1}{v} \notin L^1(0^+)$, it follows that given any fixed $\alpha \in (0, +\infty)$ there exists $t_\alpha > 0$ such that

$$\int_1^{t_\alpha} \frac{ds}{v(s)} = \frac{1}{2B} \log \alpha. \tag{6.52}$$

With the notation introduced in (6.45) define

$$y_\alpha(t) = B\frac{\alpha + V(1,t)}{\alpha - V(1,t)} \tag{6.53}$$

and observe that, because of (6.52),

$$\text{i) } y_\alpha(t_\alpha^+) = -\infty; \quad \text{ii) } y_\alpha(t_\alpha^-) = +\infty. \tag{6.54}$$

Furthermore y_α satisfies

$$y_\alpha' = \frac{y_\alpha^2 - B^2}{v(t)} \tag{6.55}$$

where defined. We claim that

$$-B \leq y(t) \leq B \quad \text{on } (0, +\infty). \tag{6.56}$$

By contradiction assume that there exists $T \in (0, +\infty)$ such that $y(T) = Y > B$, the case $Y < -B$ being similar. Let

$$\alpha = \frac{Y + B}{Y - B} V(1, T) \tag{6.57}$$

and observe that

$$y_\alpha(T) = Y = y(T) \tag{6.58}$$

and that y_α will thus be defined on $[T, T_2)$ for some maximal $T_2 \leq +\infty$. Because of (6.51), (6.55) and (6.58) we can apply the Riccati comparison lemma above with the choices $q_1 = y$, $q_2 = y_\alpha$, $G = -\frac{B^2}{v^2}$ and $T = T$. Noting that y is defined on $[T, +\infty)$ and applying the first part of the lemma we have that y_α is defined on $[T, +\infty)$. However, from (6.57) and (6.52) we have

$$t_\alpha > T$$

and we obtain a contradiction using (6.54) ii) (in case $Y < -B$ we would have obtained a contradiction via (6.54) i)); thus we have the validity of (6.56). From (6.50) we obtain

$$y' \geq \bar{a}(t)v(t). \tag{6.59}$$

Fix any $0 \leq a < b$ as in the statement of the theorem and integrate (6.59) on $[a, b]$. Using (6.56) we have

$$\int_a^b \bar{a}(s)v(s)\, ds \leq y(b) - y(a) \leq 2B. \tag{6.60}$$

Suppose now $\frac{1}{v} \in L^1(+\infty)$. In this case, there exists $t_\alpha > 0$ such that (6.52) holds only for

$$\alpha \in (0, V(1, +\infty)). \tag{6.61}$$

We claim that

$$-B \leq y(t) \leq B \frac{V(t, +\infty) + 1}{V(t, +\infty) - 1} \quad \text{on } (0, +\infty). \tag{6.62}$$

As above, we only prove the right-hand side inequality, the left-hand inequality being proved in a similar way. Assume by contradiction that there exists $T \in (0, +\infty)$ such that

$$y(T) = Y > B \frac{V(T, +\infty) + 1}{V(T, +\infty) - 1}.$$

Note that, since $V(1, +\infty) > 1$, if we let

$$\alpha = \frac{Y + C}{Y - C},$$

then, for $C > 0$ sufficiently small,

$$\alpha < V(1, +\infty).$$

It follows that the function

$$y_\alpha = C \frac{\alpha + V(1, t)}{\alpha - V(1, t)}$$

satisfies the Riccati equation (6.55), $y_\alpha(T) = Y$ and there exists $t\alpha > T$ such that $y_\alpha(t_\alpha^-) = +\infty$. Applying the Riccati comparison lemma and arguing as above we obtain a contradiction which proves that $y(t)$ satisfies the required upper estimate.

Integrating (6.59) as before we get

$$\int_a^b \bar{a}(s)v(s)\,ds \le \frac{2BV(b, +\infty)}{V(b, +\infty) - 1}. \tag{6.63}$$

The estimate of Theorem 6.12 now follows from (6.60) and (6.63). $\qquad\square$

The proof of the next result can be obtained by iterating the technique of the proof above; for details we refer to the original proof of Theorem 10 in [MRV].

Theorem 6.14. *Let $v, \bar{a}$ and z be defined as in Theorem 6.12. Then z is oscillating provided either $\frac{1}{v} \in L^1(+\infty)$ and*

$$\limsup_{t \to \infty} \left\{ \int_R^t \bar{a}(s)v(s)ds \int_t^\infty \frac{ds}{v(s)} \right\} > 1 \tag{6.64}$$

for some $R > 0$, or $v^{-1} \notin L^1(+\infty)$ and

$$\lim_{t \to \infty} \left\{ \sup_{t \le q_1 < q_2 \le \infty} \int_{q_1}^{q_2} \bar{a}(s)v(s)ds \right\} > 2B. \tag{6.65}$$

6.2.3 A more general case

We continue with the general strategy of applying Theorem 6.5 to guarantee the existence of solutions. In the next result we replace the spectral assumption $\lambda_1^F(M) < 0$ with a pointwise condition. It should be pointed out that, as the case of hyperbolic space shows, the two assumptions are related but independent.

Theorem 6.15. *Let $(M, \langle\, , \rangle)$ be a complete manifold, $a(x), b(x) \in C^{0,\mu}(M)$, $0 < \mu \le 1$. Assume that $a(x) > 0$ on M and $b(x) \ge 0$. Assume also that the Ricci tensor satisfies*

$$\mathrm{Ric} \ge -(m-1)H^2\big(1 + r(x)^2\big)^{\delta/2} \quad \text{on } M, \tag{6.66}$$

for some $\delta \ge -2$ and $H > 0$. Let $B_0 = \{x \in M : b(x) = 0\}$ and assume that

$$\lambda_1^L(B_0) > 0 \tag{6.67}$$

and

$$\text{i) } a(x) \geq Ar(x)^\alpha; \quad \text{ii) } b(x) \leq Br(x)^\beta e^{Dr(x)^\theta}, \quad \text{for } r \gg 1, \tag{6.68}$$

for some $\alpha > \frac{\delta}{2} - 1$, $0 < \theta < \min\left\{1 + \alpha - \frac{\delta}{2}, 1 + \frac{\alpha}{2}\right\}$, *and* $A, B, D > 0$, $\beta \in \mathbb{R}$. *Then there exists a positive maximal solution* u *on* M *of*

$$\Delta u + a(x)u - b(x)u^\sigma = 0, \quad \sigma > 1. \tag{6.69}$$

Remark. Consider the case of the hyperbolic m-dimensional space $\mathbb{H}^m_{-H^2}$ of constant negative curvature $-H^2$. In this case the parameter α must satisfy $\alpha > -1$, so that a behavior of $a(x) \geq 0$ of the type

$$Ar(x)^{-1+\varepsilon} \leq a(x) \leq C(1 + r(x))^{-1+2\varepsilon}$$

for $r(x) \gg 1$ and some $\varepsilon > 0$ is admissible. In this case, for $a(x)$ sufficiently small, $\lambda_1^L(M) \geq 0$, with $L = \Delta + a(x)$. This shows that Theorem 6.15 is not contained in Theorem 6.7 when $\lambda_1^L(M) < 0$.

Remark. The upper bound for θ is sharp in the sense that, in the case of equality, existence may fail; see for instance Theorem 4.8 where $\sigma = \frac{m+2}{m-2}$, $\alpha = 0$, $\delta = 0$ and $\theta = 1$, and $D = 2B, \beta = \frac{2}{m-1} + \gamma, \gamma > 0$.

Proof of Theorem 6.15. According to Theorem 6.5 we only need to construct a global subsolution $u_- \geq 0$, $u_- \not\equiv 0$.

We observe first of all that we may assume that

$$a(x) \geq A(1 + r(x))^\alpha \quad b(x) \leq B(1 + r(x))^\beta e^{Dr(x)^\theta}$$

on M.

Because of (6.66), Proposition 1.15 and Theorem 1.11 imply that

$$\Delta r \leq \widetilde{H} r^{\delta/2}$$

for some constant $\widetilde{H}$ and $r \geq R$. On the other hand, by the asymptotic behavior of the metric coefficients in geodesic polar coordinates we have

$$\Delta r = \frac{m-1}{r} + o(1) \text{ as } r \to 0+,$$

and there exists a constant C such that

$$\Delta r \leq C\left(\frac{1}{r} + r^{\delta/2}\right) \text{ weakly on } M.$$

It follows that, if $w(r)$ is a nonincreasing positive C^2 function with $w'(0) = 0$ satisfying

$$w'' + C\left(\frac{1}{r} + r^{\delta/2}\right)w'(r) + A(1 + r)^\alpha w - B(1 + r)^\beta w^\sigma \geq 0, \tag{6.70}$$

then the function $u_-(x) = w(r(x))$ is the required (weak) subsolution.

We look for a function w of the form

$$w(r) = (\mu + P(r))^\xi$$

with $P(r) \geq 0$, $P'(0) = 0$, $P'(r) \geq 0$ and $P(r) = e^{Dr^\theta}$ for $r \geq R$, where $\mu > 0$ is a constant to be chosen later, $D > 0$ and θ are the constants in (6.68) and $\xi < \frac{1}{1-\sigma} < 0$.

Letting, for ease of notation

$$H_w = w'' + C(\frac{1}{r} + r^{\delta/2})w'(r) + A(1 + r)^\alpha w,$$

we claim that, for an appropriate choice of $\mu > 0$, we have

$$H_w(r) > 0 \quad \text{on } [0, +\infty) \tag{6.71}$$

and

$$\frac{H_w(r)}{w(r)^\sigma} \geq Cr^\beta e^{Dr^\theta} \quad \text{for some } C > 0 \text{ and } r \gg 1, \tag{6.72}$$

It follows that if E is a sufficiently small positive constant, then Ew satisfies (6.70). Indeed, with obvious notation,

$$\frac{H_{Ew}(r)}{(Ew(r))^\sigma} = E^{1-\sigma}\frac{H_w(r)}{(w(r))^\sigma}$$

and using (6.71) and (6.72) we may choose E small enough that

$$\frac{H_{Ew}(r)}{(Ew(r))^\sigma} \geq B(1 + r)^\beta e^{Dr^\theta}$$

on $[0, +\infty)$. We compute

$$w'(r) = \xi(\mu + P(r))^{\xi-1}P'(r) \quad w''(r)$$
$$= \xi(\xi - 1)(\mu + P(r))^{\xi-2}(P'(r))^2 + \xi(\mu + P(r))^{\xi-2}P''(r)$$

and inserting this in the expression of H_w it is easy to see that we may choose μ large enough that $H_w > 0$ on any chosen interval $[0, R]$. On the other hand, using the expression of $P(r)$ for r large, a straightforward computation shows that

$$\frac{H_w(r)}{w(r)^\sigma} \sim Ar^\alpha\left(\mu + e^{Dr^\theta}\right)^{\xi(1-\theta)},$$

so that the conditions $\xi < \frac{1}{1-\sigma} < 0$ and $D, \theta > 0$ easily imply (6.72). Thus we may first find R such that (6.72) holds in $[R, +\infty)$ with a constant $C > 0$. Next, if necessary we may increase μ in order that (6.71) holds on $[0, R]$. It is clear that (6.72) will continue to hold with a possibly smaller C. $\quad\square$

Remark. We note that if we replace the upper bound on the coefficient $b(x)$, then the same procedure works with different choices of $w(r)$. For instance, if we assume that

$$b(x) \leq B(1+r)^\beta \quad \text{on } M,$$

then the same arguments show that one may find a subsolution of the form $u_-(x) = w(r(x))$ with $w(r) = (\mu + r^2)^{-\gamma}$ provided $\beta + \alpha \leq 2\gamma(\sigma - 1)$. Clearly the solution u supplied by the theorem satisfies the lower bound

$$u(x) \geq C(\mu + r^2)^{-\gamma}.$$

In applications to the Yamabe problem this is relevant in connection to the completeness of the conformal metric.

6.3   Global sub- and supersolutions

We now give another version of Theorem 6.15, where it is no longer assumed that b is nonnegative. This will require the explicit construction of a global supersolution, and the procedure will prove to be rather delicate. The weakening of the assumptions on b is also reflected in the stronger curvature conditions and an upper bound imposed on the coefficient a.

Theorem 6.16. *Let* $(M, \langle\,,\,\rangle)$ *be a complete manifold of dimension* $m \geq 4$ *satisfying*

$$\text{Ric} \geq -(m-1)H^2 \tag{6.73}$$

for some constant $H > 0$. *Let* $a(x), b(x) \in C^{0,\mu}(M)$, $0 < \mu \leq 1$, $\sigma \geq \frac{m+2}{m-2}$ *and suppose that, for some* $R_0 > 0$ *we have*

$$a(x) \leq \frac{(m-1)^2}{m-2} \frac{H^2}{\sigma - 1} \quad \text{on } M \setminus B_{R_0}, \tag{6.74}$$

$$b(x) > 0 \qquad \text{on } M \setminus B_{R_0}, \tag{6.75}$$

$$\lambda_1^{\Delta + \frac{m-2}{4}(\sigma-1)(a(x))}(B_{R_0}) > 0, \tag{6.76}$$

$$a(x) \geq A(1 + r(x))^\alpha \quad \text{on } M, \tag{6.77}$$

$$b(x) \leq Cr(x)^\beta e^{Dr(x)^\theta} \quad \text{for } r(x) \gg 1 \tag{6.78}$$

and for some constants $A, C, D > 0$, $\beta \in \mathbb{R}$, $-1 < \alpha \leq 0$, $\theta < 1 + \alpha$. *Then, there exists* $\eta > 0$ *such that, if* $b(x) \geq -\eta$ *on* M, *the equation*

$$\Delta u + a(x)u - b(x)u^\sigma = 0 \tag{6.79}$$

admits a positive solution u *on* M.

Proof. As usual, it suffices to construct sub- and supersolutions v_- and v_+ in Lip_{loc} such that $0 < v_- \leq v_+$. The construction of the subsolution is done as in the proof of Theorem 6.15. Note that in this respect allowing b to take negative values only helps matters. We recall that the function v_- constructed in Theorem 6.15 is radial and satisfies

$$\lim_{r \to +\infty} v_-(r) = 0.$$

Recalling that for every $E < 1$, Ev_- is still a subsolution shows that the condition

$$v_+ \geq v_-$$

can always be satisfied whenever

$$v_+ > 0 \text{ on } M \quad \text{and} \quad \lim_{r+\infty} v_+(r(x)) = +\infty.$$

We therefore concentrate on the construction of the supersolution. We first reduce the analysis to the case where $\sigma = \frac{m+2}{m-2}$. Let $\tilde{a}(x) = \frac{(m-2)(\sigma-1)}{4}a(x)$ and $\tilde{b} = \frac{(m-2)(\sigma-1)}{4}b(x)$. Then a simple computation shows that, if u_+ is a positive supersolution of

$$\Delta u + \tilde{a}(x)u - \tilde{b}(x)u^{\frac{m+2}{m-2}} = 0 \tag{6.80}$$

on M, then

$$v_+ = u_+^{\frac{4}{(m-2)(\sigma-1)}} \tag{6.81}$$

is a positive supersolution of (6.79) provided $\sigma \geq \frac{m+2}{m-2}$. Note that, because of (6.76), we have

$$\lambda_1^{\Delta+\tilde{a}(x)}(B_{R_0}) > 0. \tag{6.82}$$

Furthermore,

$$\tilde{b}(x) > 0 \quad \text{on } M \setminus B_{R_0} \tag{6.83}$$

and if $b(x) \geq -B$ on M for some $B > 0$, then $\tilde{b} \geq -\tilde{B}$ where $\tilde{B}$ is the appropriate multiple of B.

We now proceed to construct a supersolution of (6.80) in B_{R_0} and in $M \setminus B_{R_0}$, and then glue them together to form a global supersolution. We divide the argument, which is adapted from the proof of Theorem 0.1 in [RRV97], into several steps.

Step 1. Construction of a supersolution of (6.80) inside B_{R_0}.

If there exists $x_0 \in B_{R_0}$ such that

$$\tilde{a}(x_0) > 0, \tag{6.84}$$

it follows by results of J. Escobar (see [Esc87], Theorems 3.2 and 4.2), that there exists a solution $u > 0$ of the problem

$$\begin{cases} \Delta u + \tilde{a}(x)u + u^{\frac{m+2}{m-2}} = 0 & \text{on } B_{R_0}, \\ u = 0 & \text{on } \partial B_{R_0}, \ u > 0 \text{ on } B_{R_0}. \end{cases} \tag{6.85}$$

On the other hand, if $\tilde{a} \le 0$ on B_{R_0}, then, using the fact that the $\lambda_1^{\Delta}(B_{R_0}) >$, we can choose a small enough constant $\tilde{a}_1 > 0$ such that $\lambda_1^{\Delta + \tilde{a}_1}(B_{R_0})$, and apply the above argument to obtain a solution $u > 0$ of problem (6.85) with $\tilde{a}_1$ instead of $\tilde{a}(x)$. Then

$$\Delta u + \tilde{a}(x)u + u^{\frac{m+2}{m-2}} \le \Delta u + \tilde{a}_1(x)u + u^{\frac{m+2}{m-2}} = 0$$

and we conclude that (6.85) always admits a positive supersolution u vanishing on ∂B_{R_0}.

Step 2. Construction of a radial supersolution on $M \setminus B_{R_0}$.

Using the Ricci curvature assumption (6.73) and the Laplacian comparison theorem we have

$$\Delta r \le (m-1)H \coth(Hr) \quad \text{on } M. \tag{6.86}$$

Moreover, it follows from (6.74) that

$$\tilde{a} \le \hat{a} = \frac{(m-1)^2}{4}H^2 \quad \text{on } M \setminus B_{R_0}. \tag{6.87}$$

Finally, we choose a positive, decreasing function $\hat{b}(t)$ on $[0, +\infty)$ of the form

$$\hat{b}(t) = e^{-P(t)} \tag{6.88}$$

with $P \in C^2([0, +\infty))$ satisfying

$$P(0) = P'(0) = 0, P' \ge 0, P'' \ge 0, P(t) \to +\infty \text{ as } t \to +\infty \tag{6.89}$$

and such that

$$\tilde{b} \ge \hat{b}(r(x)) \quad \text{on } M \setminus B_{R_0}. \tag{6.90}$$

We look for a supersolution of (6.80) on $M \setminus B_{R_0}$ of the form $v(x) = \beta(r(x))$ with $\beta : [0, +\infty) \to \mathbb{R}$, where β solves the problem

$$\begin{cases} \beta'' + (m-1)\coth(Hr)\beta' + \hat{a}\beta - \hat{b}(r)\beta^{\frac{m+2}{m-2}} \le 0, \\ \beta'(0) = 0, \ \beta' \ge 0, \ \beta(r) > 0. \end{cases} \tag{6.91}$$

Furthermore, we require

$$\liminf_{r \to +\infty} \beta(r) > 0. \tag{6.92}$$

Indeed, if β solves (6.91), it follows from (6.86), (6.87) and (6.88) that v is positive and satisfies

$$\Delta v + \tilde{a}(x)v - \tilde{b}(x)v^{\frac{m+2}{m-2}} = \beta'' + \Delta r \beta' + \tilde{a}(x)\beta - \tilde{b}(x)\beta^{\frac{m+2}{m-2}}$$

$$\le \beta'' + (m-1)H \coth(Hr)\beta' + \hat{a}\beta - \hat{b}(r)\beta^{\frac{m+2}{m-2}} \le 0$$

on $M \setminus B_{R_0}$ and

$$\liminf_{r \to +\infty} v(x) > 0. \tag{6.93}$$

The previous computation makes it clear that the requirement $\beta' \geq 0$ in (6.91) is vital for our approach to work. As we shall see this will be a consequence of the definition (6.88) of $\hat{b}$.

To solve (6.91) we choose an increasing divergent sequence $\{T_n\} \subset (0, +\infty)$. For a given n we look for a positive solution on $[0, T_n]$ of

$$\begin{cases} \beta'' + (m-1)\coth(Hr)\beta' + \hat{a}(r)\beta - \hat{b}(r)\beta^{\frac{m+2}{m-2}} = 0, \\ \beta'(0) = 0, \ \beta(T_n) = i \in \mathbb{N}. \end{cases} \qquad (6.94)$$

To motivate the argument that follows assume that $w > 0$ is a solution of the problem

$$\Delta w + \check{a}w - \check{b}w^{\frac{m+2}{m-2}} = 0 \qquad (6.95)$$

on an open set Ω of a manifold (N, g), and let $\varphi > 0$ be a C^2 solution of

$$\Delta\varphi + \check{a}\varphi + \lambda_1\varphi = 0 \quad \text{on } \Omega. \qquad (6.96)$$

If we define the conformally related metric

$$\tilde{g} = \varphi^{\frac{4}{m-2}}g, \qquad (6.97)$$

and use the transformation law (3.81) and equations (6.95), (6.96), an elementary computation shows that the function

$$u = \varphi^{-1}w, \qquad (6.98)$$

satisfies

$$\tilde{\Delta}u - \lambda_1\varphi^{-\frac{4}{m-2}}u - \check{b}u^{\frac{m+2}{m-2}} = 0 \text{ on } \Omega, \qquad (6.99)$$

where $\tilde{\Delta}$ denotes the Laplace-Beltrami operator of $\tilde{g}$. Hence, if u is a positive solution of (6.99), then

$$w = \varphi u$$

is a positive solution of (6.95).

Note now that (6.94) is precisely the expression of the equation

$$\Delta_{\mathbb{H}^m_{-H^2}}u + \hat{a}u - \hat{b}u^{\frac{m+2}{m-2}} = 0$$

on the m-dimensional hyperbolic space $\mathbb{H}^m_{-H^2}$ of constant curvature $-H^2$ when $u(x) = \beta(r(x))$ is radial.

We also observe that the first eigenvalue of the Dirichlet Laplacian on the ball B_R in $\mathbb{H}^m_{-H^2}$ satisfies

$$\lambda_1^{\Delta_{\mathbb{H}^m_{-H^2}}}(B_R) \geq \frac{(m-1)^2}{4}H^2[\coth(HR)]^2. \qquad (6.100)$$

Indeed,

$$\Delta_{\mathbb{H}^m_{-H^2}}r = (m-1)H\coth(Hr)$$

and a standard application of the divergence theorem shows that if ϕ is a smooth function with compact support in B_R, then

$$(m-1)H\coth(HR)\int \phi^2 \leq \int \phi^2 \Delta_{\mathbb{H}^m_{-H^2}} r$$

$$= -2\int \phi\langle \nabla\phi, \nabla r\rangle \leq 2\left(\int \phi^2\right)^{1/2}\left(\int |\nabla\phi|^2\right)^{1/2},$$

and the claim follows.

It follows that for some $\lambda_1 > 0$ we have a solution φ of

$$\begin{cases} \Delta_{\mathbb{H}^m_{-H^2}} + \widehat{a}\varphi + \lambda_1\varphi = 0 & \text{on } B_{T_{n+1}}, \\ \varphi = 0 & \text{on } \partial B_{T_{n+1}}, \ \varphi > 0 \text{ on } B_{T_{n+1}} \end{cases} \tag{6.101}$$

and the radial symmetry of the problem implies that the eigenfunction φ is radial.

The above discussion shows that in order to find a solution of (6.94) we may let (N, g) be the hyperbolic space $\mathbb{H}^m_{-H^2}$ and look for a radial solution of

$$\begin{cases} \widetilde{\Delta}u - \lambda_1\varphi^{-\frac{4}{m-2}}u - \widehat{b}u^{\frac{m+2}{m-2}} = 0 & \text{on } B_{T_n}, \ u > 0, \\ u \equiv \frac{i}{\varphi(T_n)} & \text{on } \partial B_{T_n}. \end{cases} \tag{6.102}$$

Since $\lambda_1 > 0$, the functional associated to (6.102) is strictly convex and unbounded at infinity. Thus problem (6.102) admits a radial solution u. It follows that problem (6.94) admits a radial solution $\beta_{n,i} > 0$ on $[0, T_n]$. In order to determine the sign of $\beta'_{n,i}$ we proceed as follows: we consider the function

$$v_-(r) = \Lambda e^{\frac{m-2}{4}P(r)}, \quad \Lambda = \left(\frac{m-1}{2}H\right)^{\frac{m-2}{2}}.$$

Using (6.88) and the definition of $\widehat{a}$ we compute

$$\widehat{a}v_- - \widehat{b}(r)v_-^{\frac{m+2}{m-2}} = v_-\left\{\frac{(m-1)^2}{4}H^2 - \Lambda^{\frac{4}{m-2}}\right\} \equiv 0. \tag{6.103}$$

With the aid of (6.89) we see that v_- solves

$$\begin{cases} \Delta_{\mathbb{H}^m_{-H^2}}v_- + \widehat{a}v_- - \widehat{b}(r)v_-^{\frac{m+2}{m-2}} \geq 0 & \text{on } B_{T_n}, \text{ (in fact on } \mathbb{H}^m_{-H^2}) \\ v_-(x) < i & \text{on } \partial B_{T_n} \end{cases} \tag{6.104}$$

provided i is sufficiently large. On the other hand, having set

$$z(x) = \beta_{n,i}(r(x)),$$

because of (6.94) we have

$$\begin{cases} \Delta_{\mathbb{H}^m_{-H^2}} z + \widehat{a}z - \widehat{b}(r)z^{\frac{m+2}{m-2}} = 0 & \text{on } B_{T_n}, \\ z(x) = i & \text{on } \partial B_{T_n}. \end{cases} \tag{6.105}$$

Applying Proposition 6.1,

$$\beta_{n,i}(r) \geq v_-(r) \quad \text{on } [0, T_n]. \tag{6.106}$$

We fix $r_0 \in [0, T_n]$ and we consider the function

$$y(t) = \widehat{b}(r_0)t^{\frac{m+2}{m-2}} - \widehat{a}t \quad \text{on } (0, +\infty).$$

We have

$$y'(t) = \frac{m+2}{m-2}\widehat{b}(r_0)t^{\frac{4}{m-2}} - \widehat{a} \geq \frac{1}{t}y(t).$$

We let $t_0 = v_-(r_0)$. Then, $y(t_0) = 0$ because of (6.103); thus $y'(t_0) \geq 0$. On the other hand

$$y''(t) = 4\frac{m+2}{(m-2)^2}\widehat{b}(r_0)t^{\frac{4}{m-2}-1} > 0$$

and we deduce that

$$y'(t) \geq 0 \quad \forall t \geq t_0.$$

Thus, the function y is nondecreasing on $[t_0, +\infty)$. By (6.106)

$$\beta_{n,i}(r_0) \geq v_-(r_0) = t_0 \quad \text{and } y(t_0) = 0.$$

It follows that

$$\widehat{b}(r_0)\beta_{n,i}(r_0)^{\frac{m+2}{m-2}} - \widehat{a}\beta_{n,i}(r_0) \geq 0.$$

Since $r_0 \in [0, T_n]$ was chosen arbitrarily, it follows that

$$\begin{cases} \beta''_{n,i} + (m-1)H\coth(Hr)\beta' \geq 0 & \text{on } [0, T_n], \\ \beta'_{n,i}(0) = 0. \end{cases}$$

Integration of the above immediately yields

$$\beta'_{n,i}(r) \geq 0 \quad \text{on } [0, T_n]. \tag{6.107}$$

Now let $i \to \infty$. Using Proposition 6.1 we see that the sequence of solutions $\{\beta_{n,i}\}$ of (6.94) is nondecreasing. Furthermore, it is uniformly bounded on compact subsets of $[0, T_n)$ because of Lemma 4.2. Thus, $\beta_{n,i}$ converges as $i \to +\infty$ to a solution β_n of the problem

$$\begin{cases} \beta''_n + (m-1)H\coth(Hr)\beta'_n + \widehat{a}\beta_n - \widehat{b}(r)\beta_n^{\frac{m+2}{m-2}} = 0 & \text{on } [0, T_n), \\ \beta'_n(0) = 0, \ \beta'_n(r) \geq 0, \ \beta_n(r) > 0,, \beta_n(r) \to +\infty \text{ as } r \to T_n^- \end{cases}$$

and, moreover,

$$\beta_n(r) \geq v_-(r). \tag{6.108}$$

A second application of Proposition 6.1 shows that the sequence $\{\beta_n\}$ is decreasing and therefore it converges to a solution β of

$$\begin{cases} \beta'' + (m-1)H\coth(Hr)\beta' + \widehat{a}\beta - \widehat{b}(r)\beta^{\frac{m+2}{m-2}} = 0 & \text{on } [0,+\infty), \\ \beta'(0) = 0,\ \beta' \geq 0,\ \beta(r) > 0. \end{cases} \tag{6.109}$$

Furthermore, because of (6.108), which is independent of n,

$$\beta(r) \geq v_-(r) \to +\infty \quad \text{as } r \to +\infty,$$

so that (6.92) is certainly satisfied. Thus $v(x) = \beta(r(x))$ satisfies (6.93) and

$$\begin{cases} \Delta v + \widetilde{a}(x)v - \widetilde{b}(x)v^{\frac{m+2}{m-2}} \leq 0 & \text{on } M \setminus B_{R_0}, \\ v(x) > 0, \end{cases} \tag{6.110}$$

so is the required radial supersolution on $M \setminus B_{R_0}$.

Step 3. Gluing the supersolutions.

First, given $w \in C^0(M) \cap W^{1,2}_{loc}$, for $\sigma = \frac{m+2}{m-2}$ to simplify the writing we define

$$b_w = w^{-\sigma}[\Delta w + \widetilde{a}(x)w] \tag{6.111}$$

in the weak sense. Note that for $E > 0$ constant

$$b_{Ew} = E^{1-\sigma}b_w. \tag{6.112}$$

Next, we let $\widetilde{u}_+$ be a positive function in $C^0(M) \cap H^{1,2}_{loc}$ such that

$$\widetilde{u}_+(x) = \begin{cases} u(x) & \text{on } B_{R_0-\varepsilon}, \\ v(x) & \text{on } M \setminus B_{R_0} \end{cases} \tag{6.113}$$

for some $\varepsilon > 0$ sufficiently small and where $u(x)$ has been defined in (6.85) while $v(x)$ has been defined in (6.110). We then set

$$u_+(x) = E\widetilde{u}_+(x). \tag{6.114}$$

Having chosen $\varepsilon > 0$ sufficiently small we can suppose by (6.75) that $\widetilde{b}(x) > 0$ on $M \setminus B_{R_0-\varepsilon}$. Next we note that w is a supersolution of (6.80) if and only if $b_w(x) \leq \widetilde{b}(x)$ on M. Using (6.112) and (6.113) we have:

i) On $M \setminus B_{R_0}$,

$$b_{u_+}(x) = E^{1-\sigma}b_{\widetilde{u}_+}(x) = E^{1-\sigma}b_v(x) \leq E^{1-\sigma}\widetilde{b}(x) \leq \widetilde{b}(x);$$

therefore u is a supersolution of (6.80) on $M \setminus B_{R_0}$ $\forall E \geq 1$, since $\widetilde{b}(x) > 0$ on $M \setminus B_{R_0}$.

ii) On $B_{R_0+\epsilon} \setminus \overline{B}_{R_0-\epsilon}$,

$$b_{u_+}(x) = E^{1-\sigma} b_{\tilde{u}_+}(x) \leq E^{1-\sigma} \cdot \sup_{\overline{B}_{R_0} \setminus B_{R_0-\epsilon}} b_{\tilde{u}_+}(x) \leq \inf_{\overline{B}_{R_0} \setminus B_{R_0-\epsilon}} \tilde{b}(x) \leq \tilde{b}(x),$$

provided $E \geq E_0 \geq 1$ is sufficiently large. This is possible since $\tilde{b}(x) > 0$ on $\overline{B}_{R_0} \setminus B_{R_0-\epsilon}$.

iii) On $B_{R_0-\epsilon}$, since u is a supersolution of (6.85) we have

$$b_{u_+}(x) = E^{1-\sigma} b_{\tilde{u}_+}(x) = E^{1-\sigma} b_u(x) = \leq -E^{1-\sigma}.$$

To obtain a supersolution on $B_{R_0-\epsilon}$ we need

$$-E^{1-\sigma} \leq \tilde{b}(x). \tag{6.115}$$

Thus u_+ is a subsolution if and only if $\tilde{b} \geq -E^{1-\sigma}$ where E is determined in ii) that is $b(x) \geq -\eta$ where $\eta = c_m E^{1-\sigma}$.

If this condition is satisfied, we have therefore found the required positive supersolution u_+ to (6.79) with the property that

$$\lim_{r(x) \to +\infty} u_+(x) = +\infty. \qquad \square$$

Remark. It should be noted that the upper bound on $b(x)$ affects only the behavior of the subsolution, and, as observed after the proof of Theorem 6.15, different upper bounds for $b(x)$ give rise to subsolutions with different asymptotic behaviour, and therefore to different asymptotic estimates for the solution. See the remark after the proof of Theorem 6.18

More generally, the positivity of $a(x)$ together with (6.77) and (6.78) may by replaced by any other set guaranteeing the existence of a subsolution of (6.79). For instance we can assume that

$$\lambda_1^{\Delta+a(x)}(M) < 0$$

to construct a subsolution v_- on a sufficiently large ball B_R as in the proof of Theorem 6.7. In this case v_- satisfies

$$\begin{cases} \Delta v_- + a(x)v_- - b_+(x)v_-^\sigma \geq 0, & v_- > 0 \text{ on } B_R, \\ v_- \equiv 0 \text{ on } \partial B_R. \end{cases}$$

Therefore

$$u_-(x) = \begin{cases} Dv_-(x) & \text{on } B_R, \\ 0 & \text{on } M \setminus B_R \end{cases}$$

yields the desired subsolution for $D \in (0, 1]$ sufficiently small.

6.4   The case of the Yamabe problem

It is clear that we can specialize the conclusions of the previous theorems to the case of the Yamabe equation

$$c_m \Delta u - S(x)u + K(x)u^{\frac{m+2}{m-2}} = 0$$

simply replacing $a(x) = -c_m^{-1}S(x)$ and $b(x) = -c^{-1}K(x)$. We will state such a result explicitly only in the case of Theorem 6.16, leaving the other cases to the interested reader.

It should also be pointed out that, in application of the theorems of previous sections to the Yamabe equation, it is often possible to replace the condition that $S(x)$ be everywhere negative with the assumption that it is nonpositive and strictly negative off a compact set. This is based on the following proposition (compare with Proposition 3.10).

Proposition 6.17. *Let $\Omega_0 \subset\subset \Omega_1 \subset (M, \langle\, , \rangle)$ be relatively compact domains with smooth boundaries. Assume that the scalar curvature $S(x)$ verifies*

$$S(x) \le c^2 h(x) \quad on\ M \setminus \Omega_0 \tag{6.116}$$

for some $c > 0$, $h \in C^0(M)$ nonpositive, $h(x) < 0$ on $\overline{\Omega}_1 \setminus \Omega_0$. Then there exists $\eta > 0$ such that, if

$$S(x) \le \eta \quad on\ M, \tag{6.117}$$

then there exists a complete conformal metric $\widetilde{\langle\, , \rangle}$ homothetic to $\langle\, , \rangle$ on $M \setminus \Omega_1$, whose scalar curvature $\widetilde{S}(x)$ satisfies

$$\widetilde{S}(x) \le \widetilde{c}^2 h(x) \quad on\ M, \tag{6.118}$$

for some $\widetilde{c} > 0$.

Proof. We elaborate some ideas of [AM88]. It is clear that, if $\widetilde{\langle\, , \rangle}$ is a metric conformal to $\langle\, , \rangle$ and homothetic to $\langle\, , \rangle$ on $M \setminus \Omega_1$, then (6.116) implies (6.118) on $M \setminus \Omega_1$ for some suitable $\widetilde{c} > 0$. Therefore, up to modifying $\widetilde{c}$, it is enough to show that $\widetilde{\langle\, , \rangle}$ can be chosen so that $\widetilde{S}(x) < 0$ on $\overline{\Omega}_1$. We fix a constant $\delta > 0$ and solve the Dirichlet problem

$$\begin{cases} c_m \Delta \psi = \delta & on\ \Omega_1, \\ \psi = 0 & on\ \partial\Omega_1. \end{cases} \tag{6.119}$$

Next, we choose a domain Ω_2 such that $\Omega_0 \subset\subset \Omega_2 \subset\subset \Omega_1$ and a smooth cut-off function $\xi : M \to [0,1]$ such that

$$\xi \equiv 1\ on\ \Omega_0 \quad and \quad \xi \equiv 0\ on\ M \setminus \Omega_2.$$

We define $\varphi = \xi\psi$ on M; now we choose a positive constant φ_0 so that

$$\varphi(x) + \varphi_0 > 0 \quad \text{on } M \tag{6.120}$$

and

$$S(x)(\varphi(x) + \varphi_0) - c_m \Delta\varphi \le -\frac{\delta}{2} \quad \text{on } \overline{\Omega}_1 \tag{6.121}$$

(note that, in order to achieve (6.121), we use $h(x) < 0$ on $\overline{\Omega}_1 \setminus \Omega_0$ and $S(x) \le \eta$). Finally, we set

$$\widetilde{\langle\,,\,\rangle} = (\varphi(x) + \varphi_0)^{\frac{4}{m-2}} \langle\,,\,\rangle.$$

Then $\widetilde{S}(x)$ is given by

$$\widetilde{S}(x) = (\varphi(x) + \varphi_0)^{-\frac{m+2}{m-2}} [S(x)(\varphi(x) + \varphi_0) - c_m\Delta\varphi(x)].$$

Now (6.120) and (6.121) imply $\widetilde{S}(x) < 0$ on $\overline{\Omega}_1$. $\qquad\square$

We now give the announced version of Theorem 6.16 for the Yamabe equation. It is a variation of Theorem 0.1 in [RRV97].

Theorem 6.18. *Let* $(M, \langle\,,\,\rangle)$ *be a complete manifold of dimension* $m \ge 3$, *scalar curvature* $S(x)$, *sectional curvature* $Sect_M$ *and let* $K(x) \in C^\infty(M)$. *Let* $R > 0$ *and suppose that the ball* B_R *centered at* o *does not intersect the cut locus of* o. *Suppose that*

> i) $\mathrm{Ric} \ge -(m-1)H^2 \quad$ *on* M,
>
> ii) $Sect_M \le -A^2 \quad$ *on* B_R, $\tag{6.122}$
>
> iii) $S(x) \le -m(m-1)C^2(1 + r(x))^\alpha \quad$ *on* M,

for some $0 \le A, C \le H$ *and* $-1 < \alpha \le 0$ *and* $0 < A \le B$ *such that*

$$H^2 < \left(1 + \frac{1}{m(m-2)}\right) A^2 [\coth(AR)]^2. \tag{6.123}$$

Assume also that

$$K(x) \ge -C_1 r(x)^\beta e^{Dr(x)^\theta} \tag{6.124}$$

on $M \setminus B_R$ *for some constants* $C_1, D > 0$, $\beta \in \mathbb{R}$, *and* $\theta < 1 + \alpha$. *Then, there exists* $\eta > 0$ *such that, if*

$$K(x) \le \eta \quad \text{on } M,$$

the metric $\langle\,,\,\rangle$ *can be pointwise conformally deformed to a new metric of scalar curvature* $K(x)$.

Proof. We only need to verify that the assumptions of Theorem 6.16 are satisfied with $a(x) = -c_m^{-1} S(x)$ and $b(x) = -c^{-1} K(x)$. First of all, the bound on the Ricci curvature implies

$$s(x) \ge -m(m-1)H^2,$$

so that, recalling that $c_m = 4\frac{m-1}{m-2}$,

$$-c_m^{-1}s(x) \le \frac{m(m-2)}{4}H^2 \le \frac{(m-1)^2}{4}H^2, \qquad (6.125)$$

and condition (6.74) in the statement of Theorem 6.16 is automatically satisfied.

Observe next that the sectional curvature condition and the fact that the ball B_R does not intersect the cut locus of o implies, by the Laplacian comparison theorem, that

$$\Delta r \ge (m-1)A\coth(Ar) \quad \text{on } B_R. \qquad (6.126)$$

Arguing as in the proof of Theorem 6.16, this in turn implies that the first eigenvalue of the Dirichlet Laplacian on B_R satisfies

$$\lambda_1^\Delta(B_R) \ge \frac{(m-1)^2}{4}A^2[\coth(AR)]^2,$$

whence, using (6.125),

$$\lambda_1^{\Delta - c_m^{-1}S(x)}(B_R) \ge \frac{(m-1)^2}{4}A^2[\coth(AR)]^2 - \frac{m(m-2)}{4}H^2 > 0$$

by condition (6.123).   □

Remark. The conclusion of the theorem in fact continues to hold even if equality holds in (6.123). This requires a different argument for the construction of the supersolution inside the ball B_R. We refer to [RRV97] for the details.

Note also that the upper bound (6.122) iii) for $S(x)$ and the lower bound (6.78) for $K(x)$ are used to construct the subsolution and determine a lower estimate for the solution u. As noted after the proof of Theorem 6.15, replacing (6.78) with

$$K(x) \ge -C_1 r(x)^\beta \quad r(x) \gg 1,$$

with $\beta < \alpha$ allows us to produce a subsolution, and therefore a solution which satisfy the estimate

$$u(x) \ge Cr(x)^\gamma, \quad r(x) \gg 1,$$

with $\gamma = \frac{m-2}{4}(\beta + \alpha)$. In particular if $\beta + \alpha = 2$, then

$$u(x) \ge Cr^{-\frac{m-2}{2}},$$

and we may guarantee that the conformal metric

$$\widetilde{\langle\,,\,\rangle} = u^{\frac{4}{m-2}}\langle\,,\,\rangle$$

is complete.

6.5 Appendix: the Monotone Iteration Scheme

In this section we describe the monotone iteration scheme, due to H. Amann (see [Ama76]), that has been one of the main tools to obtain our existence results. Our presentation follows closely that of D. H. Sattinger, [Sat73].

Let Ω be an open domain, and let $f(x, s)$ be a continuous function on $\Omega \times \mathbb{R}$, and consider the differential equation

$$\Delta u + f(x, u) = 0, \quad \text{in } \Omega. \tag{6.127}$$

We say that a function $u_+ \in W_{loc}^{1,2}(\Omega) \cap L_{loc}^\infty(\Omega)$ is a *supersolution* of (6.127) if it satisfies the differential inequality

$$\Delta u_+ + f(x, u_+) \le 0, \quad \text{weakly in } \Omega,$$

that is, for every test compactly supported function $\rho \ge 0$ in $W^{1,2}(\Omega)$,

$$-\int \langle \nabla u_+, \nabla \rho \rangle + f(x, u_+)\rho \le 0.$$

A function u_- is a *subsolution* if it satisfies the reverse differential inequality. Similarly, if Ω is bounded, and g is a continuous function on $\partial\Omega$, $u_+ \in W_{loc}^{1,2}(\Omega) \cap C(\overline{\Omega})$ is a *supersolution of the boundary value problem*

$$\begin{cases} \Delta u + f(x, u) = 0 & \text{in } \Omega, \\ u = g & \text{on } \partial\Omega, \end{cases} \tag{6.128}$$

if

$$\begin{cases} \Delta u_+ + f(x, u_+) \le 0 & \text{in } \Omega, \\ u_+ \ge g & \text{on } \partial\Omega. \end{cases}$$

The definition of a subsolution is obtained reversing the inequalities.

Theorem 6.19. *Let Ω be a relatively compact open domain with smooth boundary $\partial\Omega$ and let $f : \overline{\Omega} \times \mathbb{R} \to \mathbb{R}$ be a locally Hölder function such that $s \to f(x, s)$ is locally Lipschitz with respect to s uniformly with respect to x. Suppose that φ and $\psi \in C^{0,1}(\overline{\Omega})$ are respectively a subsolution and a supersolution of (6.128) satisfying*

$$\varphi \le \psi \text{ on } \overline{\Omega} \text{ and } \varphi \le g \le \psi \text{ on } \partial\Omega.$$

Suppose moreover that g extends to a $C^{2,\alpha}(\overline{\Omega})$ function, which we denote with the same letter. Then the boundary value problem (6.128) has a solution $u \in C^{2,\alpha}(\overline{\Omega})$ satisfying $\varphi \le u \le \psi$.

Proof. Since $f(x, s)$ is locally Lipschitz in s uniformly with respect to x in Ω, there exists $\lambda > 0$ such that

$$s \to f(x, s) + \lambda s = F(x, s)$$

is monotone increasing for every x in Ω and every s satisfying $\min_{\overline{\Omega}} \varphi \leq s \leq \max_{\overline{\Omega}} \psi$. For every function $w \in C^\alpha(\overline{\Omega})$ we let $v = Tw \in C^{2,\alpha}(\overline{\Omega})$ be the solution of the boundary value problem

$$\begin{cases} (\Delta - \lambda)v = -[f(x,w) + \lambda w] & \text{in } \Omega, \\ v = g & \text{on } \partial\Omega, \end{cases} \tag{6.129}$$

which exists and is unique by standard elliptic theory, (see, e.g., [GT01] Theorem 6.14). We claim that the operator T is monotone, that is, if $w_1 \leq w_2$ on Ω, then $Tw_1 \leq Tw_2$. Indeed, by the monotonicity of the function $F(x, s)$ the difference $\widetilde{v} = Tw_2 - Tw_1$ satisfies

$$\begin{cases} (\Delta - \lambda)\widetilde{v} = -[F(x,w_1) - F(x,w_2)] \leq 0 & \text{in } \Omega, \\ \widetilde{v} = 0 & \text{on } \partial\Omega, \end{cases}$$

and therefore by the minimum principle, $\widetilde{v} \geq 0$ in Ω, i.e., $v_2 = Tw_2 \geq v_1 = Tw_1$, as claimed.

Now we set $u_1^- = T\varphi$, $u_1^+ = T\psi$ and for every $k \geq 1$, $u_{k+1}^- = Tu_k^-$ and $u_{k+1}^+ = Tu_k^+$. Recalling that φ and ψ are respectively sub- and supersolutions of the boundary value problem (6.128) and arguing as above, the maximum principle and an induction argument show that

$$\varphi \leq u_1^- \leq u_2^- \leq \cdots \leq u_k^- \leq \cdots \leq u_k^+ \leq u_2^+ \leq u_1^+ \leq \psi.$$

Thus, as k tends to infinity $u_k^- \to u^-$ and $u_k^+ \to u^+$.

We claim that u^- and u^+ are in $C^{2,\alpha}(\overline{\Omega})$ and solve the boundary value (6.128). Indeed, the solution v of (6.129) satisfies L^p Sobolev estimates up to the boundary of the form

$$\|v\|_{W^{2,p}} \leq C\big(\|g\|_{W^{2,p}} + \|F(x,w)\|_{L^p}\big),$$

(see, e.g., [ADN59], Theorem 15.2) and it follows easily that for every p, the sequences $\{u_k^\pm\}$ converge in $W^{2,p}$ to a solution $(u^\pm)$ of the differential equation

$$\Delta u = f(x,u).$$

By Morrey's embedding lemma, see [GT01], Theorem 7.26, for every $p > n$ the space $W^{1,p}$ embeds continuously into $C^\beta(\overline{\Omega})$ and therefore, by choosing p large enough, the above convergence takes place in $C^\alpha(\overline{\Omega})$. Finally, by [ADN59], Theorem 7.3, the solution v of (6.129) satisfies Schauder's estimates up to the boundary

$$\|v\|_{C^{2,\alpha}} \leq C\big(\big(\|g\|_{C^{2,\alpha}} + \|F(x,w)\|_{C^\alpha}\big),$$

and therefore the sequences $u_k^\pm$ converge in $C^{2,\alpha}(\overline{\Omega})$ to solutions $u^\pm$ of (6.128), as required.   □

We conclude this section with the following result that extends to the situation at hand the well-known fact that the maximum of two subharmonic functions is subharmonic.

Theorem 6.20. *Let f be a continuous function on $\Omega \times \mathbb{R}$ and suppose that $u_i \in L^\infty_{loc}(\Omega) \cap W^{1,2}_{loc}(\Omega)$, $i = 1, 2$ are subsolutions of (6.127) on Ω. Then $= \max\{u_1, u_2\}$ is also a subsolution. Similarly if v_i are supersolutions of (6.127), then so is $v = \min\{v_1, v_2\}$.*

Proof. The proof is adapted from [Le98]. We only show that the maximum of subsolutions is a subsolution. The argument in the case of supersolutions is similar. Set $u = \max\{u_1, u_2\}$ and note that $u \in L^\infty_{loc}(\Omega) \cap W^{1,2}_{loc}(\Omega)$ and

$$\nabla u = \begin{cases} \nabla u_1 & \text{if } u_1 \geq u_2, \\ \nabla u_2 & \text{if } u_1 \leq u_2. \end{cases}$$

By assumption, for every $0 \leq \varphi \in C_0^\infty(\Omega)$ we have

$$\int \langle \nabla u_i, \nabla \varphi \rangle - f(x, u_i) \leq 0, \quad i = 1, 2.$$

Let $D \subset \Omega$ be a relatively compact open set with smooth boundary such that $\operatorname{supp} \varphi \subset D$, and set $D_1 = \{x \in D : u_1(x) > u_2(x)\}$, $D_o = \{x \in D : u_1(x) = u_2(x)\}$ and $D_2 = \{x \in D : u_1(x) < u_2(x)\}$.

Let also $\gamma : \mathbb{R} \to \mathbb{R}$ be such that $\gamma' \geq 0$, $\gamma(t) = 0$ if $t \leq 0$ and $= 1$ is $t \geq 1$, and let $\gamma_n(t) = \gamma(nt)$, so that $\gamma_n(t) = 0$ if $t \leq 0$, $= 1$ if $t \geq 1/n$ and $0 \leq \gamma'_n(t) \leq nM$ with $M = \sup \gamma'$. Let $w = u_2 - u_1$. Since $w \in W^{1,2}_{loc}(\Omega)$ there exists $w_n \in C_0^\infty(\Omega)$ such that $w_n \to w$ in $W^{1,2}(D)$. By passing to a subsequence, if necessary, we may also assume that $w_n \to w$ a.e. in D and $\|w_n - w\|_{W^{1,2}(D)} \leq \frac{1}{n^2}$.

Finally, set $\varphi_1 = (1 - \gamma_n(w_n))\varphi$ and $\varphi_2 = \gamma_n(w_n)\varphi$, so that $0 \leq \varphi_i \in C_0^\infty(D)$. We claim that for a.e. $x \in D_2$ there exists $n_0 = n_0(x)$ such that $w_n(x) > 1/n$ for every $n \geq n_0$. Indeed, since $w > 0$ on D_2, there exists $n_1 = n_1(x)$ such that $w(x) > 2/n_1$. Moreover, $w_n \to w$ a.e., for a.e. x there exists $n_0 = n_0(x) \geq n_1$ such that for every $n \geq n_0$ $|w_n(x) - w(x)| \leq 1/n_1$, and therefore, for $n \geq n_0 \geq n_1$,

$$w_n(x) \geq w(x) - |w_n(x) - w(x)| \geq \frac{2}{n_1} - \frac{1}{n_1} = \frac{1}{n_1} \geq \frac{1}{n}.$$

Thus, for a.e. x and every $n \geq n_0$ we have $\gamma_n(w_n(x)) = 1$ and we conclude that $\gamma_n(w_n) \to 1$ a.e. in D_2. Similarly, since $w(x) < 0$ in D_1, we see that $\gamma_n(w_n(x)) \to 0$ a.e. in D_1.

Since $0 \leq \varphi_i \in C_0^\infty(\Omega)$, and u_i are subsolutions we have

$$\int \langle \nabla u_i, \nabla \varphi_i \rangle - f(x, u_i) \leq 0,$$

whence, adding the two inequalities, and using the definition of φ_i, a computation yields

$$
\begin{aligned}
0 \geq &\int \langle \nabla u_1, (1 - \gamma_n(w_n)) \nabla \varphi - \gamma_n'(w_n) \varphi \nabla w_n \rangle \\
&+ \int \langle \nabla u_2, \gamma_n(w_n) \nabla \varphi + \gamma_n'(w_n) \varphi \nabla w_n \rangle \\
&- \int f(x, u_1)(1 - \gamma_n(w_n)) \nabla \varphi + f(x, u_2) \gamma_n(w_n) \nabla \varphi \\
= &\int \langle \nabla u_2 - \nabla u_1, \gamma_n(w_n) \nabla \varphi \rangle + \int \langle \nabla u_2 - \nabla u_1, \gamma_n'(w_n) \varphi \nabla w_n \rangle \\
&+ \int [f(xu_1) - f(x, u_2)] \gamma_n(w_n) \varphi + \int \langle \nabla u_1, \nabla \varphi \rangle - f(x, u_1) \varphi \\
= &\, I_1(n) + I_2(n) + I_3(n) + I_4.
\end{aligned}
\tag{6.130}
$$

Observe now that, since $\nabla u_i \in L^2(D)$, $\nabla \varphi \in L^\infty(D)$, $\nabla u_1 = \nabla u_2$ on D_o, $0 \leq \gamma_n(w_n) \leq 1$, and $\gamma_n(w_n)$ tends to 1 a.e. on D_2 and to 0 a.e. on D_1, , letting $n \to \infty$, by dominated convergence we have

$$
I_1(n) = \int_{D_1 \cup D_2} \langle \nabla u_2 - \nabla u_1, \gamma_n(w_n) \rangle \nabla \varphi \to \int_{D_2} \langle \nabla u_2 - \nabla u_1, \nabla \varphi \rangle.
$$

Similarly, since $f(x, u_i)$ is bounded on D, $f(x, u_1) = f(x, u_2)$ on D_o and φ is compactly supported

$$
I_4(n) = \int_{D_1 \cup D_2} [-f(x, u_2) + f(x, u_1)] \gamma_n(w_n)) \varphi \to \int_{D_2} [-f(x, u_2) + f(x, u_1)] \gamma_n(w_n)) \varphi.
$$

To deal with I_2 note that

$$
\begin{aligned}
I_2 &= \int \langle \nabla u_2 - \nabla u_1, \nabla w_n \rangle \gamma_n'(w_n) \varphi = \int \langle \nabla w, \nabla w + (\langle \nabla w_n - \nabla w \rangle) \gamma_n'(w_n) \varphi \\
&\geq \int \langle \nabla w, \nabla w_n - \nabla w \rangle \gamma_n'(w_n) \varphi \geq - \int |\nabla w| |\nabla w_n - \nabla w| \gamma_n(w_n) \varphi.
\end{aligned}
$$

Since

$$
\gamma_n'(w_n) \leq Mn, \quad \text{and} \quad \|w_n - w\|_{W^{1,2}(D)} \leq \frac{1}{n^2},
$$

the integral on the rightmost side is bounded above by $Mn \|\varphi\|_\infty \|\nabla \varphi\|_{L^2(D)} \frac{1}{n^2} = \frac{C}{n}$ and we deduce that

$$
\liminf_n I_2(n) \geq 0.
$$

Thus, letting $n \to +\infty$ in (6.130) we obtain

$$0 \geq \lim_n I_1(n) + \liminf_n I_2(n) + \lim_n I_3(n) + I_4$$

$$\geq \int_{D_2} \langle \nabla u_2 - \nabla u_1, \nabla \varphi \rangle + [-f(x, u_2) + f(x, u_1)\varphi] + \int_D \langle \nabla u_1, \nabla \varphi \rangle - f(x, u_1)\varphi$$

$$= \int_{D_2} \langle \nabla u_2, \nabla \varphi \rangle - f(x, u_2)\varphi + \int_{D \setminus D_2} \langle \nabla u_1, \nabla \varphi \rangle - f(x, u_1)\varphi.$$

Recalling that

$$u = \max\{u_1, u_2\} = \begin{cases} u_2 & \text{on } D_2, \\ u_1 & \text{on } D \setminus D_2, \end{cases} \quad \text{and that} \quad \nabla u = \begin{cases} \nabla u_2 & \text{on } D_2, \\ \nabla u_1 & \text{on } D \setminus D_2, \end{cases}$$

the last inequality amounts to

$$0 \geq \int \langle \nabla u, \nabla \varphi \rangle - f(x, u)\varphi,$$

as required to show that u is a subsolution of (6.127). $\qquad\qquad \square$

Chapter 7

Some special cases

In this final chapter we present some special cases where one can use particular techniques to relax assumptions. Typically this happens in Euclidean and hyperbolic spaces, and more generally in the case of *models*, or manifolds with some special symmetry. For the models, the main advantage is to have a precise expression for Δr which is available, on general manifolds, only in the form of an upper or a lower bound under certain assumptions on the curvature and the cut locus.

We provide here refined techniques and a few results that, as a side product, show the degree of sharpness of the general theory and of the methods that we have developed dealing with generic complete Riemannian manifolds.

As a final remark we underline that, here and there, in this chapter we explicitly link our geometric point of view to more familiar Euclidean procedures and tools (for instance, the Rellich-Pohozaev formula), as shown below in detail.

7.1 A nonexistence result

We begin by showing that under appropriate assumptions on $a(x)$ and $b(x)$ there are no positive solutions u on M of

$$\Delta u + a(x)u - b(x)u^\sigma \geq 0.$$

To prove this fact we shall use a technique that does not require the determination of *a priori* estimates on u. The idea is to perform a sort of radialization and then apply ODE methods. However, as we will explain later, this has some geometrical heavy side effects. We need a collection of preliminary technical results; the next lemma is motivated by the work of K.S. Cheng and J. T. Lin, [CL87].

Lemma 7.1. *Let $0 < R_0 \leq R_1 \leq R_2 \leq +\infty$, $\sigma > 1$ and assume that the functions $l, \psi, b \in C^0([R_0, +\infty))$ satisfy*

$$\begin{cases} \text{i) } l(t) > 0 \text{ on } [R_0, +\infty); \quad \text{ii) } \psi(t) > 0 \text{ on } [R_0, +\infty); \\ \text{iii) } tb(t) \notin L^1(+\infty); \quad \text{iv) } tb(t) > 0 \text{ and nonincreasing on } [R_1, +\infty). \end{cases} \tag{7.1}$$

Suppose that

$$\liminf_{t \to +\infty} \frac{l(t)\psi(t)^\sigma}{tb(t)} > 0. \tag{7.2}$$

If α is a positive solution of

$$\alpha(s) \geq \psi(s)\left\{ c_1 + c_2 \int_{R_o}^{s} \left(1 - \frac{t}{s} \right) l(t)\alpha(t)^\sigma \, dt \right\} \tag{7.3}$$

on $[R_0, R_2)$ for some constant $c_1, c_2 > 0$, then $R_2 < +\infty$.

Proof. We argue by contradiction and let $R_2 = +\infty$. We set

$$\beta(s) = \frac{\alpha(s)}{\psi(s)}$$

so that (7.3) implies that β satisfies

$$\beta(s) \geq \left\{ c_1 + c_2 \int_{R_o}^{s} \left(1 - \frac{t}{s} \right) l(t)\psi(t)^\sigma \beta(t)^\sigma \, dt \right\} \quad \forall s \geq R_0.$$

Because of (7.2), there exist $c_3 > 0$ and $R_3 \geq R_0$ such that

$$\beta(s) \geq c_1 + c_3 \int_{R}^{s} \left(1 - \frac{t}{s} \right) tb(t)\beta(t)^\sigma \, dt \quad \forall s \geq R \geq R_3. \tag{7.4}$$

We choose $R_4 > \max\{R_1, R_3\}$ and we define

$$\eta(t) = \int_{R_4}^{t} vb(v) \, dv + R_4^2 \, b(R_4). \tag{7.5}$$

We observe that

$$\eta(t) \geq 0$$

and

$$i) \, \eta'(t) = t \, b(t) > 0; \quad ii) \, \eta(t) \to +\infty \text{ as } t \to +\infty \tag{7.6}$$

because of assumptions (7.1) iii), iv). We let $R \geq R_4$ and we consider

$$\eta : [R, s] \to [\nu = \eta(R), \xi = \eta(s)].$$

We indicate with $t(\eta)$ its inverse. We set

$$\gamma(\eta) = \beta(t(\eta))$$

and we perform a change of variable in (7.4) to obtain

$$\gamma(\xi) \geq c_1 + c_3 \int_{\nu}^{\xi} \left[1 - \frac{t(\eta)}{t(\xi)} \right] \gamma(\eta)^\sigma \, d\eta. \tag{7.7}$$

Next, we observe that for η sufficiently large, say $\eta \geq \eta_0$, $\eta^{-1}t(\eta)$ is nondecreasing. Indeed, it is enough to show that

$$\eta t'(\eta) - t(\eta) \geq 0, \quad \eta \geq \eta_0.$$

This, in turn, is equivalent to showing that

$$\int_{R_4}^{t} vb(v)\, dv + R_4^2 b(R_4) - t^2 b(t) \geq 0 \tag{7.8}$$

for t sufficiently large. But (7.1) iv) and our choice of R_4 imply (7.8) for $t \geq R_4$. From (7.7) we deduce

$$\gamma(\xi) \geq c_1 + c_3 \int_{\nu}^{\xi} \left(1 - \frac{\eta}{\xi}\right) \gamma(\eta)^\sigma\, d\eta \tag{7.9}$$

for all $\xi > \nu \geq \eta(R_4)$. With the aid of (7.6) ii) and (7.1) we then have: $\forall \nu > \eta(R_4)$ and $\forall \xi \in [\nu, 2\nu]$,

$$\gamma(\xi) \geq c_1 + c_3 \frac{1}{2\nu} \int_{\nu}^{\xi} (\xi - \eta)\gamma(\eta)^\sigma\, d\eta. \tag{7.10}$$

To complete the proof we define

$$g(\xi) = c_1 + c_3 \frac{1}{2\nu} \int_{\nu}^{\xi} (\xi - \eta)\gamma(\eta)^\sigma\, d\eta \tag{7.11}$$

and observe that g satisfies

$$\begin{cases} \text{i) } g(\nu) = c_1, \quad \text{ii) } g'(\nu) = 0 \\ \text{iii) } g'(\xi) = \frac{c_3}{2\nu} \int_{\nu}^{\xi} \gamma(\eta)^\sigma\, d\eta, \quad \text{iv) } g''(\xi) = \frac{c_3}{2\nu}\gamma(\xi)^\sigma \geq \frac{c_3}{2\nu} g(\xi)^\sigma \end{cases} \tag{7.12}$$

where in (7.12) iv) we have been using (7.10) and (7.11). From (7.12) we get

$$\left([g'(\xi)]^2\right)' = 2g'(\xi)g''(\xi) \geq \frac{c_3}{\nu} g'(\xi)g(\xi)^\sigma = \frac{c_3}{(\sigma+1)\nu}\left(g(\xi)^{\sigma+1}\right)'. \tag{7.13}$$

Therefore, integrating (7.13) over $[\nu, \xi]$, with the aid of (7.12) ii) we deduce

$$g'(\xi) \geq \left[\frac{c_3}{(\sigma+1)\nu}\right]^{\frac{1}{2}} \left[g(\xi)^{\sigma+1} - g(\nu)^{\sigma+1}\right]^{\frac{1}{2}}. \tag{7.14}$$

Integrating (7.14) again over $[\nu, \xi]$ we have

$$\int_{\nu}^{\xi} \frac{g'(x)\, dx}{[g(x)^{\sigma+1} - g(\nu)^{\sigma+1}]^{1/2}} \geq \left[\frac{c_3}{(\sigma+1)\nu}\right]^{1/2} (\xi - \nu). \tag{7.15}$$

Next, we perform the change of variable

$$u = \frac{g(x)}{g(\nu)}$$

and we rewrite (7.15) in the form

$$\int_1^{\frac{g(\xi)}{c_1}} \frac{du}{\sqrt{u^{\sigma+1} - 1}} \geq \sqrt{\frac{c_1^{\sigma-1} c_3}{(\sigma+1)\nu}} (\xi - \nu). \tag{7.16}$$

But $\sigma > 1$; hence $(u^{\sigma+1} - 1)^{-\frac{1}{2}} \in L^1([1, +\infty))$. On the other hand, choosing $\xi = 2\nu$ in (7.16), we immediately obtain a contradiction by taking ν sufficiently large. $\qquad\square$

To state the next result we introduce some notation. Given $\sigma > 1$ and a nonnegative $f \in C^0(M)$, we set

$$\bar{f}(r) = \frac{1}{\mathrm{vol}(\partial B_r)} \int_{\partial B_r} f, \quad r \in [0, +\infty); \tag{7.17}$$

$$\bar{f}_\sigma(r) = \left\{ \frac{1}{\mathrm{vol}(\partial B_r)} \int_{\partial B_r} f^{\frac{1}{1-\sigma}} \right\}^{1-\sigma}, \quad r \in [0, +\infty) \tag{7.18}$$

respectively for the spherical mean (as in the previous chapters) and the weighted spherical mean of f, with the convention that $\bar{f}_\sigma(r) = 0$ in case the integral in (7.18) is infinite. Furthermore $B_r = B_r(o)$ for some fixed origin $o \in M$ as before. We prove

Lemma 7.2. *Let $(M, \langle\,,\,\rangle)$ be a complete manifold of dimension $m \geq 2$ and assume*

$$\Delta r \geq (m-1)z(r) \tag{7.19}$$

for some $z(r) \in C^0([0, +\infty))$ in the weak sense on M. Let $a(x), b(x) \in C^0(M)$ satisfy

$$a(x) \leq p(r(x)) \quad on\ M\backslash\{o\} \tag{7.20}$$

for some $p(t) \in C^0((0, +\infty))$,

$$b(x) \geq 0 \quad on\ M. \tag{7.21}$$

Let $u \in C^2(M)$ be a positive solution of the differential inequality

$$\Delta u + a(x)u - b(x)u^\sigma \geq 0, \quad \sigma > 1, \quad on\ M. \tag{7.22}$$

Assume that, for any fixed $0 < \delta < s$, there exists a function $\varphi \in C^2([\delta, s])$ with the following properties:

$$\varphi'' + (m-1)z(t)\varphi' + p(t)\varphi \geq 0 \quad on\ [\delta, s], \tag{7.23}$$

$$\text{i) } \varphi'(s) = \frac{1}{\text{vol}(\partial B_s)}; \quad \text{ii) } \varphi'(t) \geq 0 \ \text{on } [\delta, s] \tag{7.24}$$

$$\varphi(s) = 0 \tag{7.25}$$

$$-\frac{\varphi(\delta)}{\varphi'(\delta)} \leq C \tag{7.26}$$

for some constant $C > 0$ independent of δ and s. Then, up to choosing δ sufficiently small (and consequently φ), there exists a constant $C_1 > 0$, independent of δ and s, such that for all $R \in [\delta, s]$ $\bar{u}(s)$ satisfies

$$\bar{u}(s) \geq C_1 \, \varphi'(\delta) \delta^{m-1} - \int_R^s \varphi(t) \, \text{vol}(\partial B_t) \, \bar{b}_\sigma(t) \, \bar{u}(t)^\sigma \, dt. \tag{7.27}$$

Proof. We fix $0 < \delta < s$, φ and $R \in [\delta, s]$. We consider the function $\varphi(r(x))$ on M; because of (7.19), (7.20) and (7.24) ii) we have

$$\Delta\varphi(r)(x) \geq -a(x)\varphi(r(x)) \tag{7.28}$$

in the weak sense on $\bar{B}_s \backslash B_\delta$. The second Green's identity, (7.24) i), (7.25) together with the definition (7.17) of $\bar{u}$ give

$$\bar{u}(s) = \int_{B_s \backslash B_R} u\Delta\varphi - \varphi\Delta u + \int_{B_R \backslash B_\delta} u\Delta\varphi - \varphi\Delta u + \int_{\partial B_\delta} u\langle\nabla\varphi, \nabla r\rangle - \varphi\langle\nabla u, \nabla r\rangle. \tag{7.29}$$

Observe that (7.22), (7.28), (7.21) and positivity of u yield

$$\int_{B_R \backslash B_\delta} u\Delta\varphi - \varphi\Delta u \geq 0. \tag{7.30}$$

Similarly

$$\int_{B_s \backslash B_R} u\Delta\varphi - \varphi\Delta u \geq \int_{B_s \backslash B_R} \varphi(r)b(x)u^\sigma. \tag{7.31}$$

Next, a straightforward application of Hölder's inequality shows

$$\int_{\partial B_t} b(x)u^\sigma \geq \bar{b}_\sigma(t)\bar{u}(t)^\sigma \, \text{vol}(\partial B_t) \tag{7.32}$$

where $\bar{b}_\sigma(t)$ has been defined in (7.18). Putting together (7.31), (7.32) and using the co-area formula (1.87), we finally get

$$\int_{B_s \backslash B_R} u\Delta\varphi - \varphi\Delta u \geq -\int_R^s \varphi(t)\bar{b}_\sigma(t)\bar{u}(t)^\sigma \, \text{vol}(\partial B_t) \, dt. \tag{7.33}$$

We now take care of the boundary term in (7.29). On the one hand,

$$\int_{\partial B_\delta} u\langle\nabla\varphi, \nabla r\rangle - \varphi\langle\nabla u, \nabla r\rangle = \varphi'(\delta)\bar{u}(\delta) \, \text{vol}(\partial b_t) - \varphi(\delta)\int_{B_\delta} \Delta u. \tag{7.34}$$

On the other hand,

$$\operatorname{vol}(\partial B_\delta) \asymp \delta^{m-1}; \quad \operatorname{vol}(B_\delta) \asymp \delta^m \quad \text{as } \delta \downarrow 0^+. \tag{7.35}$$

Using positivity of u, (7.35), and (7.34) it is easy to see that, up to choosing $0 < \delta \leq \delta_0$ sufficiently small, there exists a constant $C_1 > 0$ independent of δ and s such that

$$\int_{\partial B_\delta} u \langle \nabla\varphi, \nabla r \rangle - \varphi \langle \nabla u, \nabla r \rangle \geq C_1 \varphi'(\delta)\delta^{m-1}. \tag{7.36}$$

Estimate (7.27) now follows from (7.29), (7.30), (7.33) and (7.36).   □

Remark. In case u is a radial solution of

$$\begin{cases} u'' + (m-1)z(r)u' + a(r)u - b(r)u^\sigma \geq 0, \ \sigma > 1 \quad \text{on } [0, +\infty), \\ u'(0) = 0, \ u(0) = u_0 > 0 \end{cases}$$

with $z(r) = \frac{h'(r)}{h(r)}$, (M, h) a model (see Chapter 4), (7.27) becomes

$$u(s) \geq C_1 \varphi'(\delta)\delta^{m-1} - \int_R^s \varphi(t) h(t)^{m-1} b(t) u(t)^\sigma \, dt. \tag{7.37}$$

We apply the above results to produce a nonexistence theorem for complete Riemannian manifolds with sectional curvature bounded above by a negative constant $-B^2$, $B > 0$. The case $B = 0$ can be dealt with similarly and later, in the special case of Euclidean space, we shall apply this technique to differential inequalities of a certain type.

Theorem 7.3. *Let $(M, \langle\,,\,\rangle)$ be a complete manifold of dimension $m \geq 2$ with a pole $o \in M$ and suppose*

$$\operatorname{Riem} \leq -B^2 \tag{7.38}$$

for some constant $B > 0$. Let $a(x), b(x) \in C^0(M)$ satisfy

$$a(x) \leq AB^2 \coth(Br(x)) \text{ on } M, \text{ with } A \leq \frac{(m-1)^2}{4}, \tag{7.39}$$

$$b(x) \geq 0 \quad \text{on } M. \tag{7.40}$$

Suppose that for some constant $\sigma > 1$,

$$\bar{b}_\sigma(t) \geq \begin{cases} C \dfrac{b(t)\operatorname{vol}(\partial B_t)^{\sigma-1}}{t^{\sigma-1}e^{\frac{m-1}{2}B(\sigma-1)t}} & \text{in case } A = \frac{(m-1)^2}{4}, \\[4mm] C \dfrac{tb(t)\operatorname{vol}(\partial B_t)^{\sigma-1}}{e^{B\left(\frac{m-1}{2}+\sqrt{\frac{(m-1)^2}{4}-A^*}\right)(\sigma-1)t}} & \text{in case } A < \frac{(m-1)^2}{4} \end{cases} \tag{7.41}$$

for $t \gg 1$, some constant $C > 0$ and with

$$A^* = \max\{0, A\}, \tag{7.42}$$

where

i) $tb(t) \notin L^1(+\infty)$, ii) $tb(t) > 0$ *and nonincreasing at infinity.* (7.43)

Then the differential inequality

$$\Delta u + a(x)u - b(x)u^\sigma \geq 0 \tag{7.44}$$

has no positive C^2-solution on M.

Remark. In the proof of Theorem 7.3 we shall make use of the Hessian comparison theorem (see Chapter 1) which in the above assumptions yields

$$\Delta r \geq (m-1)B \coth Br \tag{7.45}$$

pointwise on $M \backslash \{o\}$.

Remark. We would like to comment on the case of hyperbolic space $\mathbb{H}^m_{-B^2}$ and Yamabe equation

$$c_m \Delta u + m(m-1)B^2 u + k(x)u^{\frac{m+2}{m-2}} = 0,$$

where $c_m = 4\frac{m-1}{m-2}$. Here,

$$\mathrm{vol}(\partial B_t) \sim C_1 e^{(m-1)Bt} \quad \text{as } t \to +\infty$$

for some $C_1 > 0$; furthermore,

$$a(x) = \frac{m(m-2)}{4}B^2, \quad b(x) = -\frac{1}{c_m}k(x).$$

Thus,

$$a(x) \leq AB^2 \coth(Br(x)), \quad \text{with } A = \frac{m(m-2)}{4} < \frac{(m-1)^2}{4}.$$

It follows that the requirement (7.41) on $\bar{b}_\sigma(t)$ becomes

$$\bar{b}_\sigma(t) \geq C t b(t)e^{2Bt}, \ t \gg 1, \ C > 0.$$

Now admissible choices of $b(t)$ are

$$b(t) = \frac{1}{t^2}, \quad b(t) = \frac{1}{t^2 \log t}, \quad b(t) = \frac{1}{t^2 \log t(\log \log t)}, \quad \dots, t \gg 1.$$

Thus, for instance, the above becomes

$$\bar{b}_\sigma(t) \geq \frac{e^{2Bt}}{t}, \quad t \gg 1, C > 0.$$

This is implied by

$$b(x) \geq C\frac{e^{2Br(x)}}{r(x)} \quad \text{for } r(x) \gg 1,\, C > 0. \tag{7.46}$$

This is expressed, in terms of $k(x)$, as

$$k(x) \leq -C\frac{e^{2Br(x)}}{r(x)} \quad \text{for } r(x) \gg 1,\, C > 0. \tag{7.47}$$

Inequality (7.47) relaxes the second of (4.86) in the case of m-dimensional hyperbolic space $\mathbb{H}^m_{-B^2}$. However, assumption (7.38) is definitely a stronger requirement than (4.84) of Theorem 4.8.

Note that (7.47) is independent of the dimension of $\mathbb{H}^m_{-B^2}$.

Proof of Theorem 7.3. We reason by contradiction and we assume that (7.44) has a positive C^2-solution u on M. Note that, by the first remark after the statement of the theorem, (7.45) holds on M. Next, we begin by considering the case $A = \frac{(m-1)^2}{4}$. We fix $0 < \delta \leq R < s$ and define

$$\varphi(t) = \frac{t - s}{\mathrm{vol}(\partial B_s)} e^{\frac{m-1}{2}B(s-t)} \quad \text{on } [\delta, s].$$

Then φ satisfies (7.23) of Lemma 7.2 with $z(t) = B \coth Bt$. (7.24), (7.25) and (7.26) are verified up to choosing in (7.26) $C = \frac{m-1}{2}B$. From Lemma 7.2 it follows that $\bar{u}(s)$ satisfies (7.27), that is, for some constant $C_1 > 0$ independent of δ and s,

$$\bar{u}(s) \geq C_1\varphi'(s)\delta^{m-1} - \int_R^s \varphi(t)\,\mathrm{vol}(\partial B_t)\bar{b}_\sigma(t)\bar{u}(t)^\sigma \, dt \tag{7.48}$$

for all $R \in [\delta, s)$. On the other hand,

$$\varphi'(\delta) = \frac{e^{\frac{m-1}{2}B(s-\delta)}}{\mathrm{vol}(\partial B_s)}\left(1 + (s - \delta)\frac{m-1}{2}B\right) \geq C_2\frac{s\,e^{\frac{m-1}{2}Bs}}{\mathrm{vol}(\partial B_s)}$$

with $C_2 > 0$ independent of s and $\delta \in (0, \delta_0]$ for some $\delta_0 > 0$ fixed. Furthermore,

$$-\varphi(t) = \frac{s\,e^{\frac{m-1}{2}Bs}}{\mathrm{vol}(\partial B_s)}\left(1 - \frac{t}{s}\right)e^{-\frac{m-1}{2}Bt}.$$

Therefore we choose

$$\psi(s) = \frac{s\,e^{\frac{m-1}{2}Bs}}{\mathrm{vol}(\partial B_s)}, \quad l(t) = e^{-\frac{m-1}{2}Bt}\,\mathrm{vol}(\partial B_t)\bar{b}_\sigma(t)$$

to obtain

$$\bar{u}(s) \geq \psi(s)\left\{C_3 + C_4\int_R^s \left(1 - \frac{t}{s}l(t)\bar{u}(t)^\sigma \, dt\right)\right\}$$

that is, (7.3) of Lemma 7.1. Next we observe that $\psi(t) > 0$, $l(t) > 0$ for t sufficiently large (so that we shall choose R sufficiently large in the above). Furthermore assumption (7.41) implies that (7.2) is satisfied. From Lemma 7.1 it follows that $\bar{u}(s)$ is defined on a finite interval; contradiction.

The case $A < \frac{(m-1)^2}{4}$ is treated similarly choosing on $[\delta, s]$, $0 < \delta < s$, the function φ as follows:

$$\varphi(t) = -\frac{e^{\alpha B(s-t)}\left[1 - e^{-\alpha B(s-t)}\right]}{\alpha\,\mathrm{vol}(\partial B_s)}$$

with

$$\alpha = \frac{m-1}{2} + \sqrt{\frac{(m-1)^2}{4} - A^*}. \qquad\qquad \square$$

Remark. The proof above shows how (7.41) is sensitive to the coefficient A of (7.39). One can also explore the dependence of (7.41) on the possible decay of the right-hand side of (7.39) at infinity. This is a very delicate question to which we shall return later. Note that, given (7.39), (7.41) is sharp for $A \in \left[0, \frac{(m-1)^2}{4}\right]$.

We shall now apply the above result in a special geometrical setting.

Consider the complete manifold $\mathbb{S}^{m-p-1} \times \mathbb{H}^{p+1}$ with the product metric (recall that $\mathbb{S}^m$ is the standard sphere of dimension m and that $\mathbb{H}^m$ is the standard hyperbolic space of dimension m: see also Chapter 2); a computation shows that its scalar curvature $S(x)$ is

$$S(x) = (m-1)(m-2-2p). \tag{7.49}$$

Therefore, $S(x)$ is

$$\text{positive for} \quad 0 \le p < \frac{m-2}{2}, \tag{7.50}$$

$$\text{zero for} \quad p = \frac{m-2}{2}, \tag{7.51}$$

$$\text{negative for} \quad \frac{m-2}{2} < p \le m-2. \tag{7.52}$$

Let (θ, x) denote the variables in $\mathbb{S}^{m-p-1} \times \mathbb{H}^{p+1}$; we want to study the Yamabe equation

$$c_m \Delta u - (m-1)(m-2-2p)u + k(\theta, x)u^{\frac{m+2}{m-2}} = 0, \tag{7.53}$$

$u > 0$ on $\mathbb{S}^{m-p-1} \times \mathbb{H}^{p+1}$. We restrict our attention to the special case $k(\theta, x) = k(x)$.

Given $w = w(\theta, x)$ define

$$\widehat{w}(x) = \frac{1}{\omega_{m-p-1}} \int_{\mathbb{S}^{m-p-1}} w(\theta, x)\, d\theta, \tag{7.54}$$

where $d\theta$ is the volume element on $\left(\mathbb{S}^{m-p-1}, \mathrm{can}\right)$ and $\omega_{m-p-1} = \mathrm{vol}\left(\mathbb{S}^{m-p-1}\right)$. Note that, for all $\sigma > 1$, using Hölder's inequality we have

$$\widehat{w}(x)^\sigma \le \frac{1}{\omega_{m-p-1}}\widehat{w^\sigma}(x). \tag{7.55}$$

The following result is straightforward.

Lemma 7.4. *Let u be a positive solution of (7.53) on $\mathbb{S}^{m-p-1} \times \mathbb{H}^{p+1}$. If $k(x) \le 0$, then $\widehat{u}$ satisfies*

$$c_m \Delta_{\mathbb{H}^{p+1}} \widehat{u} - (m-1)(m-2-2p)\widehat{u} + \omega_{m-p-1}k(x)\widehat{u}^{\frac{m+2}{m-2}} \ge 0. \tag{7.56}$$

Proof. We observe, as one can easily check, that

$$\Delta = \Delta_{\mathbb{S}^{m-p-1}} + \Delta_{\mathbb{H}^{p+1}}.$$

Therefore, using Stokes' theorem we have

$$\widehat{\Delta u} = \widehat{\Delta_{\mathbb{S}^{m-p-1}}u} + \widehat{\Delta_{\mathbb{H}^{p+1}}u} = \frac{1}{\omega_{m-p-1}} \int_{\mathbb{S}^{m-p-1}} \Delta_{\mathbb{S}^{m-p-1}} u(\theta, x)\, d\theta$$

$$+ \widehat{\Delta_{\mathbb{H}^{p+1}}u} = \Delta_{\mathbb{H}^{p+1}}\widehat{u(\theta,x)} = \Delta_{\mathbb{H}^{p+1}}\widehat{u}.$$

Hence,

$$c_m \Delta_{\mathbb{H}^{p+1}}\widehat{u} = c_m\widehat{\Delta u} = (m-1)(m-2-2p)\widehat{u} - \widehat{k(x)u^{\frac{m+2}{m-2}}}$$

$$= (m-1)(m-2-2p)\widehat{u} - k(x)\widehat{u^{\frac{m+2}{m-2}}}$$

and using $k(x) \le 0$ and (7.55) we obtain (7.56). $\qquad\square$

We are now ready to prove the following

Theorem 7.5. *Consider the complete manifold $\mathbb{S}^{m-p-1} \times \mathbb{H}^{p+1}$ with the product metric for $\frac{m-2}{2} < p \le m-2$. Then its metric cannot be conformally deformed to a metric of scalar curvature $k(\theta,x) = k(x) \in C^\infty(\mathbb{H}^{p+1})$ satisfying $k(x) \le 0$ and*

$$k(x) \le -C\frac{e^{2\frac{2p-m+2}{m-2}r(x)}}{r(x)} \quad \text{for } r(x) \gg 1, \tag{7.57}$$

where $r(x)$ is the hyperbolic distance from a fixed origin in $\mathbb{H}^{p+1}$ and $C > 0$ constant.

Proof. Simply apply Theorem 7.3 to the differential inequality (7.56) reasoning by contradiction. $\qquad\square$

Note the special case $p = \frac{m-2}{2}$ in which (7.57) reduces to

$$k(x) \le -\frac{C}{r(x)} \quad \text{for } r(x) \gg 1.$$

7.1.1 A Rellich-Pohozaev formula

The aim of this section is to provide a Pohozaev formula, a tool well-known to analysts in Euclidean space, which is used in order to obtain nonexistence results. The formula can be readily generalized from $\mathbb{R}^m$ to a general Riemannian manifold M, but its effectiveness in this wider context depends on the possibility of choosing a conformal vector field X on M appropriately related to the problem at hand. For Euclidean space, one has the canonical simple choice $X = r\nabla r$, where $r(x)$ is the distance function from the origin o. Similar choices are naturally available for spaces of constant sectional curvature or for models in the sense of Greene and Wu, [GW79]. However, for general Riemannian manifolds the mere existence of a globally defined conformal vector field X implies subtle geometrical restrictions, and the behavior of X and its divergence (or, alternatively, the Lie derivative of the metric in the direction of X) is difficult to understand. This is the reason why in this section we shall limit ourselves to special cases, and, in the next section, to hyperbolic space, and more specifically to its Poincaré disc $B_1(0)$ realization.

What follows can be considered as a geometric motivation of the Pohozaev formula in the particular case of the Yamabe equation in $\mathbb{R}^m$: indeed, this special case relates the formula to the Kazdan-Warner obstruction obtained by integration of (2.17) (see Chapter 2, Theorem 2.3; see also the discussion in Section 5 of [PRS03a]).

Let $(\mathbb{R}^m, \langle\,,\,\rangle)$ be the Euclidean space with its flat canonical metric. For the sake of simplicity let us suppose $m \geq 3$. We consider the conformally related metric

$$\widetilde{\langle\,,\,\rangle} = u^{\frac{4}{m-2}} \langle\,,\,\rangle$$

and, as in the previous chapters, we denote by T and $\widetilde{T}$ the trace-free Ricci tensors of $\langle\,,\,\rangle$ and $\widetilde{\langle\,,\,\rangle}$ respectively. Then clearly $T \equiv 0$ and equation (2.26) yields

$$\widetilde{T} = 2u^{-1}\left[\frac{1}{m}\Delta u\langle\,,\,\rangle - \mathrm{Hess}(u)\right] + \frac{2m}{m-2}u^{-2}\left[du \otimes du - \frac{1}{m}|\nabla u|^2\langle\,,\,\rangle\right]. \quad (7.58)$$

Let now $X = r\nabla r$. Since $\mathcal{L}_{r\nabla r}\langle\,,\,\rangle = 2\langle\,,\,\rangle$, X is conformal with respect to $\langle\,,\,\rangle$, and therefore also with respect to $\widetilde{\langle\,,\,\rangle}$. Recalling (2.17) we have

$$\widetilde{\mathrm{div}}\widetilde{W} = \frac{m-2}{2m}X(\widetilde{S}).$$

Applying the divergence theorem on B_r with respect to the metric $\widetilde{\langle\,,\,\rangle}$ to the above equation, and expressing the result in terms of the original metric we obtain

$$\frac{m-2}{2m}\int_{B_r} r\left\langle\nabla r, \nabla\widetilde{S}\right\rangle u^{\frac{2m}{m-2}}\,\mathrm{dvol}_m = \int_{\partial B_r} ru^2\widetilde{T}(\nabla r, \nabla r)\,\mathrm{dvol}_{m-1}, \quad (7.59)$$

where we have used the fact that the outward unit normal to ∂B_r with respect to $\widetilde{\langle\,,\,\rangle}$ is $u^{-2/(m-2)}\nabla r$, that the m- and $(m-1)$-dimensional volume elements are given by $\widetilde{dvol}_m = u^{2m/(m-2)} dvol_m$ and $\widetilde{dvol}_{m-1} = u^{2(m-1)/(m-2)} dvol_{m-1}$ respectively, and that, according to (2.14), $\widetilde{W}$ is defined by $\langle \widetilde{W}, Y \rangle = \widetilde{T}(X, Y)$. Now, from the expression of $\widetilde{T}$ and $X = r\nabla r$, setting $\frac{\partial u}{\partial r}$ for $\langle \nabla r, \nabla u \rangle$ we obtain

$$ru^2\widetilde{T}(\nabla r, \nabla r) = 2ru\left[\frac{1}{m}\Delta u - \text{Hess}(u)(\nabla r, \nabla r)\right] + \frac{2m}{m-2}r\left[\left(\frac{\partial u}{\partial r}\right)^2 - \frac{1}{m}|\nabla u|^2\right].$$

We insert this formula into (7.59) and observe that, using the expression for Δ in spherical coordinates,

$$\text{Hess}(u)(\nabla r, \nabla r) = \frac{\partial^2 u}{\partial r^2} = \Delta u - \frac{m-1}{r}\frac{\partial u}{\partial r} - \Delta_{\partial B_r} u.$$

Now we apply the divergence theorem on ∂B_r and use

$$|\nabla u|^2 = \left(\frac{\partial u}{\partial r}\right)^2 + |\nabla_{\partial B_r} u|^2_{\partial B_r},$$

to deal with the term $\Delta_{\partial B_r} u$, and finally use

$$c_m \Delta u = -\widetilde{S}u^{\frac{m+2}{m-2}}, \quad c_m = 4\frac{m-1}{m-2}$$

to obtain, after some simplifications,

$$\int_{B_r} r\langle \nabla r, \nabla \widetilde{S}\rangle u^{\frac{2m}{m-2}} dvol_m = \int_{\partial B_r} r\widetilde{S}u^{\frac{2m}{m-2}} dvol_{m-1}$$

$$+ mc_m \int_{\partial B_r}\left[u\frac{\partial u}{\partial r} + \frac{1}{m-2}r\left(2\left(\frac{\partial u}{\partial r}\right)^2 - |\nabla u|^2\right)\right] dvol_{m-1}. \quad (7.60)$$

The above formula is in the typical form of a Rellich-Pohozaev identity. We are going to describe a simpler and more general way to obtain it. Let $u \in C^2(M)$ be a solution of the equation

$$\Delta u = h(x)f(u) \qquad (7.61)$$

on a complete Riemannian manifold $(M, \langle , \rangle)$ with $h \in C^0(M)$ and $f \in C^0(\mathbb{R})$. Let

$$F(u) = \int_0^u f(s)\, ds,$$

and define the vector field

$$W = \left[h(x)F(u) + \frac{1}{2}|\nabla u|^2\right]X - \langle X, \nabla u\rangle\nabla u - \alpha u\nabla u, \qquad (7.62)$$

where X is a vector field on M and $\alpha \in \mathbb{R}$. Using the identities

$$\langle \nabla|\nabla u|^2, X \rangle = 2\operatorname{Hess}(u)(\nabla u, X),$$

$$\nabla u \langle \nabla u, X \rangle = \operatorname{Hess}(u)(\nabla u, X) + \frac{1}{2}\mathcal{L}_X \langle , \rangle (\nabla u, \nabla u),$$

we compute

$$\operatorname{div} W = h(x)(F(u)\operatorname{div} X - \alpha u f(u))$$
$$+ \left(\frac{1}{2}\operatorname{div} X - \alpha\right)|\nabla u|^2 + F(u)\langle \nabla h, X \rangle - \frac{1}{2}\mathcal{L}_X \langle , \rangle(\nabla u, \nabla u).$$

If we assume that X is a conformal vector field, the above inequality becomes

$$\operatorname{div} W = h(x)\big(F(u)\operatorname{div} X - \alpha u f(u)\big)$$
$$+ \left(\frac{m-2}{2m}\operatorname{div} X - \alpha\right)|\nabla u|^2 + F(u)\langle \nabla h, X \rangle.$$

Integrating over the annulus $B_r \setminus B_{r_o}$, $0 \le r_0 < r$, and using the divergence theorem we obtain

$$\int_{B_r} \operatorname{div} W = \int_{\partial B_t} \left[(h(x)F(u) + \frac{1}{2}|\nabla u|^2)\langle X, \nabla r \rangle - \langle X, \nabla u \rangle \frac{\partial u}{\partial r} - \alpha u \frac{\partial u}{\partial r}\right]\Big|_{r_o}^{r}.$$

In particular, if $(M, \langle , \rangle)$ is $\mathbb{R}^m$ with its canonical metric, and we choose $X = r\nabla r$, $f(u) = u^{\frac{m+2}{m-2}}$, $h(x) = c_m^{-1}\widetilde{S}(x)$ and $\alpha = \frac{m-2}{2}$, the above formula yields

$$\frac{m-2}{2m\,c_m}\int_{B_r} r\langle \nabla \widetilde{S}, \nabla r \rangle u^{\frac{2m}{m-2}}$$
$$= \int_{\partial B_r} \left[\frac{m-2}{2m\,c_m}r\widetilde{S}u^{\frac{2m}{m-2}} - \frac{r}{2}|\nabla u|^2 + r\left(\frac{\partial u}{\partial r}\right)^2 + \frac{m-2}{2}u\frac{\partial u}{\partial r}\right]\Big|_{r_o}^{r}, \quad (7.63)$$

and, letting $r_o = 0$, we recover (7.60).

As an application of this identity, we consider the prescribed scalar curvature equation on $\mathbb{S}^m$,

$$c_m\Delta_{\mathbb{S}^m}u - m(m-1)u + K(x)u^{\frac{m+2}{m-2}} = 0.$$

We use (the inverse of) stereographic projection to pull back the standard metric of $\mathbb{S}^m$ onto $\mathbb{R}^m$. The pull-back metric is conformal to Euclidean metric with conformality factor $4(1 + |x|^2)^{-2}$. Thus, the original conformal change of metric on $\mathbb{S}^m$ may be viewed as a conformal change of metric on $\mathbb{R}^m$ and we are led to investigating the equation

$$c_m\Delta u + b(x)u^{\frac{m+2}{m-2}} = 0 \qquad (7.64)$$

on $\mathbb{R}^m$, where $b(x)$ is bounded and admits a finite limit at infinity, and to look for positive solution $u \in C^\infty(\mathbb{R}^m)$ such that

$$u(x) \sim |x|^{2-m} \quad \text{as } |x| \to +\infty. \tag{7.65}$$

We claim that if b is radial, somewhere positive and monotone increasing where positive, then there are no positive radial solutions of (7.65) satisfying (7.65).

Indeed, assume by contradiction that such a solution exists, and identify radial functions on $\mathbb{R}^m$ with functions on $[0, +\infty)$. Since $b(x)$ is bounded, and $u(r) \sim r^{2-m}$ as $r \to +\infty$, it follows from (7.64) that $u'(r) = O(r^{1-m})$ as $r \to +\infty$. To prove this, note that since u is a radial solution of (7.64) it satisfies the ode

$$\left(r^{m-1}u'\right)' = -c_m^{-1} r^{m-1} b(r) u^{\frac{m+2}{m-2}}$$

so that, integrating between ε and r, we obtain

$$r^{m-1}u'(r) = \varepsilon^{m-1}u'(\varepsilon) - c_m^{-1} \int_\varepsilon^r b(t) t^{m-1} u(t)^{\frac{m+2}{m-2}} dt.$$

Since $b(t)$ is bounded, and $u(t) \sim t^{2-m}$, we have

$$b(t) t^{m-1} u(t)^{\frac{m+2}{m-2}} = O\left(t^{-3}\right),$$

so that the integral on the right-hand side is absolutely convergent at infinity, and we deduce that

$$|u'(r)| \le r^{1-m}\left(\varepsilon^{m-1}|u'(\varepsilon)| + c_m^{-1} \int_\varepsilon^r |b(t)| t^{m-1} u(t)^{\frac{m+2}{m-2}} dt\right) \le E r^{1-m} \quad \text{as } r \to +\infty$$

as required. Next, by the assumption on b, there exists $r_o \ge 0$ such that

$$b(r) \le 0 \text{ for } 0 \le r \le r_o \quad \text{and} \quad b(r) > 0, \ b' \ge 0, \ b' \not\equiv 0 \text{ for } r > r_o. \tag{7.66}$$

We apply equation (7.63) with $b(r)$ instead of $\widetilde{S}$ and with the value of r_o specified above, and let $r \to +\infty$. Since $b' \ge 0$, by monotone convergence, the integral over $B_r \setminus B_{r_o}$ converges to the integral over $\mathbb{R}^m \setminus B_{r_o}$, while, by the asymptotic behavior of u and u' the boundary integral over ∂B_r tends to zero, and we obtain

$$\int_{B_{r_o}} \frac{r_o}{2} u'(r_o)^2 + \frac{m-2}{2} u(r_o)u'(r_o) = -\frac{m-2}{2m\,c_m} \int_{\mathbb{R}^m \setminus B_{r_o}} r b'(r) u^{\frac{2m}{m-2}}. \tag{7.67}$$

Now, if $r_o = 0$, the left-hand side of the above equation vanishes, while the right-hand side does not, and we immediately deduce the required contradiction. Otherwise, $b \le 0$ on B_{r_o} and it follows from (7.64) that

$$\Delta u \ge 0 \quad \text{on } B_{r_o},$$

so that, by the maximum principle $\max_{B_{r_o}} u = u(r_o)$ and $u'(r_o) \ge 0$, and, again, (7.67) yields a contradiction.

We have therefore proved the following result due to W. Chen and C. Li ([CL95], Theorem 3)

Theorem 7.6. *Let K be a nonconstant rotationally symmetric function on $\mathbb{S}^m$, $m \geq 3$,, and assume that K is somewhere positive, and monotone increasing in the region where it is positive. Then the equation*

$$c_m \Delta u - m(m-1)u + K u^{\frac{m+2}{m-2}} = 0$$

has no positive smooth rotationally symmetric solutions on $\mathbb{S}^m$.

7.1.2 A nonexistence result for hyperbolic space

As already mentioned in the introduction, in this section we will work in the Poincaré model for hyperbolic space. The main reason is that the Rellich-Pohozaev formula has a particularly nice expression in this setting, and allows us to give a more transparent proof of our main nonexistence result. We leave it to the interested reader to translate the result in the standard model $\mathbb{H}^m = \mathbb{R}^m$ with metric in polar coordinates on $\mathbb{H}^m \setminus \{0\} = (0, +\infty) \times \mathbb{S}^{m-1}$ given by

$$\langle\, , \,\rangle_{\mathbb{H}^m} = dr^2 + \sinh^2 d\theta^2,$$

which has been used throughout the book.

We recall that the Poincaré model of $\mathbb{H}^m$ is

$$(\mathbb{B}^m, g_\mathbb{H}),$$

where $\mathbb{B}^m$ is the unit ball of $\mathbb{R}^m$ and $g_\mathbb{H}$ is the metric conformally related to the canonical Euclidean metric $\langle\, , \,\rangle$ given by

$$g_\mathbb{H} = \frac{4}{(1-|x|^2)^2}\langle\, , \,\rangle, \tag{7.68}$$

where, using standard notation, $|x| = \text{dist}_{\mathbb{R}^m}(x, 0)$.

We state the following problem: *Given $K \in C^\infty(\mathbb{B}^m)$, $m \geq 3$, does there exist a positive function v on $\mathbb{B}^m$ such that*

(i) $g_v = v^{\frac{4}{m-2}} g_\mathbb{H}$ *has scalar curvature K and*

(ii) g_v *is a complete Riemannian metric on $\mathbb{B}^m$?*

Setting

$$u(x) = v(x)\left[\frac{4}{(1-|x|^2)^2}\right]^{\frac{m-2}{4}},$$

we have that the above problem admits an affirmative answer if and only if the equation

$$c_m \Delta u + K(x)u^{\frac{m+2}{m-2}} = 0, \tag{7.69}$$

where $c_m = 4\frac{m-1}{m-2}$ (as in the previous chapters) and Δ is the Euclidean Laplacian, admits a positive solution u such that the conformal metric

$$g_u = u^{\frac{4}{m-2}} \langle\, , \,\rangle$$

is a complete Riemannian metric on $\mathbb{B}^m$. We note that the requirement of completeness implies that u must diverge at the boundary of $\mathbb{B}^m$, while equation (7.69) of course means that K is the scalar curvature of g_u. We also recall that, letting $r = |x|$, then the Riemannian distance ρ from the origin of $\mathbb{B}^m$ with respect to the hyperbolic metric (7.68) is given by (see [RRV94b])

$$\rho = \log\left(\frac{1+r}{1-r}\right),\qquad(7.70)$$

while

$$r = \tanh\left(\frac{\rho}{2}\right) = \frac{e^\rho - 1}{e^\rho + 1}.\qquad(7.71)$$

For the developments below, we will use the Rellich-Pohozaev-type formula which is obtained in the following way. Fix $x_o \in \mathbb{B}^m$, let $X = x - x_o$ and define a vector field by the formula

$$W = \left\{\frac{c_m}{2}|\nabla u|^2 - \frac{m-2}{2m}Ku^{\frac{2m}{m-2}}\right\}(x-x_o) - c_m\langle x-x_o, \nabla u\rangle\nabla u + c_m\alpha u\nabla u. \quad (7.72)$$

Using the fact that $\mathcal{L}_{x-x_o}\langle\, , \,\rangle = \langle\, , \,\rangle$, a computation similar to the one carried out in the previous section shows that

$$\mathrm{div}\, W = c_m\left(\frac{m-2}{2} - \alpha\right)|\nabla u|^2 + \left[\left(\alpha - \frac{m-2}{2}\right)K - \frac{m-2}{m}\langle x - x_o, \nabla K\rangle\right]u^{\frac{2m}{m-2}},$$
$$(7.73)$$

and we obtain a Rellich-Pohozaev formula integrating and using the divergence theorem.

We are now ready to state (see [RRV94b], Theorem 1.2)

Theorem 7.7. *Let $K \in C^\infty(\mathbb{B}^m)$ and suppose that*

$$\limsup_{|x|\to 1^-}(1 - |x|^2)K(x) < 0.\qquad(7.74)$$

Assume also that either

(i) *$K \le 0$ in $\mathbb{B}^m$, or*

(ii) *there exists $\alpha \le \frac{m-2}{2}$ and $x_o \in \mathbb{B}^m$ such that*

$$\left(\alpha - \frac{m-2}{2}\right)K(x) - \frac{m-2}{2m}\langle x - x_o, \nabla K(x)\rangle \ge 0 \ \ on \ \mathbb{B}^m,$$

with strict inequality in at least one point.

Then equation (7.69) does not admit any positive solutions on $\mathbb{B}^m$.

Note that if we assume that u is nonnegative and satisfies (7.69), then, by the strong maximum principle (see, [GT01], p. 35), either $u > 0$ or $u \equiv 0$.

We split the proof into several steps. We note that steps 1 to 3 are devoted to showing that the solution $u(x)$ must tend to zero in a specific way as the point x tends to the boundary of $\mathbb{B}^m$. This fact could be obtained as an application of Theorem 4.4 in Chapter 4. However we follow the alternative method presented below to describe a technique of common use in Analysis.

Proof. We assume that (7.69) admits a positive solution u and derive a contradiction.

Step 1. Assumption (7.74) implies that u is bounded.

To prove this fact we first establish the following

Proposition 7.8. *Let $B_R \subset \mathbb{R}^m$ be the ball of radius R centered at 0, let $\sigma > 1$, and let f be a positive function in $C^0(\overline{B}_R)$. Then there exists a constant $C = C(\sigma, m) > 0$ such that, if $u \in C^2(B_R)$ satisfies the differential inequality*

$$\Delta u \geq f(x)u^\sigma \quad on \ B_R, \tag{7.75}$$

then

$$u(o) \leq C(R^2 \min_{\overline{B}_R} f)^{-\frac{1}{\sigma-1}}. \tag{7.76}$$

Proof. Set $A = \min_{\overline{B}_R} f$, so that u satisfies

$$\Delta u \geq Au^\sigma. \tag{7.77}$$

For $0 < R_1 < R$ define

$$v(x) = \lambda(R_1^2 - |x|^2)^{-\frac{2}{\sigma-1}},$$

where

$$\lambda = \left(\frac{B}{A}R_1^2\right)^{\frac{1}{\sigma-1}} \qquad B = \max\left\{\frac{2m}{\sigma-1}, 4\frac{\sigma+1}{(\sigma-1)^2}\right\}.$$

A straightforward computation shows that

$$\Delta v \leq Av^\sigma \quad on \ B_{R_1}, \tag{7.78}$$

and $v(x) \to +\infty$ as $|x| \to R_1$. Thus the function $w = u - v$ attains an absolute maximum at a point $x_o \in B_{R_1}$, and satisfies the inequality

$$\Delta w \geq A(u^\sigma - v^\sigma) = h(x)w,$$

where $h \in C^0(B_{R_1})$ is the nonnegative function defined by

$$h(x) = \begin{cases} \frac{\sigma A}{u(x)-v(x)} \int_{v(x)}^{u(x)} t^{\sigma-1}dt & if \ u(x) \neq v(x), \\ \sigma Au(x)^{\sigma-1} & if \ u(x) = v(x). \end{cases} \tag{7.79}$$

If $w(x_o) > 0$, then $h(x_o) > 0$, and

$$\Delta w(x_o) \geq h(x_o)w(x_o) > 0,$$

which is impossible since w attains a maximum at x_o. Thus for every $x \in B_{R_1}$, $w(x) \leq w(x_o) \leq 0$, that is

$$u(x) \leq v(x) \text{ in } B_{R_1},$$

and in particular,

$$u(0) \leq v(0) = \left(\frac{B}{A}\right)^{\frac{1}{\sigma-1}} R_1^{-\frac{2}{\sigma-1}} = B^{\frac{1}{\sigma-1}} \left(AR_1^2\right)^{-\frac{1}{\sigma-1}},$$

and letting $R_1 \to R$ we obtain the desired estimate. $\qquad\square$

To prove the claim in Step 1, let u be a positive solution of (7.69). We fix γ in $(0,1)$ and applying Proposition 7.8 in the ball

$$B^\gamma(x) = \{y \in \mathbb{R}^m : |y - x| \leq \gamma(1 - |x|)\},$$

we deduce that

$$u(x) \leq C \left\{ (1 - |x|)^2 \max_{B^\gamma(x)} (-K(z)) \right\}^{-\frac{m-2}{4}}.$$

Using (7.74) we deduce that u is bounded above as $|x| \to 1^-$ and, it is therefore bounded.

Step 2. We show that $\lim_{|x| \to 1^-} u(x) = 0$.

By (7.74) there exists constants $0 \leq \eta < 1$ and $c > 0$ such that

$$K(x) \leq -\frac{c}{2}(1 - |x|)^{-2} \qquad\qquad (7.80)$$

on the set $\text{Ann}_{\eta,1} = B_1 \setminus \overline{B}_\eta = \{x \in \mathbb{B}^m : \eta < |x| < 1\}$. Let $\delta \in (\eta, 1)$ and let v_δ be the maximal solution of the problem

$$\begin{cases} c_m \Delta v = \frac{c}{2}(1 - |x|)^{-2} v^{\frac{m+2}{m-2}} & \text{on } \text{Ann}_{\eta,\delta}, \\ \lim_{|x| \to \eta} v(x) = \lim_{|x| \to \delta} v(x) = +\infty, \end{cases}$$

which exists by Lemma 6.4 in Chapter 6, and is radially symmetric since it must coincide with the maximal radial solution to (7.1.2). Since $v_\delta(x)$ diverges as x tends to the boundary of $\text{Ann}_{\eta,\delta}$, by usual comparison arguments, $u < v_\delta$ on this set. Moreover, as $\delta \to 1^-$, v_δ decreases and converges to a radially symmetric function v which satisfies

$$c_m \Delta v = \frac{c}{2}(1 - |x|)^{-2} v^{\frac{m+2}{m-2}} \text{ on } \text{Ann}_{\eta,1},$$

and

$$0 \le u(x) \le v(|x|).$$

Therefore, it suffices to show that

$$\lim_{|x| \to 1^-} v(|x|) = 0.$$

To this end, we first note that, arguing as in Step 1, we can show that v is bounded in a neighborhood of $\partial \mathbb{B}^m$.

Next, since v is radial, it satisfies the ode

$$c_m (r^{m-1} v')' = \frac{c}{2} r^{m-1} (1-r)^{-2} v^{\frac{m+2}{m-2}},$$

for $\eta < r < 1$. It follows that the function

$$r^{m-1} v'(r)$$

is increasing in $\eta < r < 1$, hence it has constant sign in a left neighborhood of 1. Thus v itself tends to a nonnegative limit β as $r \to 1^-$. Assume by contradiction that $\beta > 0$. Then there exists $\eta < \tau < 1$ and a constant $\theta > 0$ such that

$$(r^{m-1} v')' \ge \theta r^{m-1} (1-r)^{-2},$$

and integrating twice we deduce that $v(r) \to +\infty$ as $r \to 1^-$ contradicting the fact that v is bounded. The contradiction shows that $\beta = 0$ and therefore $\lim_{r \to 1^-} u(x) = 0$.

Step 3. There exists a constant $M > 0$ such that

$$u(x) \le M(1 - |x|). \tag{7.81}$$

Indeed, the function
$$\psi(x) = M(1 - |x|)$$
is a supersolution of (7.69) in $\mathrm{Ann}_{\eta,1}$ and, by taking M large enough we can arrange that $u(x) \le \psi(x)$ on $\partial \mathrm{Ann}_{\eta,1}$. Thus, the usual comparison principle shows that $u \le \psi$ on $\mathrm{Ann}_{\eta,1}$, and by increasing M if necessary, the inequality holds on $\mathbb{B}^m$.

Step 4. Suppose that $K \le 0$ on $\mathbb{B}^m$. Then u is a nonnegative subharmonic function which vanishes on $\partial \mathbb{B}^m$, and by the maximum principle $u \equiv 0$.

Step 5. Using the Pohozaev-type formula (7.73), we prove that, if K satisfies condition (ii) in the statement of the theorem, then (7.69) has no positive solution on $\mathbb{B}^m$.

Suppose first that $u \in C^1(\overline{\mathbb{B}}^m)$. We fix $\varepsilon > 0$, integrate (7.73) over $B_{1-\varepsilon}$ and let $\varepsilon \to 1^-$. Using the fact that $u = 0$ on $\partial \mathbb{B}^{m-1}$, so that $\nabla u = \frac{\partial u}{\partial r} \nabla r$, we obtain

$$-\frac{c_m}{2} \int_{\partial \mathbb{B}^m} |\langle \nabla u, \nabla r \rangle|^2 \langle x - x_o, \nabla u \rangle = \int_{\mathbb{B}^m} c_m \left(\frac{m-2}{2} - \alpha \right) |\nabla u|^2$$

$$+ \int_{\mathbb{B}^m} \left[\left(\alpha - \frac{m-2}{2} \right) K - \frac{m-2}{2m} \langle x - x_o, \nabla K \rangle \right] u^{\frac{2m}{m-2}}.$$

Since the left-hand side of the equality is nonpositive, and $\alpha \le \frac{m-2}{2}$ we conclude that

$$\int_{\mathbb{B}^m} \left[\left(\alpha - \frac{m-2}{2} \right) K - \frac{m-2}{2m} \langle x - x_o, \nabla K \rangle \right] u^{\frac{2m}{m-2}} \le 0. \tag{7.82}$$

On the other hand, condition (ii) in the statement of the theorem and $u > 0$ imply that the left-hand side is strictly positive, and the required contradiction follows.

However, in general we do not know that u is C^1 up to the boundary, and we need to justify that one may take the limit as $\varepsilon \to 0$ to obtain (7.82).

This can be done by the following approximation technique. Write $K = K^+ - K^-$, and, for a fixed $\delta > 0$, let

$$K_\delta^-(x) = \min\{\frac{1}{\delta}, K^-(x)\}, \tag{7.83}$$

and let $v = v_\delta$ be the solution of the problem

$$\begin{cases} c_m \Delta v - K_\delta^- v^{\frac{m+2}{m-2}} = -K^+ u^{\frac{m+2}{m-2}} & \text{in } \mathbb{B}^m, \\ v = 0 & \text{on } \partial \mathbb{B}^m. \end{cases} \tag{7.84}$$

By the asymptotic behavior of K, the function $K^+ u^{\frac{m+2}{m-2}}$ is Lipschitz and compactly supported in $\mathbb{B}^m$, and since K_δ^- is Lipschitz and bounded on $\overline{\mathbb{B}}^m$, the existence of v is guaranteed by a minimization argument, and by standard regularity, v is C^2 in $\overline{\mathbb{B}}^m$. Moreover, if $\delta_1 > \delta_2$, then $K_{\delta_1}^- \le K_{\delta_2}^-$, and by comparison,

$$v_{\delta_2} \le v_{\delta_1} \quad \text{in } \mathbb{B}^m. \tag{7.85}$$

Recalling the equations satisfied by v_δ and u and noting that $K_\delta^- \le K^-$, we deduce that the function $w = u - v$ satisfies

$$c_m \Delta w \ge K_\delta^- (u^{\frac{m+2}{m-2}} - v^{\frac{m+2}{m-2}}) = h(x) K_\delta^- w,$$

with $h(x) \ge 0$ defined by (7.79) and $\sigma = \frac{m+2}{m-2}$. Since $K_\delta^- \ge 0$ in $\mathbb{B}^m$, and $u = v = 0$ on $\partial \mathbb{B}^m$, by the maximum principle $w \ge 0$ and therefore $u \le v_\delta$ for every $\delta > 0$.

Thus as $\delta \to 0$, v_δ converges to a function v_o which vanishes on $\partial \mathbb{B}^m$ and satisfies $v_o \ge u$.

Now, by Step 3, v_δ is a solution of (7.84) and is uniformly bounded on $\mathbb{B}^m$. Moreover, if $r < 1$, $K_\delta^- = K$ on B_r for δ small enough, so that, by standard elliptic estimates (see, e.g., [Aub98], Theorem 4.40), we deduce that v_δ is uniformly bounded in $C^1(B_r)$ and therefore it converges to v_o uniformly on B_r. On the other hand, again by Step 3, v_δ satisfies the estimate (7.81) and therefore $\{|\langle \nabla r, \nabla v_\delta \rangle|\}$ is bounded and decreasing, and it follows that v_δ tends to zero as $x \to \partial \mathbb{B}^m$ uniformly in δ. Combining these two facts we conclude easily that $v_\delta \to v_o$ uniformly on $\overline{\mathbb{B}^m}$.

Integrating equation (7.74) against a test function and letting $\delta \to 0$ shows that v_o belongs to $W^{1,2}(\mathbb{B}^m) \cap C^0(\overline{\mathbb{B}^m})$ and it is a weak solution of

$$c_m \Delta v_o = K^- v_o^{\frac{m+2}{m-2}} + K^+ u^{\frac{m+2}{m-2}}.$$

Thus $w = v_o - u$ is a nonnegative weak solution of

$$\Delta w = K^- \left(v_o^{\frac{m+2}{m-2}} - u^{\frac{m+2}{m-2}} \right) = h(x)w,$$

where $h \geq 0$ is defined in (7.79). Since $w = 0$ on $\partial \mathbb{B}^m$, by comparison $w \leq 0$ and therefore $w \equiv 0$, that is $v_o = u$.

Next we fix $\delta > 0$ and we consider the vector field

$$W_\delta = \left\{ \frac{c_m}{2} |\nabla v_\delta|^2 + \frac{m-2}{2m} \left[K_\delta^- v_\delta^{\frac{2m}{m-2}} - K^+ u^{\frac{m+2}{m-2}} v_\delta \right] \right\} (x - x_o)$$
$$- c_m \langle x - x_o, \nabla v_\delta \rangle \nabla v_\delta + c_m \alpha v_\delta \nabla v_\delta.$$

Writing v instead of v_δ, a computation similar to that leading to (7.73) shows that

$$\mathrm{div} W_\delta = c_m \left(\frac{m-2}{2} - \alpha \right) |\nabla v|^2 + \frac{m-2}{2m} \langle x - x_o, \nabla K_\delta^- v_\delta^{\frac{2m}{m-2}} - \nabla K^+ u^{\frac{m+2}{m-2}} v_\delta \rangle$$
$$+ \left(\alpha - \frac{m-2}{2} \right) \left(K_\delta^- v_\delta^{\frac{2m}{m-2}} - K^+ u^{\frac{m+2}{m-2}} v_\delta \right) - L(v), \qquad (7.86)$$

where

$$L(v) = L(v_\delta) = \frac{m+2}{2} K^+ vu^{\frac{m+2}{m-2}} + \frac{m+2}{m-2} K^+ vu^{\frac{4}{m-2}} \langle x - x_o, \nabla u \rangle$$
$$+ \frac{m+2}{2m} vu^{\frac{m+2}{m-2}} \langle x - x_o, \nabla K^+ \rangle.$$

Integrating (7.86) over $\mathbb{B}^m$, noting that v is C^1 on $\overline{\mathbb{B}^m}$ and vanishes on $\partial \mathbb{B}^m$, and using the divergence theorem we obtain

$$- \frac{c_m}{2} \int_{\partial \mathbb{B}^m} \langle x - x_o, \nabla r \rangle \langle \nabla v, \nabla r \rangle^2$$
$$= \int_{\mathbb{B}^m} c_m \left(\frac{m-2}{2} - \alpha \right) |\nabla v|^2 + \frac{m-2}{2m} \langle x - x_o, \nabla K_\delta^- v^{\frac{2m}{m-2}} - \nabla K^+ u^{\frac{m+2}{m-2}} v \rangle$$
$$+ \left(\frac{m-2}{2} - \alpha \right) \int_{\mathbb{B}^m} \left(\nabla K_\delta^- v^{\frac{2m}{m-2}} - \nabla K^+ u^{\frac{m+2}{m-2}} v \right) - \int_{\mathbb{B}^m} L(v).$$

Since K^+ has compact support in $\mathbb{B}^m$, as $\delta \to 0$, the integral $L(v_\delta)$ converges to $L(u)$. Moreover,

$$
L(u) = \frac{m+2}{2} K^+ u^{\frac{2m}{m-2}} + \frac{m+2}{m-2} K^+ u^{\frac{m+2}{m-2}} \langle x - x_o, \nabla u \rangle
$$
$$
+ \frac{m+2}{2m} u^{\frac{2m}{m-2}} \langle x - x_o, \nabla K^+ \rangle = \frac{m+2}{2m} \mathrm{div}\left(K^+ u^{\frac{2m}{m-2}} (x - x_o) \right),
$$

so that $L(u) = 0$ and $L(v_\delta) = o(\delta)$ as $\delta \to 0$.

Summing up, recalling that $\alpha \le \frac{m-2}{2}$, we have obtained that

$$
o(\delta) + \int_{\mathbb{B}^m} \left[\left(\alpha - \frac{m-2}{2} \right) \left(K^+ v u^{\frac{m+2}{m-2}} - K_\delta^- v^{\frac{2m}{m-2}} \right) \right.
$$
$$
\left. - \frac{m-2}{2m} \langle x - x_o, v u^{\frac{m+2}{m-2}} \nabla K^+ - v^{\frac{2m}{m-2}} \nabla K_\delta^- \rangle, \right] < 0
\tag{7.87}
$$

with $o(\delta) \to 0$ as $\delta \to 0$.

Next, we rewrite the integrand in the above formula as

$$
\left[\left(\alpha - \frac{m-2}{2} \right) (K^+ - K_\delta^-) - \frac{m-2}{2m} \langle x - x_o, \nabla (K^+ - K_\delta^-) \rangle \right] v^{\frac{2m}{m-2}}
$$
$$
+ \left[\left(\alpha - \frac{m-2}{2} \right) K^+ - \frac{m-2}{2m} \langle x - x_o, \nabla K^+ \rangle \right] \left(v u^{\frac{m+2}{m-2}} - v^{\frac{2m}{m-2}} \right),
$$

and note that since K^+ is compactly supported in $\mathbb{B}^m$ and $v_\delta \to u$ uniformly, the integral of the second summand in tends to zero as $\delta \to 0$. Hence, (7.87) yields

$$
\limsup_{\delta \to 0} \int_{\mathbb{B}^m} \left[\left(\alpha - \frac{m-2}{2} \right) (K^+ - K_\delta^-) - \frac{m-2}{2m} \langle x - x_o, \nabla (K^+ - K_\delta^-) \rangle \right] v^{\frac{2m}{m-2}} \le 0.
\tag{7.88}
$$

On the other hand, let $G_\delta = \{ x : K(x) < -1/\delta \}$, so that G_δ is open and $K = -K_\delta^- = -1/\delta$ on G_δ. It follows that $\nabla K = -\nabla K_\delta^- = 0$ on G_δ, and, using again $\alpha \le \frac{m-2}{2}$, the fact that $K_\delta^- = K^-$ in the set where $K \ge -1/\delta$ and using condition (ii) in the statement we deduce that

$$
\left(\alpha - \frac{m-2}{2} \right) (K^+ - K_\delta^-)(x) - \frac{m-2}{2m} \langle x - x_o, \nabla (K^+ - K_\delta^-)(x) \rangle \ge 0
\tag{7.89}
$$

almost everywhere on $\mathbb{B}^m$, so that the limit in (7.88) exists and is equal to zero.

Finally, since the integrand in (7.88) is nonnegative, and, as $\delta \to 0$, tends to

$$
\left(\alpha - \frac{m-2}{2} \right) (K^+ - K^-)(x) - \frac{m-2}{2m} \langle x - x_o, \nabla (K^+ - K^-)(x) \rangle u^{\frac{2m}{m-2}}
$$

almost everywhere in $\mathbb{B}^m$, we conclude from Fatou's lemma that

$$
\int_{\mathbb{B}^m} \left[\left(\alpha - \frac{m-2}{2} \right) K - \frac{m-2}{2m} \langle x - x_o, \nabla K \rangle \right] u^{\frac{2m}{m-2}} \le 0,
\tag{7.90}
$$

and, using (ii) and $u > 0$ a contradiction is reached as above. $\square$

7.1.3 An integral obstruction

The aim of this subsection is to give a necessary condition for the existence of a metric conformally related to the Poincaré metric (7.68) on $\mathbb{B}^m$ with assigned scalar curvature $K(x)$; we recall that, as observed before, this problem consists in finding a positive solution of equation (7.69). The nature of the obstruction that we are going to describe is manifestly different from that of the Kazdan-Warner condition (see Chapter 2) which, basically, succesfully applies only in the compact setting. However, its generalization to a generic complete manifold is still missing. We have the following (see [RRV94b], Proposition 1.1)

Proposition 7.9. *Suppose that $K \in L^1(\mathbb{B}^m)$. If equation (7.69) admits a positive solution u such that*

$$\lim_{|x| \to 1^-} u(x) = +\infty, \tag{7.91}$$

then the following inequalities hold:

$$\int_{\mathbb{B}^m} K(x) < 0; \tag{7.92}$$

$$\int_s^1 r^{1-m} \left(\int_{|x|<r} K(x) \right) dr < 0 \tag{7.93}$$

for all $s \in [0,1)$.

Proof. **Step 1.** First we prove (7.92). Let $\gamma = \frac{m+2}{m-2}$. For $1 > \varepsilon > 0$ we consider a family of nonnegative, nondecreasing functions $\rho_\varepsilon(r)$, $r \in [0,1)$, such that

$$\rho_\varepsilon(r) = \begin{cases} 0 & \text{if } 0 \le r \le \frac{\varepsilon}{2}, \\ r - \frac{\varepsilon}{2} & \text{if } \varepsilon \le r < 1. \end{cases} \tag{7.94}$$

We multiply equation (7.69) by $\rho_\varepsilon\left(\frac{1}{u^\gamma}\right)$ to obtain

$$c_m \Delta u \, \rho_\varepsilon\left(\frac{1}{u^\gamma}\right) + K u^\gamma \rho_\varepsilon\left(\frac{1}{u^\gamma}\right) = 0. \tag{7.95}$$

The function $\rho_\varepsilon\left(\frac{1}{u^\gamma}\right)(x)$ is compactly supported in $\mathbb{B}^m$, because of (7.91); thus we can apply the first Green's formula in (7.95) to obtain

$$-c_m \int_{\mathbb{B}^m} \left\langle \nabla u, \nabla\left(\rho_\varepsilon\left(\frac{1}{u^\gamma}\right)\right) \right\rangle + \int_{\mathbb{B}^m} K u^\gamma \rho_\varepsilon\left(\frac{1}{u^\gamma}\right) = 0. \tag{7.96}$$

An easy computation shows that the first integral in (7.96) is equal to

$$c_m \gamma \int_{\mathbb{B}^m} u^{-\gamma-1} \rho_\varepsilon'\left(\frac{1}{u^\gamma}\right) |\nabla u|^2.$$

Since $\rho_\varepsilon' > 0$ if $u < \frac{1}{\varepsilon^\gamma}$, we deduce that, for $\varepsilon > 0$ small enough,

$$\int_{\mathbb{B}^m} K u^\gamma \rho_\varepsilon\left(\frac{1}{u^\gamma}\right) = -\gamma \int_{\mathbb{B}^m} u^{-\gamma-1} \rho_\varepsilon'\left(\frac{1}{u^\gamma}\right) |\nabla u|^2 < 0. \qquad (7.97)$$

We let $\varepsilon \to 0$; since $0 < \frac{\rho_\varepsilon(r)}{r} < 1$, by Lebesgue's theorem the left-hand side term in (7.97) admits a limit, which is

$$\int_{\mathbb{B}^m} K(x) \le 0. \qquad (7.98)$$

To be more precise, we can arrange to have that the function $\varepsilon \to \rho_\varepsilon'(r)$ is nondecreasing for all $r \ge 0$; in this way, we conclude that the right-hand side in (7.97) decreases to its limit as ε tends to 0. Thus we find

$$\int_{\mathbb{B}^m} K(x) = -\gamma \int_{\mathbb{B}^m} u^{-\alpha-1} |\nabla u|^2 < 0. \qquad (7.99)$$

Step 2. We prove now (7.93). First we observe that

$$\operatorname{div}\left(u^{-\alpha}\nabla u\right) = u^{-\alpha}\Delta u - \alpha u^{-\alpha-1}|\nabla u|^2. \qquad (7.100)$$

Next, we rewrite equation (7.69) as

$$-c_m u^{-\alpha}\Delta u = K. \qquad (7.101)$$

We use (7.100) in (7.101) and integrate over $|x| < r$ to obtain

$$\frac{1}{c_m} \int_{|x|<r} K(x) = -\gamma \int_{|x|<r} u^{-\gamma-1}|\nabla u|^2 - \int_{|x|=r} u^{-\alpha} \langle \nabla u, \nu \rangle, \qquad (7.102)$$

where ν is the exterior unit normal vector to $\mathbb{S}_r = r\mathbb{S}^{m-1}$. Using polar coordinates $(r, \sigma) \in (0,1) \times \mathbb{S}^{m-1}$, we rewrite the last integral in (7.102) as

$$\int_{|x|=r} u^{-\alpha} \langle \nabla u, \nu \rangle = \frac{r^{m-1}}{1-\gamma} \int_{\mathbb{S}^{m-1}} \frac{\partial}{\partial r}\left(u^{-\gamma+1}\right)(r, \sigma)\, d\sigma, \qquad (7.103)$$

where $d\sigma$ is the canonical measure on $\mathbb{S}^{m-1}$. We also observe that

$$\int_{\mathbb{S}^{m-1}} u^{-\gamma+1}(1, \sigma)\, d\sigma = 0 \qquad (7.104)$$

by (7.91). We multiply both sides of (7.102) by $c_m r^{1-m}$ and integrate over the interval $(s, 1)$ using (7.103) and (7.104); we deduce

$$\int_s^1 \left(r^{1-m} \int_{|x|<r} K(x)\right) dr = -c_m \left[\gamma \int_s^1 \left(r^{1-m} \int_{|x|<r} u^{-\gamma-1}|\nabla u|^2\right) dr \right.$$

$$\left. + \frac{1}{\gamma-1} \int_{\mathbb{S}^{m-1}} u^{1-\gamma}(s, \sigma)\, d\sigma\right] \qquad (7.105)$$

from which (7.93) follows immediately. $\square$

7.2 Special symmetries and existence

In this section we prove an existence result (see Theorem 7.12 below) where we allow the coefficient $a(x)$ to be nonpositive everywhere on M. However, this is obtained by assuming the strong curvature assumption (7.116). A geometrical application of this result is contained in Corollary 7.13.

In what follows we shall use the following version of Gronwall inequality.

Lemma 7.10. *Let φ and ψ be nonnegative, locally integrable functions on $[0, +\infty)$ and let η be a solution of the differential inequality*

$$\eta'(r) \le \varphi(r)\eta(r) + \psi(r) \quad on \; [0, +\infty). \tag{7.106}$$

Then, $\forall r \in [0, +\infty)$,

$$\eta(r) \le \eta(0)e^{\int_0^r \varphi(t)\,dt} + \int_0^r \psi(t)e^{\int_t^r \varphi(\xi)\,d\xi}\,dt. \tag{7.107}$$

Proof. First of all we observe that, $\forall s \in [0, +\infty)$,

$$\left(\eta(r)e^{-\int_s^r \varphi(t)\,dt}\right)' = e^{\int_r^s \varphi(t)\,dt}\{\eta'(r) - \varphi(r)\eta(r)\},$$

so that, using (7.106), we deduce

$$\left(\eta(r)e^{-\int_s^r \varphi(t)\,dt}\right)' \le \psi(r)e^{-\int_s^r \varphi(t)\,dt}.$$

Integrating this inequality on $[0, r]$ gives

$$\eta(r)e^{-\int_s^r \varphi(t)\,dt} \le \eta(0)e^{\int_0^s \varphi(t)\,dt} + \int_0^r \psi(t)e^{-\int_s^t \varphi(\xi)\,d\xi}\,dt$$

and therefore the validity of (7.107). $\qquad\square$

A second ingredient we shall use is the following

Proposition 7.11. *Let M_h be an m-dimensional model and suppose that*

$$\tilde{C}r^{\frac{\alpha}{2}} \le \frac{h'(r)}{h(r)} \tag{7.108}$$

for $r \gg 1$ and some constants $\tilde{C} > 0$, $\alpha > -2$. Let $\sigma > 1$; then, given $\varepsilon > 0$, there exists $\delta > 0$ such that for $\beta_0 < \delta$ the initial value problem

$$\begin{cases} \beta'' + (m-1)\frac{h'}{h}(r)\beta' = g(r)\beta^\sigma, \\ \beta'(0) = 0, \;\; \beta(0) = \beta_0 \end{cases} \tag{7.109}$$

with $g \in C^0([0,+\infty))$ *satisfying*

$$\begin{cases} \text{(i) } g > 0 & \text{on } [0,+\infty), \\ \text{(ii) } g(r) \le \dfrac{C}{r^{1-\frac{\alpha}{2}}(\log r)^{1+\varepsilon}} & \text{for } r \gg 1 \end{cases} \tag{7.110}$$

has a positive, (bounded) C^2-solution β on $[0,+\infty)$.

Proof. First, we produce a subsolution of (7.109) on $[0,+\infty)$. We follow an idea contained in the proof of Proposition 3.2 of [RRS95]. Fix $R > 1$ and define $\tilde{g} \in C^0([0,+\infty))$ by setting

$$\tilde{g}(r) = \begin{cases} \dfrac{\varepsilon(\log R)^{\sigma\varepsilon}}{r(\log r)^{1+\varepsilon}} \left\{ (m-1)\dfrac{h'(r)}{h(r)} - \dfrac{1}{r} - \dfrac{1+\varepsilon}{r\log r} \right\} & \text{for } r \ge R, \\ \dfrac{\varepsilon(\log R)^{\sigma\varepsilon}}{R(\log R)^{1+\varepsilon}} \left\{ (m-1)\dfrac{h'(R)}{h(R)} - \dfrac{1}{R} - \dfrac{1+\varepsilon}{R\log R} \right\} & \text{for } 0 \le r < R. \end{cases}$$

Next, set

$$\gamma(r) = \frac{1}{(\log R)^\varepsilon} - \frac{1}{(\log r)^\varepsilon} \quad \text{on } [R,+\infty).$$

A direct computation shows that

$$\gamma'' + (m-1)\frac{h'}{h}(r)\gamma' = \tilde{g}(r)\frac{1}{(\log R)^{\sigma\varepsilon}}$$

on $[R,+\infty)$. Furthermore, γ is increasing on $[0,+\infty)$ and

$$\lim_{r\to+\infty} \gamma(r) = \frac{1}{(\log R)^\varepsilon}.$$

It follows that

$$\gamma(r)^\sigma \le \frac{1}{(\log R)^{\varepsilon\sigma}} \quad \text{on } [R,+\infty)$$

and therefore γ satisfies

$$\gamma'' + (m-1)\frac{h'}{h}(r)\gamma' \ge \tilde{g}(r)\gamma(r)^\sigma \tag{7.111}$$

on $[R,+\infty)$. However, from (7.108), using also $\alpha > -2$ we have that, up to choosing R sufficiently large, for some absolute constant $T > 0$,

$$\tilde{g}(r) \ge \frac{T\varepsilon(\log R)^{\sigma\varepsilon}}{r^{1-\frac{\alpha}{2}}(\log r)^{1+\varepsilon}}$$

$\forall r \in [R,+\infty)$ and we can also suppose that R is large enough so that

$$T\varepsilon(\log R)^{\sigma\varepsilon} > C$$

where C is the constant appearing in (7.110) (ii). It follows that

$$\widetilde{g}(r) \geq g(r) \quad \text{on } [R, +\infty). \tag{7.112}$$

Then, (7.111) and (7.112) imply

$$\gamma'' + (m-1)\frac{h'}{h}(r)\gamma' \geq g(r)\gamma(r)^\sigma. \tag{7.113}$$

Next we choose $\delta > 0$ so small that the solution of (7.109) with $0 < \beta_0 < \delta$ is defined on a maximal interval $[0, R_1)$ with $R + 1 < R_1 \leq +\infty$ and in such a way that, having set $R_0 = R + 1$,

$$\beta(R_0) < \gamma(R_0), \quad \beta'(R_0) < \gamma'(R_0). \tag{7.114}$$

If $R_1 = +\infty$ we have nothing to prove; thus let $R_1 < +\infty$. On $[R_0, R_1)$ set $w = \beta - \gamma$ so that, by (7.109) and (7.113), w satisfies

$$w'' + (m-1)\frac{h'}{h}(r)w' - g(r)(\beta^\sigma - \gamma^\sigma) \leq 0. \tag{7.115}$$

Let $\bar{r} \in [R_0, R_1)$ be chosen; then on $[R_0, \bar{r}]$ we have the validity of (7.115). Define $z(r)$ on $[R_0, \bar{r}]$ as

$$z(r) = \begin{cases} \frac{\sigma}{\gamma(r) - \beta(r)} \int_{\beta(r)}^{\gamma(r)} t^{\sigma-1}\, dt & \text{if } \beta(r) \neq \gamma(r), \\ \sigma\beta(r)^{\sigma-1} & \text{if } \beta(r) = \gamma(r). \end{cases}$$

Then $z(r) \in C^0([R_0, \bar{r}])$, $z(r) \geq 0$ and (7.115), (7.114) can be now written as

$$\begin{cases} w'' + (m-1)\frac{h'}{h}(r)w' - c(r)w \leq 0, \\ w(R_0) < 0, \quad w'(R_0) < 0, \end{cases}$$

with $c(r) = g(r)z(r) \geq 0$. Therefore, by the maximum principle, w attains its negative minimum either at $\bar{r}$ or at R_0. The latter case cannot occur because it would imply $w'(R_0) \geq 0$; hence the negative minimum is achieved at $\bar{r}$, that is, $\beta(\bar{r}) < \gamma(\bar{r})$. Since $\bar{r} \in [R_0, R_1)$ was chosen arbitrarily, we have that

$$\beta(r) < \gamma(r) \quad \text{on } [R_0, R_1)$$

contradicting the fact that $\beta(r) \to +\infty$ as $r \to R_1$ since $[0, R_1)$ was its maximal interval of definition. $\square$

We are now ready to prove

Theorem 7.12. *Let $(M, \langle\, ,\, \rangle)$ be a complete Riemannian manifold with a pole o satisfying*

$$\mathrm{Riem}_{(M, \langle\, ,\, \rangle)} \leq -B^2, \tag{7.116}$$

for some constant $B > 0$. Let $a(x), b(x) \in C^{0,\mu}(M)$ for some $0 < \mu \le 1$. Assume that $b(x)$ is nonnegative and strictly positive outside some compact set and that

$$B_0 = \{x \in M : b(x) = 0\}$$

satisfies $\lambda_1^L(B_0) > 0$ with $L = \Delta + a(x)$. Let $\widehat{a}(r), \widehat{b}(r) \in C^0([0,+\infty))$ be such that

$$\widehat{a}(r) \le 0 \quad on \ [0,+\infty), \tag{7.117}$$

$$\widehat{b}(r) \le C \frac{e^{\frac{\sigma-1}{B(m-1)} \int_0^r \widehat{a}(\xi)\,d\xi}}{r(\log r)^{1+\varepsilon}} \tag{7.118}$$

for $r \gg 1$ and some $\sigma > 1, \varepsilon > 0$, and suppose that

$$\begin{cases} a(x) \ge \widehat{a}(r(x)), \\ b(x) \le \widehat{b}(r(x)) \end{cases} \tag{7.119}$$

on M. Then, the equation

$$\Delta u + a(x)u - b(x)u^\sigma = 0$$

admits a positive maximal C^2-solution on M.

Proof. We first consider the ODE problem

$$\begin{cases} (i) \ \alpha'' + (m-1)B\coth(Br)\alpha' + \widehat{a}(r)\alpha = 0 \ \text{on} \ [0,+\infty), \\ (ii) \ \alpha'(0) = 0, \ \alpha(0) = \alpha_0 \in (0,1]. \end{cases} \tag{7.120}$$

Standard Picard iteration (see for instance [CL55]) guarantees the existence of a solution α on $[0,+\infty)$. Then α solves the integrated equation

$$\alpha(r) = \alpha_0 - \int_0^r (\sinh Bt)^{1-m} \int_0^t (\sinh Bs)^{m-1}\widehat{a}(s)\alpha(s)\,ds \tag{7.121}$$

on $[0,+\infty)$. Thus, using (7.117), we have

$$\alpha(r) = \alpha_0 + \int_0^r B(\cosh Bt)(\sinh Bt)^{1-m} \int_0^t \frac{(\sinh Bs)^{m-1}}{B\cosh Bt}(-\widehat{a}(s))\alpha(s)\,ds$$

$$\le \alpha_0 + \int_0^r B(\cosh Bt)(\sinh Bt)^{1-m} \int_0^t \frac{(\sinh Bs)^{m-1}}{B\cosh Bs}(-\widehat{a}(s))\alpha(s)\,ds$$

$$= \alpha_0 + \left\{ \frac{1}{2-m}(\sinh Bt)^{2-m} \int_0^t \frac{(\sinh Bs)^{m-1}}{B\cosh Bs}(-\widehat{a}(s))\alpha(s)\,ds \right\}\Big|_0^r$$

$$\quad + \frac{1}{B(m-2)}\int_0^r \tanh Bt(-\widehat{a}(t))\alpha(t)\,dt$$

$$\le \alpha_0 - \frac{1}{B(m-2)}\int_0^r \widehat{a}(t)\alpha(t)\,dt,$$

that is,

$$\alpha(r) \le \alpha_0 - \frac{1}{B(m-2)} \int_0^r \widehat{a}(t)\alpha(t)\, dt. \tag{7.122}$$

Setting

$$\eta(r) = -\frac{1}{B(m-2)} \int_0^r \widehat{a}(t)\alpha(t)\, dt,$$

from (7.122) we deduce

$$\eta'(r) \le \frac{1}{B(m-2)}(-\widehat{a}(r))\eta(r) + \frac{\alpha_0}{B(m-2)}(-\widehat{a}(r)).$$

Furthermore, $\eta(0) = 0$. Applying Lemma 7.10 and (7.122) we obtain

$$\alpha(r) \le \alpha_0 + \eta(r) \le \alpha_0 + \int_0^r \frac{\alpha_0}{B(m-2)}(-\widehat{a}(t))e^{\int_t^r \frac{1}{B(m-2)}(-\widehat{a}(\xi))\, d\xi}\, dt.$$

It follows that

$$\alpha(r) \le \alpha_0 - \alpha_0 \int_0^r \frac{d}{dt}\left(e^{\int_t^r \frac{-\widehat{a}(\xi)}{B(m-2)}\, d\xi}\right)\, dt = \alpha_0 e^{\int_0^r \frac{-\widehat{a}(\xi)}{B(m-2)}\, d\xi},$$

that is,

$$\alpha(r) \le \alpha_0 e^{-\frac{1}{B(m-2)}\int_0^r \widehat{a}(t)\, dt}. \tag{7.123}$$

Note that from (7.120) and (7.117) we also deduce $\alpha' \ge 0$. We now define

$$g(r) = \widehat{b}(r)\alpha_0^{\sigma-1} e^{-\frac{\sigma-1}{B(m-2)}\int_0^r \widehat{a}(\xi)\, d\xi} > 0 \tag{7.124}$$

and observe that, because of (7.124) and (7.118),

$$g(r) \le \frac{C}{r(\log r)^{1+\varepsilon}} \tag{7.125}$$

for $r \gg 1$ and some constant $C > 0$. Applying Proposition 7.11 we deduce the existence of a positive solution β of the ODE problem

$$\begin{cases} \beta'' + (m-1)B\coth(Br)\beta' - g(r)\beta^\sigma = 0 & \text{on } [0,+\infty), \\ \beta'(0) = 0, \ \beta(0) = \beta_0 > 0. \end{cases} \tag{7.126}$$

Note that $\beta' \ge 0$ since $g(r) \ge 0$. Next we define

$$\gamma(r) = \gamma_0 + \int_0^r (\sinh Bt)^{1-m} \int_0^t (\sinh Bs)^{m-1}\left\{\widehat{b}(s)\gamma(s)^\sigma - \widehat{a}(s)\gamma(s)\right\}\, ds.$$

Choosing $\gamma_0 \in (0, \alpha_0\beta_0)$ we claim that $\gamma(r)$ is defined on all of $[0,+\infty)$. Observe that if $\gamma(r)$ is bounded above on each interval of the type $[0,T]$, then we are done. Towards this aim it is sufficient to show that defining $\bar{R}$ as

$$\bar{R} = \sup\{r_0 \in (0,+\infty) : \gamma(r) \le \alpha(r)\beta(r) \ \text{on } [0,r_0]\}$$

we have $\bar{R} = +\infty$. We suppose the converse, that is, $\bar{R} < +\infty$. Then, using our choice of γ_0 and (7.123),

$$\gamma(\bar{R}) = \gamma_0 + \int_0^{\bar{R}} (\sinh Bt)^{1-m} \int_0^t (\sinh Bs)^{m-1} \left\{ \widehat{b}(s)\gamma(s)^\sigma - \widehat{a}(s)\gamma(s) \right\} ds$$

$$< \alpha_0 \beta_0 \qquad\qquad\qquad\qquad\qquad\qquad\qquad\qquad\qquad\qquad (7.127)$$

$$+ \int_0^{\bar{R}} (\sinh Bt)^{1-m} \int_0^t (\sinh Bs)^{m-1} \{g(s)\alpha(s)\beta(s)^\sigma - \widehat{a}(s)\alpha(s)\beta(s)\} ds.$$

Since $\alpha', \beta' \geq 0$ we have

$$\Delta_{\mathbb{H}^m_{-B^2}} \alpha\beta = \alpha\Delta_{\mathbb{H}^m_{-B^2}}\beta + \beta\Delta_{\mathbb{H}^m_{-B^2}}\alpha + 2\alpha'\beta' \geq g\alpha\beta^\sigma - \widehat{a}\alpha\beta,$$

in other words,

$$\left((\sinh Br)^{m-1}(\alpha\beta)'\right)' \geq (\sinh Br)^{m-1}\{g(r)\alpha\beta^\sigma - \widehat{a}(r)\alpha\beta\}.$$

Integrating over $[0, \bar{R}]$ and using $\alpha'(0) = \beta'(0) = 0$, $\alpha(0) = \alpha_0, \beta(0) = \beta_0$ we get

$$\alpha(R)\beta(R) \geq \alpha_0\beta_0 + \int_0^{\bar{R}} (\sinh Bt)^{1-m}$$

$$\times \int_0^t (\sinh Bs)^{m-1}\{g(s)\alpha(s)\beta(s)^\sigma - \widehat{a}(s)(r)\alpha(s)\beta(s)\} ds. \qquad (7.128)$$

Since $\alpha_0 \in (0, 1]$, comparing (7.127) and (7.128) we obtain

$$\gamma(\bar{R}) < \alpha(\bar{R})\beta(\bar{R})$$

contradicting the definition of $\bar{R}$. It follows that γ is defined on $[0, +\infty)$ and it satisfies

$$\begin{cases} \gamma'' + (m-1)B\coth(Br)\gamma' + \widehat{a}(r)\gamma - \widehat{b}(r)\gamma^\sigma = 0 & \text{on } [0, +\infty), \\ \gamma'(0) = 0, \ \gamma(0) = \gamma_0 > 0. \end{cases} \qquad (7.129)$$

Furthermore, since $\widehat{a}(r) \leq 0$ and $\widehat{b}(r) \geq 0$ we have

$$\gamma'(r) \geq 0 \quad \text{on } [0, +\infty).$$

Defining

$$u_-(x) = \gamma(r(x)) \quad \text{on } M,$$

assumptions (7.116), (7.119), $\gamma' \geq 0$ and the Hessian comparison theorem (see Chapter 1) show that u_- is a positive subsolution of

$$\Delta u + a(x)u - b(x)u^\sigma = 0.$$

Applying Theorem 6.5 we obtain the desired conclusion. $\square$

Applying Theorem 7.12 we have that the equation

$$c_m \Delta_{\mathbb{H}^{p+1}} v - (m-1)(m-2-2p)v + K(x)v^{\frac{m+2}{m-2}} = 0 \qquad (7.130)$$

with $0 \leq p < \frac{m-2}{2}$ has a smooth positive solution v on $\mathbb{H}^{p+1}$ whenever $K(x) \leq 0$ on $\mathbb{H}^{p+1}$ and

$$K(r) \geq -C \frac{e^{-\frac{m-2-2p}{m-1}r(x)}}{r(\log r)^{1+\varepsilon}} \qquad (7.131)$$

for $r(x) \gg 1$ and some constants $C, \varepsilon > 0$.

Letting (θ, x) be the generic point in $\mathbb{S}^{m-p-1} \times \mathbb{H}^{p+1}$ we define

$$u(\theta, x) = v(x).$$

Then

$$c_m \Delta u - (m-1)(m-2-2p)u + K(x)u^{\frac{m+2}{m-2}} = 0$$

on $\mathbb{S}^{m-p-1} \times \mathbb{H}^{p+1}$. We have thus proved

Corollary 7.13. *Let the generic point of $\mathbb{S}^{m-p-1} \times \mathbb{H}^{p+1}$ be denoted by (θ, x) and let $\widehat{K}(\theta, x) = K(x)$ with $K(x) \leq 0$ on $\mathbb{S}^{m-p-1} \times \mathbb{H}^{p+1}$ and satisfying (7.131). Let $0 \leq p < \frac{m-2}{2}$; then, the product metric on $\mathbb{S}^{m-p-1} \times \mathbb{H}^{p+1}$ can be conformally deformed to a new metric with scalar curvature $\widehat{K}(\theta, x)$.*

7.3   The case of Euclidean space and further results

We observe that the techniques introduced above can be used to produce positive supersolutions of

$$\Delta u - a(x)u = 0 \quad \text{on } M$$

with a controlled behaviour at infinity; this of course requires again the strong geometric assumption (7.38). However, we shall later use this technique on $\mathbb{R}^m$ to produce interesting results from the point of view of the study of the qualitative behaviour of positive solutions of Yamabe type equations.

7.3.1   A linear comparison result

We begin with the following *comparison result*.

Lemma 7.14. *Let M_g be a model and let $a(r) \in C^0([0, +\infty))$ be nonnegative. Let $u = \alpha \circ r$ be a positive, radial, C^2-solution on M_g of*

$$\Delta u + a(r)u \leq 0 \qquad (7.132)$$

and let $v = \beta \circ r$ be a positive radial C^2-subsolution of (7.132) on $M \backslash B_R(o)$ for some $R > 0$ such that $\beta'(R) > 0$. Then, there exists a constant $C > 0$ such that

$$u(x) \leq Cv(x) \quad \text{on } M \backslash B_R(o). \qquad (7.133)$$

Proof. First of all we observe that existence of a solution $u > 0$ of (7.132) implies by a result of Fisher-Colbrie and Schoen, [FCS80], that the operator $L = \Delta + a(x)$ satisfies

$$\lambda_1^L(M) \geq 0. \tag{7.134}$$

Next, (7.132) yields

$$\begin{cases} \alpha'' + (m-1)\frac{g'(r)}{g(r)}\alpha' + a(r)\alpha \leq 0 & \text{on } [0, +\infty), \\ \alpha'(0) = 0 \end{cases} \tag{7.135}$$

so that nonnegativity of $a(r)$, positivity of α and (7.135) imply $\alpha'(r) \leq 0$. Having made this observation, we choose $\xi > 0$ to satisfy

$$\xi\beta'(R) > \alpha'(R) \quad \text{and} \quad \xi\beta(R) > \alpha(R). \tag{7.136}$$

Next, we fix $\bar{R} > R$ and let φ be a positive radial solution of the Dirichlet eigenvalue problem for L on $B_{\bar{R}+1}$. Because of (7.134) φ satisfies

$$\begin{cases} \varphi'' + (m-1)\frac{g'(r)}{g(r)}\varphi' + a(r)\varphi \leq 0, \\ \varphi'(0) = 0, \quad \varphi(\bar{R}+1) = 0. \end{cases} \tag{7.137}$$

In particular

$$\varphi'(t) \leq 0 \quad \text{on } [0, \bar{R}+1). \tag{7.138}$$

On the interval $[R, \bar{R}]$ we consider the function

$$w(t) = \alpha(t) - \xi\beta(t);$$

because of our choice ξ it follows that

$$\text{i) } w'(R) < 0; \quad \text{ii) } w(R) < 0. \tag{7.139}$$

On the other hand, w satisfies

$$\Delta w + a(r)w \leq 0 \quad \text{on } \overline{B_{\bar{R}}} \setminus B_R. \tag{7.140}$$

But, on $\overline{B_{\bar{R}}} \setminus B_R$, φ is positive and solves

$$\Delta\varphi + a(r)\varphi \leq 0.$$

By the generalized maximum principle (see [PW67], page 73) we have, either

$$w = c\varphi$$

for some negative constant c or $\frac{w}{\varphi}$ has its negative absolute minimum at $\bar{R}$ or R. However, in this latter case, we must have

$$\left(\frac{w}{\varphi}\right)'(R) \geq 0; \tag{7.141}$$

but from (7.138) and (7.139),

$$w'(R)\varphi(R) - \varphi'(R)w(R) < 0$$

contradicting (7.141). We therefore conclude that $w(\bar{R}) < 0$. Since $\bar{R} > R$ was chosen arbitrarily we have proved

$$u \leq \xi w \quad \text{on } M \backslash B_R.$$

This shows the validity of (7.133). □

7.3.2 Back to Corollary 5.8

We give the next result in order to comment again on Corollary 5.8. We begin with the following observation which is contained in the proof of Theorem 7.3.

Theorem 7.15. *Let $(M, \langle , \rangle)$ be a complete, connected, simply connected manifold satisfying*

$$\text{Riem} \leq -B^2$$

for some constant $B > 0$. Let u be a nonnegative C^2-solution of

$$\Delta u + a(x)u \geq 0$$

on B_R with $u(0) > 0$; suppose

$$a(x) \leq AB^2 \coth(Br(x)) \quad \text{on } M, \text{ with } A \leq \frac{(m-1)^2}{4}.$$

Then

$$\bar{u}(r) \geq C \frac{r e^{\frac{m-1}{2}Br}}{\text{vol}(\partial B_r)} \tag{7.142}$$

for some constant $C > 0$ independent of r.

A similar result holds when $B = 0$ but we shall analyze this and the next proposition later on in more details.

Proposition 7.16. *Let $(M, \langle , \rangle)$ be a complete manifold of dimension $m \geq 2$, connected and simply connected. Suppose that*

$$\text{Riem} \leq -B^2, \quad B > 0$$

and for $a(x) \in C^0(M)$,

$$a(x) \leq \frac{(m-1)^2}{4}B^2 \coth(Br(x)) \quad \text{on } M.$$

Then, there exists a positive C^2-supersolution w of

$$\Delta w + a(x)w \leq 0 \tag{7.143}$$

satisfying

$$C^{-1}r(x)e^{-\frac{m-1}{2}Br(x)} \leq w(x) \leq Cr(x)e^{-\frac{m-1}{2}Br(x)} \tag{7.144}$$

for some constant $C > 0$ and $r(x) \gg 1$.

Proof. We choose $\hat{a}(r) \in C^0([0, +\infty)$ to satisfy

$$\text{i) } \hat{a}(r) > 0 \text{ on } [0, +\infty); \quad \text{ii) } a(x) \leq \hat{a}(r(x)) \text{ on } M \tag{7.145}$$

and, for some fixed $\varepsilon > 0$,

$$\hat{a}(r) \begin{cases} = \frac{(m-1)^2}{4}B^2 \coth Br & \text{on } [\varepsilon, +\infty), \\ \leq \frac{(m-1)^2}{4}B^2 \coth Br & \text{on } [0, \varepsilon). \end{cases} \tag{7.146}$$

Next, we consider hyperbolic space $(\mathbb{H}^m, ds^2)$ with constant sectional curvature $-B^2$. On $\mathbb{H}^m \setminus \{o\} = (0, +\infty) \times \mathbb{S}^{m-1}$ we represent the metric in the form

$$ds^2 = dt^2 + \frac{1}{B}\sinh^2(Bt)\,d\theta^2 \tag{7.147}$$

where $d\theta^2$ is the canonical metric on $\mathbb{S}^{m-1}$ and $t(x) = \text{dist}_{(\mathbb{H}^m, ds^2)}(x, o)$. On $\mathbb{H}^m$ the equation

$$\Delta_{ds^2}u + \hat{a}(r)u = 0 \tag{7.148}$$

has a positive radial solution. Indeed, by (7.147) a radial solution of (7.148) is of the form $u = \alpha \circ t$ with α a solution of

$$\begin{cases} \alpha'' + (m-1)B\coth(Bt)\alpha' + \hat{a}(t)\alpha = 0 & \text{on } [0, +\infty), \\ \alpha'(0) = 0, \ \alpha(0) = \alpha_0, \end{cases} \tag{7.149}$$

and it is well known that (7.149) admits a solution. Having chosen $\alpha_0 > 0$ so that $u(0) = \alpha_0 > 0$, Theorem 7.15 implies that u is positive on $\mathbb{H}^m$. We fix $b > 0$ and consider

$$v(t(x)) = [t(x) - b][\sinh Bt(x)]^{-\frac{m-1}{2}} \quad \text{on } (b, +\infty).$$

It is immediate to verify that v is a positive radial subsolution of (7.148) on $\mathbb{H}^m \setminus \bar{B}_b$. Furthermore

$$v'(b + \varepsilon) > 0$$

for some $\varepsilon > 0$ sufficiently small. Then, according to Lemma 7.14 and Theorem 7.15, $u(x) = \alpha(t(x))$ satisfies

$$C^{-1}t(x)e^{-\frac{m-1}{2}Bt(x)} \leq u(x) \leq Ct(x)e^{-\frac{m-1}{2}Bt(x)} \tag{7.150}$$

for some constant $C > 0$ and $t(x) \gg 1$. We define, on $(M, \langle\,,\,\rangle)$,

$$w(x) = \alpha(r(x)); \tag{7.151}$$

it is then clear that w satisfies (7.144). Furthermore, from (7.145) and (7.149) we have

$$\alpha'(t) \leq 0 \quad \text{on } [0, +\infty). \tag{7.152}$$

By the Hessian comparison theorem, see (7.45), we then deduce that w satisfies (7.143) and it is C^2 and positive. $\qquad\square$

We go back to Corollary 5.8. In the assumptions of Proposition 7.16 we see that two positive solutions u and v of

$$\Delta u + a(x)u - b(x)u^\sigma = 0, \, \sigma > 1 \quad \text{on } M,$$

$b(x) \geq 0$, $b(x) \not\equiv 0$, coincide provided

$$u(x) - v(x) = o\!\left(r(x)e^{-\frac{m-1}{2}Br(x)}\right) \quad \text{as } r(x) \to +\infty \tag{7.153}$$

or, in other words,

$$(u(x) - v(x))\frac{e^{\frac{m-1}{2}Br(x)}}{r(x)} \to 0 \quad \text{as } r(x) \to +\infty. \tag{7.154}$$

This may seem to be a stronger requirement than (5.81), but this is not the case, because there

$$a(x) \equiv \frac{m(m-2)}{4} < \frac{(m-1)^2}{4}$$

and it is in this latter case that we have produced w in Proposition 7.16.

We can also use Proposition 7.16 to explicitate assumption (3.35) in the nonexistence result of Theorem 3.3.

However, we shall now focus on the case of $\mathbb{R}^m$.

7.3.3 The Euclidean space

To deal with the case of $\mathbb{R}^m$ we need a number of refined results contained in [BRS98].

Lemma 7.17. *Let* $a(t), b(t) \in C^0([0, +\infty))$ *satisfy*

$$b(t) \geq 0 \tag{7.155}$$

and

$$a(t) \leq \frac{A^2}{t^2} \quad \text{with } A \leq \frac{m-2}{2}, m \geq 3. \tag{7.156}$$

Let α *be a solution of the problem*

$$\begin{cases} \alpha'' + (m-1)\frac{1}{t}\alpha' + a(t)\alpha - b(t)\alpha|\alpha|^{\sigma-1} \geq 0, \\ \alpha(0) = \alpha_0 > 0, \, \alpha'(0) = 0 \end{cases} \tag{7.157}$$

on $[0,T)$, $T \le +\infty$, $\sigma > 1$. Then α is positive; moreover, for all $\delta < T$ sufficiently small and for all $t \in [\delta, T)$ we have

$$
\alpha(t) \ge
\begin{cases}
\frac{m-2}{4}\alpha_0\left(\frac{\delta}{t}\right)^{\frac{m-2}{2}} \log\left(\frac{t}{\delta}\right) & \text{if } A = \frac{m-2}{2}, \\[2ex]
\frac{1}{4}\frac{m-2+\sqrt{(m-2)^2-4A^2}}{\sqrt{(m-2)^2-4A^2}}\alpha_0\left(\frac{\delta}{t}\right)^{\frac{1}{2}\left(m-2-\sqrt{(m-2)^2-4A^2}\right)} & \text{if } A < \frac{m-2}{2}.
\end{cases}
\tag{7.158}
$$

Remark. In particular, any solution of the problem

$$
\begin{cases}
\beta'' + \frac{m-1}{t}\beta' + a(t)\beta \ge 0, \\
\beta(0) = \beta_0 > 0, \ \ \beta'(0) = 0
\end{cases}
\tag{7.159}
$$

with $a(t)$ as in the statement of Lemma 7.17, is defined and positive on $[0,+\infty)$ and satisfies the estimate from below given in (7.158).

Proof. We follow the lines outlined above for the general case of a Riemannian manifold with curvature bounded above. Let $[0,\widetilde{T})$ be the maximal subinterval of $[0,T)$ on which α is positive, and fix $0 < \delta < s \le \widetilde{T}$. We are going to show that $\widetilde{T} = T$ by applying the following version of Green's second identity,

$$
s^{m-1}[\varphi'(s)\alpha(s) - \varphi(s)\alpha'(s)] = \delta^{m-1}[\varphi'(\delta)\alpha(\delta) - \varphi(\delta)\alpha'(t)]
\tag{7.160}
$$
$$
+ \int_\delta^s \left\{ t^{m-1}[\varphi''(t)\alpha(t) - \varphi(t)\alpha''(t)] + (m-1)t^{m-2}[\varphi'(t)\alpha(t) - \varphi(t)\alpha'(t)] \right\} dt
$$

to the functions α and φ, where $\varphi = \varphi_s$ is the solution of the problem

$$
\begin{cases}
\varphi'' + \frac{m-1}{t}\varphi' + \frac{A^2}{t^2}\varphi = 0 & \text{on } [\delta, s], \\
\varphi(s) = 0, \ \ \varphi'(s) = s^{-(m-1)}.
\end{cases}
\tag{7.161}
$$

By (7.161), φ is negative and monotonically increasing in (δ, s), and it is easy to check that

$$
\varphi(t) =
\begin{cases}
s^{\frac{2-m}{2}} t^{\frac{2-m}{2}} \log\left(\frac{t}{s}\right) & \text{if } A = \frac{m-2}{2}, \\[2ex]
-\frac{1}{\sqrt{(m-2)^2-4A^2}} t^{\frac{1}{2}\left(2-m-\sqrt{(m-2)^2-4A^2}\right)} \\
\times s^{\frac{1}{2}\left(2-m+\sqrt{(m-2)^2-4A^2}\right)}\left(1 - \left(\frac{t}{s}\right)^{\sqrt{(m-2)^2-4A^2}}\right) & \text{if } A < \frac{m-2}{2}.
\end{cases}
$$

A straightforward computation shows that

$$
\alpha(t)\varphi''(t) - \alpha''(t)\varphi(t) \ge -\frac{m-1}{t}[\alpha(t)\varphi'(t) - \alpha'(t)\varphi(t)] - \varphi(t)b(t)\alpha(t)|\alpha(t)|^{\sigma-1},
$$

so that, substituting in Green's identity above yields

$$
\alpha(s) \ge -\int_\delta^s t^{m-1}\varphi(t)b(t)\alpha(t)|\alpha(t)|^{\sigma-1}\, dt + \delta^{m-1}[\varphi'(\delta)\alpha(\delta) - \varphi(\delta)\alpha'(\delta)].
\tag{7.162}
$$

Since $s \leq \widetilde{T}$, the above integral is negative so that

$$\alpha(s) \geq \delta^{m-1}[\varphi'(\delta)\alpha(\delta) - \varphi(\delta)\alpha'(\delta)].$$

For $A = \frac{m-2}{2}$, using $\varphi'(\delta) = s^{\frac{2-m}{2}}\delta^{-\frac{m}{2}}\left[1 + \frac{m-2}{2}\log\left(\frac{s}{\delta}\right)\right]$, we can estimate the right-hand side of (7.162) by

$$\delta^{\frac{m}{2}}\left\{\alpha(\delta)\frac{m-2}{2\delta} - |\alpha'(\delta)|\right\}\frac{1}{s^{\frac{m-2}{2}}}\log\left(\frac{s}{\delta}\right),$$

and since $\alpha(0) = \alpha_0 > 0$ and $\alpha'(0) = 0$, the quantity in braces is greater than $\frac{m-2}{4\delta}\alpha_0 > 0$ for $\delta > 0$ sufficiently small.

We therefore conclude that the first inequality of (7.158) holds in $[\delta, \widetilde{T}]$.

Similarly, if $A < \frac{m-2}{2}$, one verifies that the second inequality of (7.158) holds in $[\delta, \widetilde{T}]$; together with the definition of $\widetilde{T}$, this implies that $T = \widetilde{T}$, and the proof is complete.   □

The next lemma deals with the case where $a(t)$ decays at infinity faster than the critical decay $\frac{A^2}{t^2}$.

Lemma 7.18. *Let $a(t), b(t) \in C^0([0, +\infty))$. Assume that $b(t) \geq 0$, and that there exist constants $A > 0$ and $\varepsilon > 0$ such that*

$$a(t) \leq \min\left\{\frac{(m-2)^2}{4t^2}, \frac{A^2}{t^{2+\varepsilon}}\right\} \quad \text{for } t \geq 0.$$

If α is a solution of (7.159) defined on $[0, +\infty)$, then there is a constant $C > 0$ depending only upon $\alpha(0)$ and $a(t)$, such that

$$\alpha(t) \geq C \quad \text{for } t \in [0, +\infty). \tag{7.163}$$

Proof. The idea of the proof is the same as that of Lemma 7.17 and consists of applying Green's identity (7.160) with suitable test functions φ having the properties specified in the next

Claim. For all s sufficiently large, there exists a piecewise C^2-function φ_s with the following properties:

(i)
$$\varphi_s''(t) + \frac{m-1}{t}\varphi_s'(t) + a(t)\varphi_s(t) \geq 0 \quad \text{on } (0, s];$$

(ii)
$$\varphi_s < 0 \quad \text{in } (0, s), \quad \varphi_s(s) = 0 \quad \text{and} \quad \lim_{s \to +\infty} s^{m-1}\varphi_s'(s) = 1;$$

(iii) there exist constants c_1, c_2 independent of s such that, for δ small enough,

$$|\varphi_s(\delta)| \leq c_1\left(\log\frac{1}{\delta}\right)^{-\frac{m-2}{2}} \quad \text{and} \quad \varphi_s'(\delta) \geq c_2\left(\log\frac{1}{\delta}\right)\delta^{-\frac{m}{2}}.$$

Postponing for the moment the proof of the claim (after Proposition 7.19), we finish the proof of the lemma.

Arguing as in the proof of Lemma 7.17, we have

$$s^{m-1}\varphi'_s(s)\alpha(s) \geq - \int_\delta^s t^{m-1}\varphi_s(t)b(t)\alpha(t)|\alpha(t)|^{\sigma-1}\,dt \qquad (7.164)$$
$$+ 2^{m-1}[\varphi'_s(\delta)\alpha(\delta) - \varphi_s(\delta)\alpha'(\delta)]$$

for δ small enough. Since $s^{m-1}\varphi'_s(s) \to 1$ and $\varphi_s(t) < 0$, this yields

$$\alpha(s) \geq \frac{1}{2}s^{m-1}[\varphi'_s(\delta)\alpha(\delta) - \varphi_s(\delta)\alpha'(\delta)].$$

Using property (iii) of the claim, we see that the right-hand side is bounded from below by

$$\frac{c_2}{2}\delta^{\frac{m}{2}}\left(\log\frac{1}{\delta}\right)\left[\frac{\alpha(\delta)}{\delta} - \frac{c_1}{c_2}|\alpha'(\delta)|\right],$$

which is positive, provided $\delta > 0$ is small enough. $\square$

Using Lemma 7.18 we now proceed similarly to what we did in the proof of Proposition 7.16 to establish

Proposition 7.19. *Let* $a(t) \in C^0([0,+\infty))$, $a(t) \geq 0$ *and satisfying, for* $m \geq 3$,

$$a(t) \leq \begin{cases} \text{i)}\ \frac{A^2}{t^2} & \text{for } 0 \leq A \leq \frac{m-2}{2}, \\ \text{ii)}\ \min\left\{\frac{(m-2)^2}{4t^2}, \frac{A^2}{t^{2+\varepsilon}}\right\} & \text{for some } \varepsilon > 0, A > 0. \end{cases} \qquad (7.165)$$

Then the equation

$$\Delta u + a(|x|)u = 0$$

has a positive radial solution $u \in C^2(\mathbb{R}^m)$ *satisfying, for some constant* $C > 0$ *and* $|x| \gg 1$,

in case (7.165) i),

$$\begin{cases} C^{-1}(\log|x|)|x|^{-\frac{m-2}{2}} \leq u(x) \leq C(\log|x|)|x|^{-\frac{m-2}{2}} & \text{if } A = \frac{m-2}{2}, \\ C^{-1}|x|^{-\gamma} \leq u(x) \leq C|x|^{-\gamma} & \text{if } 0 \leq A < \frac{m-2}{2} \end{cases} \qquad (7.166)$$

where $\gamma = \frac{1}{2}\left(m - 2 - \sqrt{(m-2)^2 - 4A^2}\right)$,

while in case (7.165) ii),

$$C^{-1} \leq u(x) \leq C. \qquad (7.167)$$

We now go back to Lemma 7.18 to provide a

Proof of the claim. We divide the argument into three steps.

Step 1. We begin by considering the differential equation

$$\varphi'' + \frac{m-1}{t}\varphi' + \frac{A^2}{t^{2+\varepsilon}} = 0. \tag{7.168}$$

Assuming, without loss of generality, that $\nu = \frac{m-2}{\varepsilon}$ is not an integer, the general solution is given in terms of the Bessel function of the first kind by

$$\varphi(t) = C_1 t^{-\frac{m-2}{2}} J_\nu\left(\frac{2A}{\varepsilon}t^{-\frac{\varepsilon}{2}}\right) + C_2 t^{-\frac{m-2}{2}} J_{-\nu}\left(\frac{2A}{\varepsilon}t^{-\frac{\varepsilon}{2}}\right),$$

see [Wat66] page 97. Given $s > 0$ we define

$$\varphi_{1,s}(t) = \frac{\Gamma(\nu)\Gamma(-\nu)}{\varepsilon} s^{-\frac{m-2}{2}} t^{-\frac{m-2}{2}} \tag{7.169}$$

$$\times \left[-J_{-\nu}\left(\frac{2A}{\varepsilon}s^{-\frac{\varepsilon}{2}}\right) J_\nu\left(\frac{2A}{\varepsilon}t^{-\frac{\varepsilon}{2}}\right) + J_\nu\left(\frac{2A}{\varepsilon}s^{-\frac{\varepsilon}{2}}\right) J_{-\nu}\left(\frac{2A}{\varepsilon}t^{-\frac{\varepsilon}{2}}\right)\right]$$

so that $\varphi_{1,s}$ is a solution of (7.168) satisfying

$$\varphi_{1,s}(s) = 0.$$

Since $J_\lambda(x)$ behaves like $\sqrt{\frac{2}{\pi x}}\cos\left(x - \frac{\lambda\pi}{2} - \frac{\lambda}{4}\right)$ as $x \to +\infty$, the function $\varphi_{1,s}$ is not of definite sign in $(0, s)$; however, we claim that there exists p_1 independent of s such that, for all $s > p_1$, $\varphi_{1,s}(t) < 0$ in $[p_1, s)$. Indeed, from the power series representation of Bessel functions

$$J_\lambda(x) = \left(\frac{x}{2}\right)^\lambda \sum_0^{+\infty} \frac{(-1)^k (x/2)^k}{\Gamma(k+1)\Gamma(k+\lambda+1)},$$

it follows easily that $J_\nu(x)$ (resp. $\Gamma(1-\nu)J_{-\nu}(x)$) is positive and increasing (resp. decreasing) in an interval $(0, x_+)$ (resp. $(0, x_-)$). By rewriting the right-hand side of (7.169) as

$$-\frac{\Gamma(\nu)}{\varepsilon t^{\frac{m-2}{2}} s^{\frac{m-2}{2}}} J_\nu\left(\frac{2A}{\varepsilon}t^{-\frac{\varepsilon}{2}}\right)\Gamma(1-\nu)J_{-\nu}\left(\frac{2A}{\varepsilon}s^{-\frac{\varepsilon}{2}}\right)$$

$$\times \left[1 - \frac{J_\nu\left(\frac{2A}{\varepsilon}s^{-\frac{\varepsilon}{2}}\right)\Gamma(1-\nu)J_{-\nu}\left(\frac{2A}{\varepsilon}t^{-\frac{\varepsilon}{2}}\right)}{J_\nu\left(\frac{2A}{\varepsilon}t^{-\frac{\varepsilon}{2}}\right)\Gamma(1-\nu)J_{-\nu}\left(\frac{2A}{\varepsilon}s^{-\frac{\varepsilon}{2}}\right)}\right],$$

we may therefore conclude that there exists $p_1 > 0$, which depends only upon x_+ and x_-, such that $\varphi_{1,s}(t)$ is negative for $p_1 \le t < s$.

We will also need to know the asymptotic behavior of $\varphi_{1,s}(t)$ and $\varphi'_{1,s}(t)$ for s and t large. Using the formula

$$J_\lambda(x) = \frac{\left(\frac{x}{2}\right)^\lambda}{\Gamma(1+\lambda)} + o(x^\lambda) \quad \text{as } x \to 0,$$

which can be read off from the power series representation above, and the recurrence formulas for the derivative of J_λ,

$$\left(x^\lambda J_\lambda(x)\right)' = x^\lambda J_{\lambda-1}(x), \quad \left(x^{-\lambda} J_\lambda(x)\right)' = x^{-\lambda} J_{\lambda+1}(x),$$

a lengthy but straightforward computation shows that

$$\varphi_{1,s} = \frac{s^{2-m} - t^{2-m}}{m-2} + \left(t^{2-m} + s^{2-m}\right)(\eta_t + \eta_s),$$

$$\varphi'_{1,s}(t) = t^{-(m-1)} + \frac{\varepsilon A^2}{(m-2)(m-2-\varepsilon)} s^{2-m} t^{-\varepsilon-1} + t^{1-m}(\eta_s + \eta_t)$$
$$+ s^{2-m} t^{-\varepsilon-1}(\eta_s + \eta_t),$$

where $\lim_{t\to+\infty} \eta_t = 0$ and $\lim_{s\to+\infty} \eta_s = 0$. This immediately gives

$$\lim_{s\to+\infty} s^{m-1}\varphi'_{1,s}(s) = 1$$

which implies that there exists $p_2 \geq p_1$, independent of s, such that

$$|\varphi_{1,s}(t)| \leq \frac{2m-1}{2(m-2)^2} t^{2-m}, \quad \frac{4m-9}{4(m-2)} t^{1-m} \leq \varphi'_{1,s}(t) \leq 2t^{1-m} \qquad (7.170)$$

for all $t, p_2 \leq t \leq s$.

Step 2. Fix $p \in (p_2, s)$ and let $\varphi_2(t) = \varphi_{2,p,s}(t)$ be the solution of the differential equation

$$\varphi'' + \frac{m-1}{t}\varphi' + \frac{(m-2)^2}{4t^2}\varphi = 0 \quad \text{on } (0, +\infty)$$

satisfying $\varphi_2(p) = \varphi_{1,s}(p)$, $\varphi'_2(p) = \varphi'_{1,s}(p)$.
   It is easy to verify that φ_2 is given by the formula

$$\varphi_2(t) = p^{\frac{m-2}{2}}\left[\varphi_{1,s}(p)t^{\frac{2-m}{2}} - \left(\frac{m-2}{2}\varphi_{1,s}(p) + p\varphi'_{1,s}(p)t^{\frac{2-m}{2}} \log\left(\frac{p}{t}\right)\right)\right].$$

According to (7.170) above, we have

$$\frac{m-2}{2}\varphi_{1,s}(p) + p\varphi'_{1,s}(p) > \frac{m-3}{2(m-2)}p^{2-m},$$

which implies that

$$\varphi_2(t) < 0 \quad \text{for } 0 < t \leq p.$$

Step 3. Let t_0 be such that $a(t) \leq \frac{A^2}{t^{2+\varepsilon}}$ for $t \geq t_0$ and let $p = \max\{t_0, p_2\}$. Define

$$\varphi_s(t) = \begin{cases} \varphi_{2,p,s}(t) & \text{for } 0 < t \leq p, \\ \varphi_{1,s}(t) & \text{for } t > p. \end{cases}$$

By steps 1 and 2 above, φ_s is a piecewise C^2-solution of

$$\varphi''(t) + \frac{m-1}{t}\varphi'(t) + a(t)\varphi(t) \geq 0$$

which is negative in $(0, s]$, vanishes for $t = s$ and satisfies

$$\lim_{s \to +\infty} s^{m-1}\varphi_s'(s) = 1.$$

To conclude the proof of our claim, it remains to be shown that φ_s satisfies condition (iii), that is, there are constants c_1 and c_2 independent of s such that

$$|\varphi_s(\delta)| \leq c_1\left(\log\frac{1}{\delta}\right)^{-\frac{m-2}{2}} \quad \text{and} \quad \varphi_s'(\delta) \geq c_2\left(\log\frac{1}{\delta}\right)\delta^{-\frac{m}{2}}$$

for δ small enough.

Indeed, if $\delta < p$ the definition of φ_2 and (7.170) imply

$$|\varphi_s(\delta)| \leq p^{\frac{m-2}{2}+1}\varphi_{1,s}'(p)\delta^{\frac{m-2}{2}}\log\frac{p}{\delta} \leq 2p^{\frac{2-m}{2}}(\log p + 1)\log\left(\frac{1}{\delta}\right)\delta^{\frac{2-m}{2}}.$$

On the other hand, a simple computation gives

$$\varphi_s'(\delta) = p^{\frac{m-2}{2}}\left[-\frac{m-2}{2}\varphi_{1,s}(p) + \left(\frac{m-2}{2}\varphi_{1,s}(p) + p\varphi_{1,s}'(p)\right)\right.$$
$$\left.\times\left(1 + \frac{m-2}{2}\log\left(\frac{p}{\delta}\right)\right)\right]\delta^{-\frac{m}{2}}$$
$$\geq \frac{m-2}{2}p^{\frac{m-2}{2}}\left(m\varphi_{1,s}(p) + p\varphi_{1,s}'(p)\right)\log\left(\frac{p}{\delta}\right)\delta^{-\frac{m}{2}},$$

whence again by (7.170)

$$\varphi_s'(\delta) \geq \frac{m-3}{4}p^{\frac{2-m}{2}}\log\left(\frac{p}{\delta}\right)\delta^{-\frac{m}{2}}$$

and the proof of the claim is completed. □

As a first consequence of Proposition 7.19 we give the next uniqueness result for ground states (see Chapter 3 for the definition) of a Yamabe type equation in $\mathbb{R}^m$. This is exactly the subtle case for uniqueness and in what follows we shall apply Corollary 5.8.

Theorem 7.20. Let $m \geq 3$ and $a(x), b(x) \in C^0(\mathbb{R}^m)$, $b(x) \geq 0$, $b(x) \not\equiv 0$, with

$$a(x) \leq \begin{cases} \text{i)} \ \frac{A^2}{|x|^2} & \text{for } 0 \leq A \leq \frac{m-2}{2}, \\ \text{ii)} \ \min\left\{\frac{(m-2)^2}{4|x|^2}, \frac{A^2}{|x|^{2+\varepsilon}}\right\} & \text{for some } \varepsilon > 0, \ A > 0. \end{cases} \tag{7.171}$$

Let $u(x)$ and $v(x)$ be positive solutions of

$$\Delta u + a(x)u - b(x)u^\sigma = 0 \quad on \ \mathbb{R}^m.$$

Assume, as $|x| \to +\infty$,

in case (7.171) i),

$$\begin{cases} u(x) - v(x) = o\left((\log|x|)|x|^{-\frac{m-2}{2}} \right) & if \ A = \frac{m-2}{2}, \\ u(x) - v(x) = o\left(|x|^{-\frac{1}{2}\left(m-2-\sqrt{(m-2)^2 - 4A^2} \right)} \right) & if \ 0 \le A < \frac{m-2}{2}; \end{cases}$$

in case (7.171) ii),

$$u(x) - v(x) \to 0.$$

Then,

$$u(x) \equiv v(x) \quad on \ \mathbb{R}^m.$$

Remark. As already observed, this case is particularly significant when $u(x)$ and $v(x)$ are ground states of equation $\Delta u + a(x)u - b(x)u^\sigma = 0$ on $\mathbb{R}^m$.

As an immediate application of Proposition 7.19 we consider the nonexistence result given in Theorem 3.3. We then have

Theorem 7.21. Let $a(x), b(x) \in C^0(\mathbb{R}^m)$ and assume $b(x) \ge 0$. Let $H \ge 1$ and $A \in \mathbb{R}$ be constants such that

$$\max\{0, A\} \le H - 1.$$

Let $m \ge 3$ and suppose that $Ha(x)$ satisfy (7.171). Then the differential inequality

$$u\Delta u + a(x)u^2 - b(x)u^{\sigma+1} \ge -A|\nabla u|^2, \ \sigma \ge 1$$

has no nonnegative C^2-solutions u on $\mathbb{R}^m$ satisfying

$$\operatorname{supp} u \cap \{x \in \mathbb{R}^m : b(x) > 0\} \neq \emptyset$$

and,

in case (7.171) i),

$$\begin{cases} \left\{ \int_{\partial B_r} \left[(\log|x|)|x|^{-\frac{m-2}{2}} \right]^{\frac{\beta+1}{H}(2-\beta)} u^{2(\beta+1)} \right\}^{-1} \notin L^1(+\infty) & if \ A = \frac{m-2}{2}, \\ \left\{ \int_{\partial B_r} |x|^{-\gamma\frac{\beta+1}{H}(2-\beta)} u^{2(\beta+1)} \right\}^{-1} \notin L^1(+\infty) & if \ 0 \le A < \frac{m-2}{2}, \end{cases}$$

with $\gamma = \frac{1}{2}\left(m - 2 - \sqrt{(m-2)^2 - 4A^2} \right)$,

while in case (7.171) ii)

$$\left\{ \int_{\partial B_r} u^{2(\beta+1)} \right\}^{-1} \notin L^1(+\infty)$$

for some $\beta > 1$ and $\max{0, A} \le \beta \le H - 1$.

Bibliography

[ADN59]   S. Agmon, A. Douglis, and L. Nirenberg. Estimates near the boundary for solutions of elliptic partial differential equations satisfying general boundary conditions. I. *Comm. Pure Appl. Math.*, 12:623–727, 1959.

[Ahl38]   L. V. Ahlfors. An extension of Schwarz's lemma. *Trans. Amer. Math. Soc.*, 43:359–364, 1938.

[AM85]   P. Aviles and R. C. McOwen. Conformal deformations of complete manifolds with negative curvature. *J. Differential Geom.*, 21(2):269–281, 1985.

[AM88]   P. Aviles and R. C. McOwen. Conformal deformation to constant negative scalar curvature on noncompact Riemannian manifolds. *J. Differential Geom.*, 27(2):225–239, 1988.

[Ama76]   H. Amann. Supersolutions, monotone iterations, and stability. *J. Differential Equations*, 21(2):363–377, 1976.

[Aub76]   T. Aubin. Équations différentielles non linéaires et problème de Yamabe concernant la courbure scalaire. *J. Math. Pures Appl. (9)*, 55(3):269–296, 1976.

[Aub98]   T. Aubin. *Some nonlinear problems in Riemannian geometry.* Springer Monographs in Mathematics. Springer-Verlag, Berlin, 1998.

[BE87]   J.P. Bourguignon and J.P. Ezin. Scalar curvature functions in a conformal class of metrics and conformal transformations. *Trans. Amer. Math. Soc.*, 301:723–736, 1987.

[Bes08]   A. L. Besse. *Einstein manifolds.* Classics in Mathematics. Springer-Verlag, Berlin, 2008. Reprint of the 1987 edition.

[Bis77]   R.L. Bishop. Decompositon of cut loci. *Proc. Amer. Math. Soc.*, 65:133–137, 1977.

[BMR09]   B. Bianchini, L. Mari, and M. Rigoli. Spectral radius, index estimates for Schrödinger operators and geometric applications. *J. Funct. Anal.*, 256:1769–1820, 2009.

[BRS98]   L. Brandolini, M. Rigoli, and A. G. Setti. Positive solutions of Yamabe type equations on complete manifolds and applications. *J. Funct. Anal.*, 160:176–222, 1998.

[BVV91]   M.-F. Bidaut-Véron and L. Véron. Nonlinear elliptic equations on compact Riemannian manifolds and asymptotics of Emden equations. *Invent. Math.*, 106:489–539, 1991.

[Cal57]   E. Calabi. An extension of Hopf's maximum principle with an application to Riemannian geometry. *Duke Math. J.*, 25:45–56, 1957.

[Car88]   É. Cartan. *Leçons sur la géométrie des espaces de Riemann.* Les Grands Classiques Gauthier-Villars. [Gauthier-Villars Great Classics]. Éditions Jacques Gabay, Sceaux, 1988. Reprint of the second (1946) edition.

[CGT82]   J. Cheeger, M. Gromov, and M. Taylor. Finite propagation speed, kernel estimates for functions of the Laplace operator, and the geometry of complete Riemannian manifolds. *J. Differential Geom.*, 17(1):15–53, 1982.

[Cha84]   I. Chavel. *Eigenvalues in Riemannian geometry*, volume 115 of *Pure and Applied Mathematics*. Academic Press Inc., Orlando, FL, 1984.

[Cha06]   I. Chavel. *Riemannian geometry*, volume 98 of *Cambridge Studies in Advanced Mathematics*. Cambridge University Press, Cambridge, second edition, 2006. A modern introduction.

[Che55]   S.S. Chern. An elementary proof of the existence of isothermal parameters on a surface. *Proc. Amer. Math. Soc.*, 6:771–782, 1955.

[CL55]   E. A. Coddington and N. Levinson. *Theory of ordinary differential equations.* McGraw-Hill, New York, 1955.

[CL87]   K. S. Cheng and J.-T. Lin. On the elliptic equations $\Delta u = K(x)u^\sigma$ and $\Delta u = K(x)e^{2u}$. *Trans. Amer. Math. Soc.*, 304(2):639–668, 1987.

[CL95]   W. X. Chen and C. Li. A note on the Kazdan-Warner type conditions. *J. Differential Geom.*, 41(2):259–268, 1995.

[CN92]   K.-S. Cheng and W.-M. Ni. On the structure of the conformal scalar curvature equation on $\mathbf{R}^n$. *Indiana Univ. Math. J.*, 41(1):261–278, 1992.

[Dav95]   E. B. Davies. *Spectral theory and differential operators*, volume 42 of *Cambridge Studies in Advanced Mathematics*. Cambridge University Press, Cambridge, 1995.

[dC92]   M. P. do Carmo. *Riemannian geometry.* Mathematics: Theory & Applications. Birkhäuser Boston Inc., Boston, MA, 1992. Translated from the second Portuguese edition by Francis Flaherty.

[Eis49]   L. P. Eisenhart. *Riemannian Geometry.* Princeton University Press, Princeton, N. J., 1949. 2d printing.

[EL78]   J. Eells and L. Lemaire. A report on harmonic maps. *Bull. London Math. Soc.*, 10:1–68, 1978.

[Esc87]   J. Escobar. Positive solutions for some semilinear elliptic equations with critical Sobolev exponents. *Comm. Pure. Appl. Math.*, 40:623–657, 1987.

[Esc90]   J. Escobar. Uniqueness theorems on conformal deformations of metrics, Sobolev inequalities, and an eigenvalue estimate. *Comm. Pure. Appl. Math.*, 43:857–883, 1990.

[FC85]   D. Fischer-Colbrie. On complete minimal surfaces with finite Morse index in three-manifolds. *Invent. Math.*, 82(1):121–132, 1985.

[FCS80]   D. Fischer-Colbrie and R. Schoen. The structure of complete stable minimal surfaces in 3-manifolds of non-negative scalar curvature. *Comm. Pure. Appl. Math.*, 33:199–211, 1980.

[Fed69]   H. Federer. *Geometric measure theory.* Die Grundlehren der mathematischen Wissenschaften, Band 153. Springer-Verlag New York Inc., New York, 1969.

[For91]   O. Forster. *Lectures on Riemann surfaces*, volume 81 of *Graduate Texts in Mathematics.* Springer-Verlag, New York, 1991. Translated from the 1977 German original by Bruce Gilligan, Reprint of the 1981 English translation.

[GHV72]   W. Greub, S. Halperin, and R. Vanstone. *Connections, curvature, and cohomology. Vol. I: De Rham cohomology of manifolds and vector bundles.* Academic Press, New York, 1972. Pure and Applied Mathematics, Vol. 47.

[Gri76]   P. A. Griffiths. *Entire holomorphic mappings in one and several complex variables.* Princeton University Press, Princeton, N. J., 1976. The fifth set of Hermann Weyl Lectures, given at the Institute for Advanced Study, Princeton, N. J., October and November 1974, Annals of Mathematics Studies, No. 85.

[Gri99]   A. Grigor'yan. Analytic and geometric background of recurrence and non-explosion of the Brownian motion on Riemannian manifolds. *Bull. Amer. Math. Soc. (N.S.)*, 36(2):135–249, 1999.

[GT01]   D. Gilbarg and N. Trudinger. *Elliptic partial differential equations of second order.* Classics in Mathematics. Springer-Verlag, Berlin, 2001. Reprint of the 1998 edition.

[GW79]   R. E. Greene and H. Wu. *Function theory on manifolds which possess a pole*, volume 699 of *Lecture Notes in Mathematics*. Springer, Berlin, 1979.

[HW53]   P. Hartman and A. Wintner. On the existence of Riemannian manifolds which cannot carry non-constant analytic or harmonic functions in the small. *Amer. J. Math.*, 75:260–276, 1953.

[Ili96]   S. Ilias. Inégalités de Sobolev et résultats d'isolement pour les applications harmoniques. *J. Funct. Anal.*, 139(1):182–195, 1996.

[Jin88]   Z. R. Jin. A counterexample to the Yamabe problem for complete non-compact manifolds. In *Partial differential equations (Tianjin, 1986)*, volume 1306 of *Lecture Notes in Math.*, pages 93–101. Springer, Berlin, 1988.

[Kor14]   A. Korn. Zwei Anwendungen der Methode der sukzessiven Annäherungen. *Schwarz-Festschr.*, pages 215–229, 1914.

[Kui49]   N. Kuiper. On conformally flat spaces in the large. *Ann. of Math.*, 50:916–924, 1949.

[KW74a]   J. L. Kazdan and F. W. Warner. Curvature functions for compact 2-manifolds. *Ann. of Math. (2)*, 99:14–47, 1974.

[KW74b]   J. L. Kazdan and F. W. Warner. Curvature functions for open 2-manifolds. *Ann. of Math. (2)*, 99:203–219, 1974.

[KW75a]   J. L. Kazdan and F. W. Warner. Existence and conformal deformation of metrics with prescribed Gaussian and scalar curvatures. *Ann. of Math. (2)*, 101:317–331, 1975.

[KW75b]   J. L. Kazdan and F. W. Warner. Scalar curvature and conformal deformation of Riemannian structure. *J. Differential Geometry*, 10:113–134, 1975.

[Le98]   V. K. Le. On some equivalent properties of sub- and supersolutions in quasilinear elliptic equations. *Hiroshima Math. J.*, 28:373–380, 1998.

[Lee97]   J. M. Lee. *Riemannian manifolds*, volume 176 of *Graduate Texts in Mathematics*. Springer-Verlag, New York, 1997. An introduction to curvature.

[Lee03]   J. M. Lee. *Introduction to smooth manifolds*, volume 218 of *Graduate Texts in Mathematics*. Springer-Verlag, New York, 2003.

[Lee11]   J. M. Lee. *Introduction to topological manifolds*, volume 202 of *Graduate Texts in Mathematics*. Springer, New York, second edition, 2011.

[Li90]   P. Li. On the structure of complete Kähler manifolds with nonnegative curvature near infinity. *Invent. Math.*, 99(3):579–600, 1990.

[Lic16] L. Lichtenstein. Zur Theorie der konformen Abbildung nichtanalytischer, singularitätenfreier Flächenstücke auf ebene Gebiete. *Krak. Anz.*, 1916:192–217, 1916.

[Lic58] A. Lichnerowicz. *Geómétrie des groupes des transformation.* Travaux et Recherches Mathématiques, III. Dunod, Paris, 1958.

[LP87] J. M. Lee and T. H. Parker. The Yamabe problem. *Bull. Amer. Math. Soc.*, 17:37–91, 1987.

[LR96] P. Li and M. Ramachandran. Kähler manifolds with almost nonnegative Ricci curvature. *Amer. J. Math.*, 118(2):341–353, 1996.

[LTY98] P. Li, L. F. Tam, and D. Yang. On the elliptic equation $\Delta u + ku - Ku^p = 0$ on complete Riemannian manifolds and their geometric applications. *Trans. Amer. Math. Soc.*, 350(3):1045–1078, 1998.

[LY90] P. Li and S.-T. Yau. Curvature and holomorphic mappings of complete Kähler manifolds. *Compositio Math.*, 73(2):125–144, 1990.

[Mil63] J. Milnor. *Morse theory.* Based on lecture notes by M. Spivak and R. Wells. Annals of Mathematics Studies, No. 51. Princeton University Press, Princeton, N.J., 1963.

[MR10] P. Mastrolia and M. Rigoli. Diffusion-type operators, Liouville theorems and gradient estimates on complete manifolds. *Nonlinear Anal.*, 72:3767–3785, 2010.

[MRS10] L. Mari, M. Rigoli, and A. G. Setti. Keller-Osserman conditions for diffusion-type operators on Riemannian manifolds. *J. Funct. Anal.*, 258:665–712, 2010.

[MRV] P. Mastrolia, M. Rimoldi, and G. Veronelli. Myers-Type Theorems and Some Related Oscillation Results. *J. Geom. Anal.*, pages 1–17. 10.1007/s12220-011-9213-0.

[MS10] O. Munteanu and N. Sesum. The Poisson equation on complete manifolds with positive spectrum and applications. *Adv. Math.*, 223:198–219, 2010.

[Ni82] W. M. Ni. On the elliptic equation $\Delta u + K(x)u^{(n+2)/(n-2)} = 0$, its generalizations, and applications in geometry. *Indiana Univ. Math. J.*, 31(4):493–529, 1982.

[NST01] L. Ni, Y. Shi, and L.-F. Tam. Poisson equation, Poincaré-Lelong equation and curvature decay on complete Kähler manifolds. *J. Differential Geom.*, 57:339–388, 2001.

[Oba62a] M. Obata. Certain conditions for a Riemannian manifold to be isometric with a sphere. *J. Math. Soc. Japan*, 14:333–340, 1962.

[Oba62b] M. Obata. Conformal transformations of compact Riemannian manifolds. *Illinois J. Math*, 6:292–295, 1962.

[Pet06a] P. Petersen. *Riemannian geometry*, volume 171 of *Graduate Texts in Mathematics*. Springer-Verlag, New York, 2006.

[Pet06b] P. Petersen. *Riemannian geometry*, volume 171 of *Graduate Texts in Mathematics*. Springer, New York, second edition, 2006.

[PR]   S. Pigola and M. Rimoldi. Characterizations of Model Manifolds by Means of Certain Differential Systems. *Canad. Math. Bull.*, pages 1–14. doi:10.4153/CMB-2011-134-0.

[PRS03a] S. Pigola, M. Rigoli, and A. G. Setti. Some applications of integral formulas in Riemannian geometry and PDE's. *Milan J. Math.*, 71:219–281, 2003.

[PRS03b] S. Pigola, M. Rigoli, and A. G. Setti. Volume growth, "a priori" estimates, and geometric applications. *Geom. Funct. Anal.*, 13(6):1302–1328, 2003.

[PRS05a] S. Pigola, M. Rigoli, and A. G. Setti. A Liouville-type result for quasilinear elliptic equations on complete Riemannian manifolds. *J. Funct. Anal.*, 219:400–432, 2005.

[PRS05b] S. Pigola, M. Rigoli, and A. G. Setti. Maximum principles on Riemannian manifolds and applications. *Mem. Amer. Math. Soc.*, 174(822):x+99, 2005.

[PRS05c] S. Pigola, M. Rigoli, and A. G. Setti. Vanishing theorems on Riemannian manifolds, and geometric applications. *J. Funct. Anal.*, 229:424–461, 2005.

[PRS07]   S. Pigola, M. Rigoli, and A. G. Setti. Some characterizations of space-forms. *Trans. Amer. Math. Soc.*, 359(4):1817–1828 (electronic), 2007.

[PRS08]   S. Pigola, M. Rigoli, and A. G. Setti. *Vanishing and finiteness result in geometric analysis*, volume 266 of *Progress in Mathematics*. Birkhauser Verlag Ag, 2008.

[PRS10]   S. Pigola, M. Rigoli, and A. G. Setti. Existence and non-existence results for a logistic-type equation on manifolds. *Trans. Amer. Math. Soc.*, 362:1907–1936, 2010.

[PW67]   M. H. Protter and H. F. Weinberger. *Maximum principles in differential equations*. Englewood Cliffs Prentice-Hall, 1967.

[RRS95]   A. Ratto, M. Rigoli, and A.G. Setti. On the Omori-Yau maximum principle and its application to differential equations and geometry. *J. Funct. Anal.*, 134:486–510, 1995.

[RRV94a] A. Ratto, M. Rigoli, and Véron. Conformal immersions of complete Riemannian manifolds and extensions of the Schwarz lemma. *Duke Math. J.*, 74(1):223–236, 1994.

[RRV94b] A. Ratto, M. Rigoli, and L. Véron. Scalar curvature and conformal deformations of hyperbolic space. *J. Funct. Anal.*, 121:15–77, 1994.

[RRV97] A. Ratto, M. Rigoli, and L. Veron. Scalar curvature and conformal deformations of noncompact riemannian manifolds. *Math. Z.*, 225:395–426, 1997.

[RS01] M. Rigoli and A. G. Setti. Liouville type theorems for ϕ-subharmonic functions. *Rev. Mat. Iberoamericana*, 17:471–520, 2001.

[RSV05] M. Rigoli, M. Salvatori, and M. Vignati. Some remarks on the weak maximum principle. *Rev. Mat. Iberoamericana*, 21(2):459–481, 2005.

[RZ07] M. Rigoli and S. Zamperlin. *A priori* estimates, uniqueness and existence of positive solutions of Yamabe type equations on complete manifolds. *J. Funct. Anal.*, 245:144–176, 2007.

[Sat73] D. H. Sattinger. *Topics in stability and bifurcation theory*. Lecture Notes in Mathematics, Vol. 309. Springer-Verlag, Berlin, 1973.

[Sch84] R. Schoen. Conformal deformation of a Riemannian metric to constant scalar curvature. *J. Differential Geom.*, 20(2):479–495, 1984.

[Spi79] M. Spivak. *A comprehensive introduction to differential geometry. Vol. II*. Publish or Perish Inc., Wilmington, Del., second edition, 1979.

[Swa68] C. A. Swanson. *Comparison and oscillation theory of linear differential equations*. Academic Press, New York, 1968. Mathematics in Science and Engineering, Vol. 48.

[Swa75] C. A. Swanson. Picone's identity. *Rend. Mat. (6)*, 8(2):373–397, 1975. Collection of articles dedicated to Mauro Picone on the occasion of his ninetieth birthday, II.

[SY94] R. Schoen and S.-T. Yau. *Lectures on differential geometry*. Conference Proceedings and Lecture Notes in Geometry and Topology, I. International Press, Cambridge, 1994.

[Tru68] N. S. Trudinger. Remarks concerning the conformal deformation of Riemannian structures on compact manifolds. *Ann. Scuola Norm. Sup. Pisa (3)*, 22:265–274, 1968.

[Wat66] G.N. Watson. *A treatise on the theory of Bessel functions*. Cambridge University Press, 1966.

[Yam60] H. Yamabe. On a deformation of Riemannian structures on compact manifolds. *Osaka Math. J.*, 12:21–37, 1960.

[Yau73]   S.T. Yau. Remarks on conformal transformations. *J. Diff. Geom.*, 3:369–381, 1973.

[YN59]   K. Yano and T. Nagano. Einstein spaces admitting a one-parameter group of conformal transformations. *Ann. of Math.*, 69:451–461, 1959.

List of Symbols

$[X, Y]$	Lie bracket of two vector fields X and Y, page 12
$\|\mathrm{Hess}(u)\|$	norm of $\mathrm{Hess}(u)$, page 19
$\|X\|$	norm of the vector field X, page 19
$\|x\|$	distance of the point $x \in \mathbb{R}^m$ from the origin 0, page 211
$\bigwedge^2(U)$	space of skew-symmetric 2-forms, page 14
$C^\infty(M)$	set of smooth functions defined on M, page 16
(U, φ)	local chart, page 8
Δu	Laplacian of the function u, page 18
δ_i^j	suggestive version of the Kronecker symbol, page 8
δ_{ij}	Kronecker symbol, page 8
$\dot{\gamma}$	tangent vector of the curve γ, page 22
$x^1, \ldots, x^m$	coordinate functions, page 8
$\frac{\partial u}{\partial \nu}$	directional derivative of the function u in the direction of ν, page 62
$\mathrm{Hess}(u)$	Hessian of the function u, page 18
$\lambda_1^{L_H}(\Omega)$	first eigenvalue of the operator L_H on the bounded domain Ω, page 74
$\lambda_1^{L_H}(M)$	first eigenvalue of the operator L_H on the Riemannian manifold M, page 74
$\mathcal{L}_X \langle\, ,\, \rangle$	Lie derivative of the metric $\langle\, ,\, \rangle$ in the direction of X, page 13

$\mathcal{L}_X \omega$	Lie derivative of the 1-form ω in the direction of X, page 13
$\mathcal{L}_X f$	Lie derivative of the function f in the direction of X, page 13
$\mathcal{L}_X Y$	Lie derivative of the vector field Y in the direction of X, page 13
$\mathbb{B}^m$	unit ball of $\mathbb{R}^m$, page 71
$\mathbb{B}_R^m(0)$	open disk of radius R centered at the origin in $T_o M \approx \mathbb{R}^m$, page 54
$\mathbb{H}^m$	standard hyperbolic space of dimension m, page 47
$\mathbb{H}_{-H^2}^m$	hyperbolic space of constant sectional curvature $-H^2$, page 75
$\mathbb{S}^m$	standard sphere of dimension m, page 47
$\mathbb{S}_+^m$	standard upper hemisphere, page 53
$\mathbb{S}_+^m\left(\sqrt{k}\right)$	upper hemisphere of radius $k^{-1/2}$, page 63
$\mathbb{S}_{k^2}^m$	sphere of constant sectional curvature k^2, page 61
$\mathfrak{X}(M)$	set of smooth vector fields on M, page 10
$\text{Ann}_{\eta,1}$	annulus $B_1 \setminus \overline{B}_\eta$, page 214
$\text{cut}(o)$	cut locus of the point o, page 21
II	second fundamental tensor, page 33
$\langle\,,\,\rangle_{ij}$	(local) components of the metric, page 8
$\nabla \omega$	covariant derivative of the 1-form ω, page 11
∇f	gradient of the function f, page 11
∇X	covariant derivative of the vector field X, page 10
$\nabla_Y \omega$	covariant derivative of ω in the direction of Y, page 11
$\nabla_Y X$	covariant derivative of X in the direction of Y, page 10
$\|f\|_\infty$	L^∞-norm of the function f, page 153
ω_m	volume of the unit sphere in $\mathbb{R}^m$, page 28
ω_{ik}	covariant derivative of the coefficient ω_i, page 11

$\mathrm{Conf}(M)$	group of conformal diffeomorphisms on M, page 146
$\dim(M)$	dimension of the manifold M, page 8
$\mathrm{div}\,X$	divergence of a vector field X, page 10
$\mathrm{hess}(u)$	$(1,1)$ version of $\mathrm{Hess}(u)$, page 20
Id	identity matrix, page 23
$\mathrm{Iso}(M)$	group of isometries of M, page 146
$\mathrm{Lip}_0(M)$	set of Lipschitz functions on M with compact support, page 29
ric	$(1,1)$ version of Ric, page 20
Riem	Riemann curvature tensor of type $(1,3)$, page 15
$\mathrm{Sect}(u \wedge v)$	sectional curvature of the plane π spanned by u and v, page 16
$\mathrm{supp}\,\varphi$	support of the function φ, page 77
Tor	torsion tensor, page 12
tr	trace, page 10
$\mathrm{vol}\,\partial B_R(o)$	volume of the boundary of the geodesic ball $B_R(o)$, page 28
$\mathrm{vol}\,B_R(o)$	volume of the geodesic ball $B_R(o)$, page 28
W	Weyl tensor, page 45
$\otimes$	tensor product, page 10
$\overline{\mathbb{B}^m}$	closed unit ball of $\mathbb{R}^m$, page 71
$\oslash$	Kulkarni-Nomizu product, page 46
$\frac{\partial \cdot}{\partial r}$	derivative in the radial direction, page 28
$\partial B_R(o)$	boundary of the geodesic ball centered at o with radius R, page 21
$(\varphi_t)_*$	push-forward of the flow, page 14
ρ	Riemannian distance from the origin of $\mathbb{B}^m$ with respect to the hyperbolic metric, page 212
Ric	Ricci tensor, page 16

$\mathrm{Ric}(\nabla r, \nabla r)$	radial Ricci curvature, page 23
$(M, \langle\,,\,\rangle)$	Riemannian manifold with metric $\langle\,,\,\rangle$, page 7
$\{\theta^i\}$	local orthonormal coframe, page 8
$\{e_i\}$	local orthonormal frame, page 8
$\mathrm{sgn}(\cdot)$	signum function, page 110
$\sharp$	sharp map, page 11
$\sqrt{g}$	square root of the determinant of the metric in polar geodesic coordinates, page 23
Θ^i_j	curvature forms, page 14
θ^i_j	Levi-Civita connection forms, page 8
φ_t	local flow of a vector field, page 14
$\wedge$	wedge product, page 8
$\widetilde{\langle\,,\,\rangle}$	conformally deformed metric, page 37
A	Schouten tensor, page 45
a_+	positive part of the function a, page 89
A^f_t	t-level set of a function f, page 29
$B_R(o)$	geodesic ball centered at o with radius R, page 21
C	Cotton tensor, page 45
$C^\infty(U)$	set of smooth functions defined on the open set U, page 9
$C^{0,\alpha}(M)$	space of locally Hölder continuous functions on M with exponent α, page 165
$C^\infty_0(M)$	set of smooth function with compact support on M, page 27
c_m	constant appearing in the (geometric) Yamabe equation ($m \geq 3$), page 39
$d\omega$	exterior differential of the 1-form ω, page 12
df	exterior differential of the function f, page 11

dx^i	differential of the coordinate function x^i, page 8
exp_o	exponential map of M at o, page 21
f^*	pull-back *via* the map f, page 32
f_i	local components of the differential df, page 11
$G(x, y)$	Green kernel, page 95
H	mean curvature vector field, page 34
h^α_{ijk}	coefficients of the covariant derivative of II, page 35
h^α_{ij}	coefficients of the second fundamental tensor II, page 33
h^ν	mean curvature in the direction of ν, page 34
$K_p(\pi)$	sectional curvature of the 2-plane π, page 16
K_{rad}	radial sectional curvature, page 31
R	curvature tensor of type $(0, 4)$, page 15
r	Euclidean distance from the origin in $\mathbb{R}^m$, page 212
$r(x)$	Riemannian distance function, page 21
R^i_{jkt}	(local) components of the Riemann curvature tensor of type $(1, 3)$, page 15
$R_{ijkt,l}$	covariant derivatives of the (local) components of the curvature tensor, page 16
R_{ijkt}	(local) components of the curvature tensor of type $(0, 4)$, page 15
R_{ij}	(local) components of the Ricci tensor, page 16
S	scalar curvature, page 16
T	traceless Ricci tensor, page 18
$T^*_p M$	cotangent space at p, page 10
$T^s_r(M)$	set of tensor fields of type (r, s), page 10
$T_p M$	tangent space at p, page 10

T_{ij} (local) components of the traceless Ricci tensor, page 18

u^* supremum of the function u, page 102

u_* infimum of the function u, page 148

u_{ijkt} fourth derivatives of the function u, page 20

u_{ijk} third derivatives of the function u, page 19

u_{ij} (local) components of Hess(u), page 18

$W^{1,1}(M)$ Sobolev space of functions in $L^1(M)$ with (weak) gradient in $L^1(M)$, page 29

$W^{1,2}(M)$ Sobolev space of functions in $L^2(M)$ with (weak) gradient in $L^2(M)$, page 143

X^i_k covariant derivative of the coefficient X^i, page 10

Index

L^∞ *a priori* estimate, 149
L_H, 74
a priori estimates
    from above, 113
    from below, 106

Bessel function, 235
Bianchi identities
    first, 15
    second, 16
Bishop-Gromov comparison theorem, 28
Bochner-Weitzenböck formula, 19
boundary point lemma, 66

Cartan's lemma, 33
co-area formula, 29
Codazzi equations, 35
Codazzi tensor, 46
comparison result, 140
conformal
    deformation of the metric, 37
    diffeomorphism, 146
    vector field, 41, 207
coordinate functions, 8
cotangent space, 10
Cotton tensor, 45
covariant derivative
    of a 1-form, 11
    of a function, 11
    of a generic tensor field, 11
    of a vector field, 10
    of the metric, 12
curvature forms, 14
curvature tensor
    of type $(0,4)$, 15
    Riemann, 15
cut locus, 21
cut point, 21
    ordinary, 21
    singular, 21

Darboux
    coframe, 32
    frame, 33
    frames along f preserving orientations, 34
de Rham cohomology groups, 47
decomposition of the curvature tensor
    using the Schouten tensor, 46
differential
    of a coordinate function, 8
Dirichlet problem
    for the operator L_H, 74
divergence
    of a vector field, 10
dual orthonormal frame, 8

Einstein manifold, 16
Einstein summation convention, 8
entire
    subsolution, 166
estimate
    from above, 112
    from below, 105
exponential map, 21
exterior differential
    of a 1-form, 12

Fatou's lemma, 218

first Bianchi identities, 15
first eigenvalue
    negativity, 170
    nonnegativity, 75
first eigenvalue of the operator L_H
    on M, 74
    on bounded domains, 74
first structure equations, 8
fundamental theorem of Riemannian
    geometry, 13

Gauss equations, 35
Gauss lemma, 22
Gram-Schmidt orthonormalization pro-
    cess, 8
Green kernel, 94
Gronwall inequality, 221
ground state, 103
group
    of conformal diffeomorphisms, 146
    of isometries, 146

Hessian, 18
    $(1,1)$ version, 20
Hessian comparison theorem, 31
Hopf classification theorem, 56
Hopf-Rinow theorem, 124
hypersurface, 34

immersion, 32
    isometric, 32
    minimal, 34
    totally geodesic, 34
    totally umbilical, 34
index
    of a Schrödinger operator, 173
integral curve, 22
interior of a set, 74
isothermic coordinates, 46, 48

Künneth formula, 47
Kazdan-Warner obstruction, 18, 41
Kronecker symbol, 8
Kulkarni-Nomizu product, 46

Laplace-Beltrami operator, 18
    transformation law, 99
Laplacian, 18
Laplacian comparison theorem, 25
Levi-Civita connection forms, 8
Lie bracket
    of two vector fields, 12
Lie derivative
    geometric meaning, 14
    of a 1-form, 13
    of a function, 13
    of a vector field, 13
local chart, 8
local orthonormal coframe, 8
locally conformally flat manifold, 46
lowering indices, 15

Mayer-Vietoris argument, 47
mean curvature
    in the direction of a unit normal
        vector field, 34
    of an immersed hypersurface, 34
    vector field, 34
metric
    induced by an immersion, 32
    parallelism of the, 12
    torsion-free, 12
minimizing geodesic, 22
model manifold, 115
Monotone iteration scheme, 191
monotonicity of the first eigenvalue,
    74
Morse lemma, 56

negative part, 125
Newton inequality, 60
Newton's inequality, 23
nonparabolic manifold, 94
normal bundle, 35

Obata type vector field, 57
Obata's theorem, 53
oscillating solution, 171

parabolic manifold, 94

parallel translation, 22
Picone's identity, 74
Poincaré model
    of the hyperbolic space, 211
Poisson equation, 94
pole, 29, 202
positive part, 112
positive part of a function, 89

radial Ricci curvature, 23, 26
reference point, 22
Rellich-Pohozaev formula, 207
Rellich-Pohozaev ientity, 208
Riccati differential inequalities, 25
Ricci equations, 36
Ricci identities, 19
Ricci tensor, 16
    $(1,1)$ version, 20
    traceless, 18, 40
Riemann theorem, 46
Riemann-Köbe uniformization theorem, 48
Riemannian
    manifold, 7
    metric, 7
Riemannian distance function, 21
Riemannian product
    of locally conformally flat manifolds, 47

scalar curvature, 16
Schouten tensor, 45
Schur's theorem, 17
second fundamental tensor, 33
second Green formula, 27
second structure equations, 14
    pull-back, 34
sectional curvature, 16
    radial, 31
Seifert-Van Kampen theorem, 56
sharp map, 11
smallness
    in a spectral sense, 159

solution
    maximal, 165, 166
    minimal, 166
spectral theory
    of Schrödinger operators, 158
sphere
    m-dimensional, 43
spherical mean, 170, 173
stretching factor, 146
subsolution, 111, 191
supersolution, 111, 191
    of a boundary value problem, 191
support of a function, 77
symmetries
    of the Riemann curvature tensor, 15
    of the second derivatives, 19

tangent space, 10
tangent vector of a curve, 22
tensor field
    of type (r, s), 10
third derivatives, 19
torsion tensor, 12
trace, 10
transformation law
    for a local o.n. coframe, 37
    for the connection forms, 38
    for the curvature forms, 38
    for the curvature tensor, 38
    for the Hessian, 99
    for the Laplace-Beltrami operator, 99
    for the Ricci tensor, 38
    for the scalar curvature, 39
    for the traceless Ricci tensor, 43

umbilical point, 34

Van der Waerden-Bortolotti covariant derivation, 35
vector field, 10

volume
    of a geodesic ball, 28
    of the boundary of a geodesic
        ball, 28

weak maximum principle at infinity,
        133
weakly distance decreasing, 155
weighted spherical mean, 200
Weyl tensor, 44

Yamabe
    equation(s), 39
    invariant, 47
Yamabe equation
    on hyperbolic space, 211

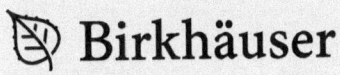
birkhauser-science.com

Progress in Mathematics (PM)

Edited by
Hyman Bass, University of Michigan, USA
Joseph Oesterlé, Institut Henri Poincaré, Université Paris VI, France
Alan Weinstein, University of California, Berkeley, USA
Yuri Tschinkel, Courant Institute of Mathematical Sciences, New York, USA

Progress in Mathematics is a series of books intended for professional mathematicians and scientists, encompassing all areas of pure mathematics. This distinguished series, which began in 1979, includes research level monographs, polished notes arising from seminars or lecture series, graduate level textbooks, and proceedings of focused and refereed conferences. It is designed as a vehicle for reporting ongoing research as well as expositions of particular subject areas.

PM 301: Ruzhansky, M.; Sugimoto, M.; Wirth, J. (Eds.)
Evolution Equations of Hyperbolic and Schrödinger Type.
Asymptotics, Estimates and Nonlinearities (2012).
ISBN 978-3-0348-0350-2

PM 300: Bump, D.; Friedberg, S.; Goldfeld, D. (Eds.)
Multiple Dirichlet Series, L-functions and Automorphic Forms
(2012).
ISBN 978-0-8176-8333-7

PM 299: Müller-Hoissen, F.; Pallo, J. M.; Stasheff, J. (Eds.)
Associahedra, Tamari Lattices and Related Structures. Tamari
Memorial Festschrift (2012).
ISBN 978-3-0348-0404-2

PM 298: Getz, J.; Goresky, M.
Hilbert Modular Forms with Coefficients in Intersection
Homology and Quadratic Base Change (2012).
ISBN 978-3-0348-0350-2

PM 297: Dai, X.; Rong, X. (Eds.)
Metric and Differential Geometry. The Jeff Cheeger
Anniversary Volume (2012).
ISBN 978-3-0348-0256-7

PM 296: Itenberg, I.; Jöricke, B.; Passare, M. (Eds.)
Perspectives in Analysis, Geometry, and Topology. On the
Occasion of the 60th Birthday of Oleg Viro (2012).
ISBN 978-0-8176-8276-7

PM 295: Joseph, A.; Melnikov, A.; Penkov, I. (Eds.)
Highlights in Lie Algebraic Methods (2012).
ISBN 978-0-8176-8273-6

PM 294: Barreira, L.
Thermodynamic Formalism and Applications to Dimension
Theory (2011).
ISBN 978-3-0348-0205-5

PM 293: Mazzucchelli, M.
Critical Point Theory for Lagrangian Systems (2011).
ISBN 978-3-0348-0162-1

PM 292: van den Ban, E. P.; Kolk, J.A.C. (Eds.)
Geometric Aspects of Analysis and Mechanics. In Honor of the
65th Birthday of Hans Duistermaat (2011).
ISBN 978-0-8176-8243-9

PM 291: Greene, R.E.; Kim, K.-T.; Krantz, S.G.
The Geometry of Complex Domains (2011).
ISBN 978-0-8176-4139-9

PM 290: Mantegazza, C.
Lecture Notes on Mean Curvature Flow (2011).
ISBN 978-3-0348-0144-7

PM 289: Colombo, F.; Sabadini, I.; Struppa, D. C.
Noncommutative Functional Calculus (2011).
ISBN 978-3-0348-0109-6

PM 288: Neeb, K.-H.; Pianzola, A. (Eds.)
Developments and Trends in Infinite-Dimensional Lie Theory
(2011).
ISBN 978-0-8176-4740-7

PM 287: Cattaneo, A.S.; Giaquinto, A.; Xu, P. (Eds.)
Higher Structures in Geometry and Physics (2011).
ISBN 978-0-8176-4734-6

PM 286: Abbes, A.
Éléments de Géométrie Rigide.
Volume 1: Construction et Étude Géométrique des Espaces
Rigides (2011).
ISBN 978-3-0348-0011-2

PM 285: Soifer, A.
Ramsey Theory. Yesterday, Today, and Tomorrow (2010).
ISBN 978-0-8176-8091-6